ENVIRONMENTAL RISK ASSESSMENT AND MANAGEMENT FROM A LANDSCAPE PERSPECTIVE

ENVIRONMENTAL RISK ASSESSMENT AND MANAGEMENT FROM A LANDSCAPE PERSPECTIVE

Edited by

Lawrence A. Kapustka
Wayne G. Landis

A JOHN WILEY & SONS, INC., PUBLICATION

Library of Congress Cataloging-in-Publication Data:

Kapustka, Lawrence.
    Environmental risk assessment and management from a landscape perspective / Lawrence A.
Kapustka, Wayne G. Landis.
        p. cm.
    Includes bibliographical references and index.
    ISBN 978-0-470-08997-2 (cloth)
    1. Environmental risk assessment. 2. Ecological assessment (Biology) 3. Environmental
management. 4. Landscape protection. I. Landis, Wayne G. II. Title.
    GE145.K376 2010
    333.71′4–dc22

                                                                                    2009037584

Printed in the United States of America

10 9 8 7 6 5 4 3 2 1

CONTENTS

PREFACE

With the rapid changes in today's world of publishing, we find this production having aspects of traditional and modern outlets. The printed version of this book will look and feel like any other technical, dare we say scholarly, book. One exception is the existence of an ftp site which contains supplemental materials, including color graphics and some models not included in the book. These related materials can be found at ftp://ftp.wiley.com/public/sci_tech_med/environmental_risk.

Terminology in the general field of risk assessment can at times be confusing, in particular because there are identical terms that have different meaning among jurisdictions and across disciplines. We have attempted to establish consistency throughout the book, but we accept that we may not have succeeded in all cases. One prominent item relates to nuanced distinctions regarding Ecological Risk Assessment (EcoRA) versus Environmental Risk Assessment (EnRA).

As the EcoRA framework was being developed in the United States in the late 1980s, there were clear separations between ecological risks and human health considerations. Much of the environmental legislation in the United States and elsewhere included some variation of the theme "to protect human health and the environment." With this enabling language, *environment* became synonymous with *ecology* in many circles. The field becomes more confusing with the often used definition of ecology being some aspect of "organisms interacting with their abiotic and biotic environment." Indeed, as chemical or physical parameters were correlated with nominal conditions and within protective boundaries for valued aquatic and terrestrial species, and to the extent that engineering disciplines came to dominate much of environmental management, connections to ecological entities were often forgotten—at least not explicitly identified. In some international arenas, EnRA was interpreted to be the coordinated of human health and ecological assessment of risks. Adding to the confusion, systems ecologists tend to view the abiotic parameters (what many others refer to as environment) and biotic parameters [plants, animals (including humans)] as being under the umbrella of ecology.

In this book, we have opted to use EcoRA and EnRA in lieu of another commonly used acronym, ERA. Throughout the chapters, we have attempted to hold to the terms introduced by the authors. However, we urge readers not to fret too much over these particular word choices, because in the end these may be largely distinctions without differences.

LAWRENCE A. KAPUSTKA
WAYNE G. LANDIS

Calgary, Alberta, Canada
Bellingham, Washington
March 2010

CONTRIBUTORS

Timothy F. H. Allen, University of Wisconsin, Madison, Wisconsin

Valerie C. Chen, Western Washington University Bellingham, Washington

Audrey M. Colnar, Western Washington University Bellingham, Washington

Mark D. Dixon, University of South Dakota Vermillion, South Dakota

Rebecca A. Efroymson, Oak Ridge National Laboratory Oak Ridge, Tennessee

Kenneth Froese, Calgary, Alberta, Canada

Hector Galbraith, Manomet Center for Conservation Sciences, Manomet, Massachusetts

Deborah A. Goeldner, Golden Geospatial Inc., Golden, Colorado

William W. Hargrove, USDA Forest Service, Asheville, North Carolina

Dale J. Hoff, United States Environmental Protection Agency, Duluth, Minnesota

Michael J. Hooper, Texas Tech University, Lubbock, Texas

Henriette I. Jager, Oak Ridge National Laboratory, Oak Ridge, Tennessee

G. Darrel Jenerette, University of California Riverside, Riverside, California

Alan R. Johnson, Clemson University, Clemson, South Carolina

Laurel Kaminski, Western Washington University, Bellingham, Washington

Lawrence A. Kapustka, LK Consultancy, Calgary, Alberta, Canada

Goro Kushima, Western Washington University, Bellingham, Washington

Wayne G. Landis, Western Washington University, Bellingham, Washington

David Lapen, Agriculture and Agri-Food Canada, Ottawa, Ontario, Canada

Greg Linder, USGS/BRD/CERC, Brooks, Oregon

Edward Little, USGS/BRD/CERC, Columbia, Missouri

Ronald J. McCormick, Sokio Systems, Calgary, Alberta, Canada

Alison McLaughlin, Health Canada, Ottawa, Ontario, Canada

S. Jannicke Moe, NIVA (Norwegian Institute for Water Research), Oslo, Norway

Sverker Molander, Chalmers University, Göteborg, Sweden

Rosana Moraes, Golder Associates Sarl, Lyon, France

Marla Orenstein, Habitat Health Impact Consulting Corp., Calgary, Alberta, Canada

Joanne Parrott, Environment Canada, Burlington, Ontario, Canada

James Pittman, Earth Economics, Seattle, Washington

Jeff T. Price, World Wildlife Fund, Washington, DC

Ananda Seebach, Western Washington University, Bellingham, Washington

Juliet C. Stromberg, Arizona State University, Tempe, Arizona

Edward Topp, Agriculture and Agri-Food Canada, London, Ontario, Canada

Sandra J. Turner, Paz Y Pasitos, Clemson, South Carolina

Jianguo Wu, Arizona State University, Tempe, Arizona

PROLOGUE

Every great advance in natural knowledge has involved the absolute rejection of
authority.

Thomas H. Huxley

Facts do not cease to exist just because they are ignored.

Aldous Huxley

Since the very early 1990s, we two have been discussing how ecology interacts with environmental toxicology, risk assessment, and environmental management. Our perspective piece "Ecology: the science versus the myth" (Kapustka and Landis 1998) was one of our early attempts to put our discussions and thoughts to paper.

We were struck at that time on the number of myths that had worked themselves into acceptable usage. Ideas such as ecosystem health, equilibrium, and recovery to an original state were being used as the basis of research agendas, environmental management, and policy-making. Our paper was designed to demonstrate just how dated these ideas were and to provide alternative approaches that would be appropriate for managing ecological systems. One of the keys to that paper was the incorporation of the ideas of Wu and Loucks (1995) and the hierarchical patch dynamics paradigm (HPDP) as an alternative to the standard models of ecological structure being used in the early 1990s. The HPDP is an approach that inherently integrates spatial and temporal interactions, even at large scales, so taking a regional or landscape approach was an easy next step.

In the intervening 11 years, we and many of our colleagues have been investigating sites or phenomena that can be thousands of square kilometers in size. This book is our attempt to create a coherent picture of how to approach risk assessment at a regional scale using a landscape perspective. Perhaps the beginnings of a useful paradigm for evaluating risks and making management decisions will be an outcome of our current effort. The fun thing about paradigms is that they are there to be broken, spurring other researchers on to a better understanding of how to estimate risks and manage ecological structures at these scales.

As the reader will notice, this book is not just about chemicals. Risk assessment has been trapped by its original focus on contaminants and pesticides and is often seen as an extension of toxicology. Risk analysis is used in many other arenas: transportation, terrorism, human health, financial forecasts, and engineering. The use in each field follows a similar framework, but with specific considerations depending upon its application. In this book other ecological topics, especially invasive species, are used to demonstrate the broad applicability of the environmental risk assessment process. It would be satisfying to see risk assessment being adopted as a general analysis framework for environmental management.

So that you do not have to read the epilogue first to see how it all turns out, the answer is yes, it is possible to do risk assessments at very large scales for a variety of environmental, ecological, and human health goals. While there are certainly challenges, there is no conceptual barrier to using risk assessment (or more properly risk analysis) as a tool for managing broad and varied landscapes over varied time horizons. The book takes a building block approach to demonstrating this thesis.

The first block provides the foundations—that is, the theory behind risk assessment, ecology, and scale. In Chapter 1, Introduction, we provide a regulatory and historical background for the beginnings of risk assessment and its use at regional and landscape scales. Particularly noteworthy is the far-reaching intent of the founding legislation in contrast to the state of the environmental sciences in the 1970s. In the next chapter, Ecological Risk Assessment Toward a Landscape Perspective, Kapustka outlines the components of the risk assessment process. Particularly noteworthy is the blending of social science and biophysical science in the process. The chapter explicitly denotes the importance of stakeholders in the risk assessment process, a role that is often usurped by regulators and many, if not most, risk assessment experts. Chapter 3, also by Kapustka, discusses the challenges and vagueness of three prominent aspects of contemporary risk assessments, namely, population, habitat, and ecological systems.

The next two chapters set the stage for thinking at the scale that this book has as its target. In Chapter 4, Relevance of Spatial and Temporal Scales to Ecological Risk Assessment, Alan R. Johnson and Sandra J. Turner discuss the importance of intrinsic scale and the pitfalls in extrapolating across scales in risk assessment. G. Darrel Jenerette and Jianguo Wu in Chapter 5, Quantitative Measures and Ecological Hierarchy, detail quantitative approaches in examining the organization of ecological systems based upon ideas of scaling and the analysis of patterns. One of the key ideas is that relatively small changes at the system level may lead to catastrophic changes to individuals.

Jannicke Moe introduces the application of Bayes Theorem in risk assessment in Chapter 6, Bayesian Models in Assessment and Management. This chapter is one of the best introductions to a variety of applications including Bayesian interference, Bayesian networks, influence diagrams, and hierarchical Bayesian networks. Examples of the use of Bayesian tools at large spatial scales are also provided.

In Chapter 7, Linking Regional and Local Risk Assessment, Rosana Moraes and Sverker Molander introduce the use of the Procedure for Ecological Tiered Assessment of Risks (PETAR) in integrating local and regional risks. The PETAR approach incorporates cause–effect chains into conceptual models, assists in the selection of sites for exposure and effects analysis, and incorporates a weight of evidence approach.

Kenneth L. Froese and Marla Orenstein in Chapter 8, Integrating Health in Environmental Risk Assessments, refute the common misconception that ecological risk assessment and human health risk assessment are separate entities. The authors demonstrate that this misconception is in part due to a narrow definition of health that is not reflective of current international practice. They posit that at regional scales, human health as defined by well-being is straightforwardly incorporated as part of the risk assessment process.

Wildlands are important to large numbers of stakeholders, yet mechanisms of measuring the value of these systems are very controversial. In Chapter 9, Rebecca Efroymson et al. review the method for evaluating total economic and other values of these common aspects of landscapes. A number of approaches are evaluated: Multimetric tools, habitat metrics, and complex models are some of the examples.

Climate change is one of the dominant social issues of this time. Chapter 10, Predicting Climate Change Risks to Riparian Ecosystems in Arid Watersheds: The Upper San Pedro as a Case Study, demonstrates that risk assessment can be used to evaluate risks due to current and upcoming alterations to climate. Hector Galbraith et al. use a series of temperature scenarios and evaluate risks to riparian endpoints under each. Supplemental figures to Chapter 10 can be found on the ftp site.

The next series of chapters form the interface between the methods and the case study segments of this book. In Chapter 11, Invasive Species and Environmental Risk Assessment, Greg Linder and Edward Little present a landmark review and synthesis paper. Linder and Little provide an encyclopedic summary of the application of risk assessment for the evaluation of invasive species. The review covers topics such as uncertainty analysis, management tools, and a series of important recommendations for furthering the field. The book now transitions to a series of case studies, with the first two being the application of risk assessment to invasive species.

Chapter 12, Landscape Nonindigenous Species Risk Assessment: Asian Oyster and Nun Moth Case Studies, covers marine and terrestrial cases based in the East Coast of the United States. Wayne G. Landis et al. demonstrate the applicability of the Hierarchical Invasive Risk Model (HIRM) to each of these scenarios. The nun moth is an insect pest that has not yet arrived to the United States from Europe. Asian oyster has been suggested as a species to be introduced to the Chesapeake Bay as a means of reestablishing the shellfish industry.

Chapter 13, Ecological Risk Assessment of the Invasive *Sargassum muticum* for the Cherry Point Reach, Washington, again demonstrates the applicability of the HIRM approach in a very different circumstance. Sargassum, although an invasive, provides habitat for a number of valued species along the coast of the Northwest United States. Ananda Seebach et al. calculated that for some endpoints, negative risk or benefit results from the invasive.

Chapter 14, Integrated Laboratory and Field Investigations: Assessing Contaminant Risk to American Badgers, takes the study of regional risk to the field. Dale J. Hoff et al. demonstrate the utility of having site-specific data on home range size, prey type, uptake, and the presence of other stressors for the assessment species. Many of the previous chapters in this book discuss the importance of quantifying and then reducing uncertainty in regional risk assessments, this is the chapter that describes how to reduce that uncertainty with site-specific data collection and analysis. Supplementary material to Chapter 14 can be found on the ftp site.

Human and veterinary pharmaceuticals can be found originating from both point and nonpoint sources. Chapter 15, Environmental Risk Assessment of Pharmaceuticals, focuses on this important issue. Joanne Parrott et al. list the issues with conducting risk assessments for this topic that covers a number of chemical classes, modes of action, and numerous routes of exposures. The authors also take into account the several regulatory environments being used to evaluate the environmental release of these biological active materials.

The third and last block ties society and regional risk together into a knot of valuation and sustainability. Chapter 16 by Ronald J. McCormick, James Pittman, and Timothy F. H. Allen analyzes the limitations of classical economic paradigms and argue for a transition to economic ecology, a perspective that embraces the dynamics of ecological systems. In Chapter 17 James Pittman and Ronald J. McCormick cover economic analyses and ecosystem valuation, and McCormick concludes in Chapter 18 with a brief discussion of metrics of sustainability within social and ecological landscapes. This section ties the human landscape to the ecological, demonstrating the deep connections between the two.

Finally, we present an Epilogue. The book demonstrates that regional risk assessment by its nature incorporates more than simply chemical toxicity. Social and ecological landscapes need to be considered. Although perhaps daunting in scope, regional risk assessment is possible to accomplish in a manner that informs long-term environmental management decision-making.

Have fun in reading this book and let us know what you think.

REFERENCES

Kapustka LA, Landis WG. 1998. Ecology: The science versus the myth. *Human Ecol Risk Assessment* **4**:829–838.

Wu J, Loucks, OL. 1995. From balance of nature to hierarchical patch dynamics: A paradigm shift in ecology. *Q Rev Biol* **70**:439–466.

1

INTRODUCTION

Lawrence A. Kapustka and Wayne G. Landis

Environmental management evolved considerably in the last quarter of the 20th century, passing from an era of human dominance over nature toward an ideal of sustainability. The earlier period was emboldened in North America by the philosophy of manifest destiny. The new era is marked by self-awareness of our limited capacity to tame nature. The vague constructs of sustainability are tested by desires to hold onto the luxuries of developed societies while striving for environmental and social equity; a struggle captured in the tension between romantic illusions and pragmatic actions.[1] New ways of thinking about resources and resource use are permeating debate among intellectuals, politicians, and common people (McKibben 2007).

The complexities of human interactions across different ethnic, social, cultural, religious, and political perspectives have shown that environmental management is fraught with truly wicked problems—situations for which there are no right answers, but there are many wrong ones.[2] If the movement embracing sustainability is to succeed, surely there will need to be greater acknowledgment and understanding of ecological systems. There also needs to be a dampening of rhetoric that erupts from

---

[1] Johann Hari. Move Over, Thoreau. Posted January 12, 2009. As viewed on January 14, 2009 at http://www.slate.com/id/2207168/pagenum/all/p2.
[2] The concept of wicked problems was introduced by Rittel and Webber (1973).

Environmental Risk and Management from a Landscape Perspective, edited by Kapustka and Landis

normative science (Lackey 2001, 2007). A landscape perspective can begin to inform this quest.

A landscape perspective requires that attention be given to the relevant temporal and spatial scales for the issues to be addressed. It also demands that humans be considered to be part of the ecological systems being interrogated and managed. But for many pragmatic reasons, as well as many short-sighted excuses, the proper scales are seldom considered. Obviously, there is a vast spread in spatial scale that is required to sustain populations across the spectrum of plants, animals, microbes, and even humans (detailed discussions of scale appear in Chapter 4). For some microbes in soils, the relevant scale to observe population dynamics or functional processes is in the realm of mm^3 to cm^3; for vertebrate populations, home ranges (excluding migratory species) span four or five orders of magnitude with some exceeding 100,000 ha. In contrast, most contaminated sites are measured in tens of hectares with the rare site exceeding 5000 ha. Similarly for temporal scales, microbial events can be measured in minutes or less, but most ecological developments affecting humans occur over decades to tens or hundreds of millennia—except of course those events that follow chaotic patterns associated with a tipping point. Most risk assessments focus on a few years; the rare risk assessment projects out to 100 years or more. Accordingly, there often is a considerable disconnect between the spatial and temporal scales relevant to ecological developments and the respective scales of investigations used in ecological risk assessments (EcoRAs) or in risk management (Fig. 1.1).

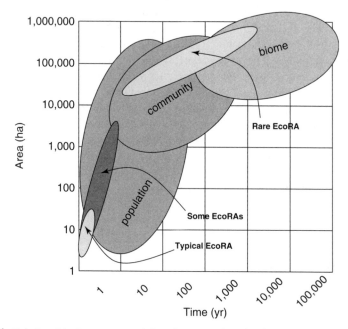

Figure 1.1. Relationship between spatial and temporal scales for ecological developments and risk assessments. [After Fig. 1 in Kapustka (2008).]

The consideration of the scales (as in Fig. 1.1) also will mandate a change in perception of rare, yet important, events. Events that are rare at the scale of a Superfund site or most risk assessments are seen as anomalies. These events, such as a fire, flood, extinction, or the introduction of an invasive species, are anomalies because the arbitrary scale of the site does not include an ecological context. For an individual Superfund site the likelihood of invasion may be low. However, within a landscape an invasion event has a high probability and, depending upon location, may impact the remediation of a contaminated site. Within a landscape, fire, floods, invasions, and extirpations will be occurring and will have important impacts at a variety of scales and to a number of ecological services.

The modern era of environmental management followed the passage of the National Environmental Protection Act (NEPA) of 1969 that was signed into law in the United States in January 1970.[3] There were other events in the arts and cultural circles, politics, and economics that set the stage for the environmental movement, which led to this remarkable document that had as its primary purpose:

> To declare a national policy which will encourage productive and enjoyable harmony between man and his environment; to promote efforts which will prevent or eliminate damage to the environment and biosphere and stimulate the health and welfare of man; to enrich the understanding of the ecological systems and natural resources important to the Nation.

The opening of Sec. 101 [42 USC § 4331] of NEPA provides a remarkable framing of the national environmental policy:

> The Congress, recognizing the profound impact of man's activity on the interrelations of all components of the natural environment, particularly the profound influences of population growth, high-density urbanization, industrial expansion, resource exploitation, and new and expanding technological advances and recognizing further the critical importance of restoring and maintaining environmental quality to the overall welfare and development of man, declares that it is the continuing policy of the Federal Government, in cooperation with State and local governments, and other concerned public and private organizations, to use all practicable means and measures, including financial and technical assistance, in a manner calculated to foster and promote the general welfare, to create and maintain conditions under which man and nature can exist in productive harmony, and fulfill the social, economic, and other requirements of present and future generations of Americans.

A series of US federal laws designed to address environmental issues followed. This included the Clean Air Act (1970), the Water Quality Improvement Act (1970), the Water Pollution and Control Act Amendments (1972), the Resource Recovery Act (1970), the Resource Conservation and Recovery Act (1976), the Toxic Substances Control Act (1976), the Occupational Safety and Health Act (1970), the Federal Environmental Pesticide Control Act (1972), the Endangered Species Act (1973), the Safe

---

[3]The full act can be viewed at http://ceq.hss.doe.gov/Nepa/regs/nepa/nepaeqia.htm (accessed July 2009).

Drinking Water Act (1974), the Federal Land Policy and Management Act (1976), and the Surface Mining Control and Reclamation Act (1977). Similar lists of legislative acts were passed in countries throughout the world.

As with many lofty political ambitions, translating policy into actions takes many twists and turns. Well-intentioned actions can often have devastating unintended consequences, in part due to imperfect knowledge or unawareness of how things really work. Efforts to suppress forest fires in the early and mid-1900s led to massive buildup of understory and litter—fuel that, once ignited, burns through the canopy and results in markedly different successional trajectories of the affected landscapes. Environmental management decisions that focus on narrowly constructed goals are just as likely to have unintended consequences. Strikingly different recommendations emerge if the objectives are to protect or enhance specific wildlife populations versus elimination of contaminants from a site. Indeed the focus on eliminating contaminants tends to lead to destruction of habitat needed to harbor the valued wildlife purportedly being protected. Thus, from a landscape perspective we ought to promote holistic examinations that are more likely to explore alternative scenarios and lead to very different decisions (a theme developed in Chapters 8 and 17).

The language in the US NEPA appeared to foster holistic approaches to environmental management. It prescribed a multidisciplinary focus to evaluate environmental impacts of projects and activities that would significantly affect the human environment. So why have so many environmental management problems emerged?

Implementation of the lofty aspirations of the US NEPA was not so easy. Challenges emerged quickly due to the very different approaches across the disciplines of engineering, ecology, economics, and sociology. Though project designs typically contained discretely bounded and quantified estimates of monetary costs for engineering, there were no equivalent metrics to characterize costs of ecological resources or the well-being of human communities. Project benefits resulting from construction of a road or a reservoir easily trumped vague concerns about the loss of habitat for an obscure fish or butterfly, or the displacement of a few farmers who's land would be inundated. Welfare economics were used or misused to justify the greater good of society over the losses of a few individuals, in part due to attention solely on readily monetized factors and assigning many factors to a category of "externalities" that conveniently (though not necessarily by being mean-spirited) could be dismissed from consideration. Sociologists in the early 1970s relied largely on generalized narratives; biologists seemed to do little more than generate lists of species expected to occupy the affected area. When pressed for answers, biologists and sociologists could not match the apparent certitude of the engineers.

Indeed, the field of ecology in the late 1960s and early 1970s was not prepared to provide quantitative answers that were useful for environmental management decision-making. The work by May (1973) on the relationship between species diversity and community dynamics occurred after the major environmental legislation was enacted, and his work on nonlinear dynamics occurred later in the decade (May and Oster 1976). Ecology and related sciences were still mired in the clutches of Clementsian ecology that emphasized stability and climax systems (Clements 1916); however, competing theories were gaining traction, at least in the academic community. The resurrection of Gleason's (1926) ideas on the role chance plays in the organization of communities resulted in the continuum concept (Curtis 1955, Curtis and McIntosh 1951, Whittaker 1953) and provided a counterpoint to Clementsian ecology, but application of the continuum concepts to address environmental management issues was quite limited. Ecology shifted into a new realm with the groundbreaking research by Wu and Loucks (1995) into new formulations of community dynamics.

Similarly, ecotoxicology was nonexistent at the onset of the modern environmental movement. It was the passage of the environmental legislation described in this chapter that provided the impetus for developing the field. Initially (through the 1970s), studies were focused largely on determining the concentrations that resulted in mortality in situations of acute exposures. Development of ecological risk assessment did not even begin until the mid-1980s.

The needs of environmental management were far ahead of the science of both ecology and environmental toxicology at the time. And so, by default the processes of environmental management came to be dominated by the linear problem-solving approach of engineers. That legacy still remains.

A decade into the modern environmental era, the disaster of Love Canal[4] in upper New York state led to additional legislation in the United States to deal with hazardous waste sites. CERCLA (aka Superfund in 1980 and amendments in 1986) created a need to document the effects of chemical releases on exposed humans and ecological resources, to identify causal linkages between the chemicals and the observed biological responses, and forecast the benefits of remediation. Though aspects of ecological risk assessment were embodied in pesticide and chemical registration as well as water quality standards for discharge permitting, it was the institutionalization of Superfund in the mid-1980s that expanded the depth and breadth of ecological risk assessments. By the end of the 1980s the general framework for EcoRAs was established (Suter 2008). Field observations pertaining to the nature and extent of contaminants were evaluated against toxicity responses to the chemicals. Forensic ecology[5] was used to establish weight-of-evidence linkage of chemicals to effects, and EcoRA was used to evaluate alternative remedies that either eliminated the offending chemicals or restricted exposure to receptors.

Risk assessment is performed in many disparate disciplines and can be expressed in qualitative or quantitative terms. It is used in engineering to evaluate the probability of structural failures and to forecast design requirements for flood control, building design, slope stability, driving a car, flying an airplane, walking in the city, or keeping foods and drinking water from being tainted with various toxic chemicals. Financial institutions and individuals employ risk assessment to evaluate investment opportunities, set insurance premiums and payouts, weigh potential liabilities, and structure

---

[4]See *The Love Canal Tragedy* at http://www.epa.gov/history/topics/lovecanal/01.htm as viewed January 14, 2009.

[5]The term "retrospective risk assessment" has been used to identify plausible causal relationships. However, as risk is inherently a forward looking attempt to describe likelihood of occurrence, the term is convoluted. Also see Taleb's (2007) discussion of retrospective distortion.

contracts. Most readers of this book will be familiar with human health and ecological risk procedures that estimate cancers occurrence, mortality rates, morbidity rates, and so on. Regardless of the type of assessment, risk procedures examine possible scenarios by estimating the magnitude of exposure to the stress element, relating effects to different magnitudes of exposure, and concluding with a description of the likelihood of the events occurring in each scenario. Ultimately, the purpose of the risk assessment is to inform decision-makers tasked with managing the risks. In assigning probability of certain scenarios, informed management decisions can be structured to avoid, minimize, or mitigate adverse consequences in a manner that is proportional to the risks and consequences.

The most important part of the risk assessment process is the Problem Formulation stage in which the questions to be addressed are refined so that the analysis and characterization stages are relevant to the management goals. More details pertaining to Problem Formulation are presented in Chapter 2.

The marriage of risk assessment and multicriteria decision analyses approaches provides a formal approach to consider tradeoffs and assist decision-makers as they navigate through challenging problems. Risk assessment procedures are sufficiently flexible to accommodate mixes of quantitative and qualitative information (illustrated in Chapters 6, 7, 12, and 13). Furthermore, the advanced methods are robust in handling varying degrees of uncertainty and natural variations (see Chapter 6), both normal attributes of biological and sociological data. Recent advances in risk assessment include alignment with comparative risk and multicriteria decision analysis tools (Linkov et al. 2006). Also, computational models that account for variability and uncertainty provide powerful insights that aid the evaluation of scenarios through identification of the most influential parameters also known as sensitivity analysis.

There are many challenges involved in making sound environmental management decisions. Environmental management issues are complex due to inherent characteristics of the environment we attempt to manage, but also due to the many competing interests expressed by various stakeholders. Unaided, humans are not particularly good in processing all the information that comes from assessment of these multifaceted issues, such as those that face regulatory bodies; what may seem perfectly logical in terms of engineering feasibility or legal/regulatory compliance can be viewed very differently by other stakeholders.

Information comes to decision-makers in many different forms (e.g., modeling/monitoring data, environmental risk analysis, benefit–cost analysis, and stakeholders' preferences). In the public sector, adherence to an agreed process can be of equal or greater importance when compared to the actual decision. The process must be understood by stakeholders, and it must be seen as being fair, equitable, and consistent. Yet, if each of these overarching aims is to be met, the process must also be flexible (that is, it must accommodate project- or site-specific information). Perhaps most importantly, the process should be constructed in such a manner that it can reward innovation and good-faith efforts. In other words, there should not be built-in disincentives that stifle creative environmental management.

Decision-makers typically receive different types of technical input. But how can individuals or groups integrate (or judge) the relative importance of information from each source? While modeling and monitoring results are usually presented as quantitative estimates, risk assessment and benefit–cost analyses incorporate higher degrees of qualitative judgment. Structured information about stakeholder preferences may not be presented to the decision-makers at all, and it may be handled in an *ad hoc* or subjective manner that exacerbates the difficulty of defending the decision process as reliable and fair [see Kiker et al. (2005)]. Moreover, where structured approaches are employed, they may be perceived as lacking the flexibility to adapt to localized concerns or faithfully represent minority viewpoints. There has been considerable activity in recent years that has examined formal approaches to guide the decision process, ones that handle both quantitative and subjective preference information equally well.

Fundamentally, the design elements of the risk framework either constrain or expand the utility of the information that will be used to make decisions. Deciding on issues of scale, both temporal and spatial, becomes one of the most important challenges. If not constructed properly to account for both ecological realities and socioeconomic interests, the resulting analysis may be irrelevant to the problems being addressed or they may be tautological—that is, circular arguments that dictate the "answers" that emerge, thereby giving a false sense of validity to the regulatory decisions that ensue. Great care must be exercised during the design phase so that equity, fairness, openness, clarity of purpose, and agreement on terminology are recognizable features of the approach. A most important consideration of the framework is to ensure that the breadth of affected stakeholders' values is captured in the explicit delineation of endpoints to be assessed.

In the interplay of policy and regulatory actions, inevitably there are varying degrees of tension that arise due to differences in stakeholders' tolerance or acceptance of environmental risks. These tensions often are created as a direct consequence of the processes followed in reaching decisions, but there is much more. Explorations from nearly two decades ago into risk perception have provided powerful illustrations into the way people handle multiple forms of information as they make decisions. In general, we can conclude that scientific or technical descriptions of a risk event or activity form only a small part of the body of information that people process as they consider accepting or rejecting the risk [see Gladwell (2005) for his discussions of the theory of thin slices]. Those science-based or technological features are largely limited to understanding the mechanisms and characterization of uncertainty. From the regulatory side, the most critical feature influencing public acceptance of decisions is trust in the responsible institution (Peters et al. 1997). Many of the remaining features relate in one form or another to communications and the degree of control that the public feels they can exercise, either directly or indirectly. Historically, public notice and public hearings/comment periods have been the primary means for public input into the environmental management regulatory process. However, all too often, at least to the affected public, it appears that the crucial decisions have been made by government officials and industry proponents well before public input is sought. As the regulatory process evolves to meet current challenges, there are opportunities to achieve the goals of public input in ways that are more satisfying to all stakeholders and simultaneously streamline the process so that efforts can be focused on issues in proportion to the importance of the issues.

Our society has shown a particular inability to consider scale, especially for events that are rare and consequential. Taleb (2007) calls such events *Black Swans*. Black Swans are events that are rare, essentially not predictable as to a specific type of event, but have high consequence. The dot com downturn, the subprime lending mortgage meltdown, World War II, and the fall of the Soviet Union are all examples of societal events that represent Black Swans. Ecological events include the influenza outbreak of 1914, the invasions of zebra mussel and sea lamprey into the Great Lakes, and the impacts of persistent organic pollutants on wildlife populations popularized by *Silent Spring* (Carson 1962).

It is not clear that our current regulatory and management structure for the environment in Western society can manage future Black Swans at the proper scales. Each type of activity or event tends to be regulated by a specific statute by a specific agency, and without a landscape context. The context of the regulations can be traced to the early 1970s when our understanding of landscapes was nascent. If landscapes are to be managed to perpetuate ecological services, sustainability, or other societal goals, then a change in perception and approach is required.

At the core of this book and as detailed in subsequent chapters, a landscape perspective is crucial in addressing contemporary environmental management concerns. The scale of time and space dictate the appropriate focus: Using the wrong scales leads to asking the wrong questions, resulting in irrelevant "answers." Choosing the proper scales for assessment seems intuitively obvious, but as we consider conflicting objectives, disparate values, and multiple receptors, one quickly realizes the enormity of the task.

REFERENCES

Carson R. 1962. *Silent Spring*. Houghton Mifflin, Boston.

Clements FE. 1916. *Plant Succession: An Analysis of the Development of Vegetation*. Carnegie Institute Publication 242, Washington, DC.

Curtis JT. 1955. A prairie continuum in Wisconsin. *Ecology* **36**:558–566.

Curtis JT, McIntosh RP. 1951. An upland forest continuum in the prairie–forest border region of Wisconsin. *Ecology* **32**:476–496.

Gladwell M. 2005. *Blink: The Power of Thinking Without Thinking*. Black Bay Books, New York.

Gleason HA. 1926. The individualistic concept of the plant association. *Bull Torrey Bot Club* **53**:7–26.

Kapustka L. 2008. Limitations of the current practices used to perform ecological risk assessment. *Integr Environ Assess Manage* **4**:290–298.

Kiker GA, Bridges TS, Varghese A, Seager TP, Linkov I. 2005. Application of multicriteria decision analyses in environmental decisions making. *Integr Environ Assess Manage* **1**:95–108.

Lackey RT. 2001. Values, policy, and ecosystem health. *BioScience* **51**:437–443.

Lackey RT. 2007. Science, scientist, and policy advocacy. *Conserv Biol* **21**:12–17.

Linkov I, Satterstrom FK, Kiker GA, Bridges TS, Benjamin SL, Belluck DA. 2006. From optimization to adaptation: Shifting paradigms in environmental management and their application to remedial decisions. *Integr Environ Assess Manage* **2**:92–98.

May RM. 1973. *Stability and Complexity in Model Ecosystems, second edition*. Princeton University Press, Princeton, NJ, 265 pp.

May RM, Oster GF. 1976. Bifurcations and dynamical complexity in simple ecological models. *Am Nat* **110**:573–599.

McKibben B. 2007. *Deep Economy: The Wealth of Communities and the Durable Future*. Henry Holt and Co., New York.

Peters RG, Covello VT, McCallum DB. 1997. The determinants of trust and credibility in environmental risk communications: An empirical study. *Risk Anal* **17**:43–54.

Rittel H, Webber M. 1973. Dilemmas in a general theory of planning. *Policy Sci* **4**:155–169.

Suter GW. 2008. Ecological risk assessment in the United States environmental protection agency: A historical overview. *Integr Environ Assess Manage* **4**:285–289.

Taleb NN. 2007. *The Black Swan: The Impact of the Highly Improbable*. Random House, New York.

Whittaker RH. 1953. A consideration of climax theory: The climax as a population and pattern. *Ecol Monogr* **23**:41–78.

Wu J, Loucks OL. 1995. From balance of nature to hierarchical patch dynamics: A paradigm shift in ecology. *Q Rev Biol* **70**:439–466.

2

ECOLOGICAL RISK ASSESSMENT TOWARD A LANDSCAPE PERSPECTIVE

Lawrence A. Kapustka

The fundamental concepts of risk assessment are imbedded in our cultural prehistory. Then of course the considerations were intimately linked to daily survival: Was it safe to cross a river, attack a woolly mammoth, eat the camas bulbs, or enter the territory of hostile neighbors in search food, shelter, and treasure? Scientific inquiry enabled improved understanding of connections between events and consequences. For example, efforts by Pasteur and Koch established a basis for managing bacterial contaminants and thereby develop safeguards against diseases.

The application of risk assessment and risk management to biological situations likely grew out of the science of epidemiology. Epidemiology, the study of diseases in populations (Schwabe et al. 1977), emerged as a formal discipline in the late 1800s following a severe cholera epidemic in London. The cause of the epidemic was traced to contaminated drinking water, and the solution was simple: Remove the pump handles from the contaminated wells. The process that was used to make the diagnosis and find the solution represented a new way of thinking, one that searched for patterns and traced linkages that could explain the observed phenomenon.

But formal constructs of risk assessment as we know them today are relatively new. Though at one level, the procedures appear to be simple and the math is generally elementary, the execution of the procedures quickly becomes complicated. The many decisions that must be made in setting up an assessment, though none individually may seem very challenging, can become daunting with the interconnectedness of the suite

Environmental Risk and Management from a Landscape Perspective, edited by Kapustka and Landis
Copyright © 2010 John Wiley & Sons, Inc.

of decisions. Perhaps as illustrated in the popular exposé on the human psyche, *Blink* (Gladwell 2005), we are ill-equipped to address such complexities. Experts are easily, often unwittingly, diverted by their prejudices and can make monumental errors in judgment. The strong focus on chemical stressors, along with simplistic and outdated explanations of how ecological systems work, may indeed lead us to develop highly precise descriptions of risks that are precisely wrong. In the collection of chapters in this book, we proffer the landscape and systems perspectives as improvements to the risk approaches, especially when integrated into tools from the field of decision science. Mistakes will still be made, but hopefully fewer and with better chance for corrections. To set the stage, I first examine the basic features of contemporary risk assessment.

RATIONALE FOR RISK ASSESSMENT

The origin of formal risk assessment can be traced to the insurance industry (Suter, 2008). Actuarial tables aided the search for patterns of mortality in specific age groups and provided clues to link behaviors and lifestyles to the patterns—the foundation of insurance premiums. The art of describing the occurrence of death or morbidity linked to putative causal factors (epidemiology) could be used to forecast events (a form of quantitative risk assessment) and design remedial actions. With the dawn of the age of chemicals following World War II, efforts were made to link chemical contamination in food and adverse health consequences. Approaches established by the early 1980s form the core of risk assessment as it is currently practiced (NRC 1983, US EPA 1992, 1998).

Our basic aversion to external risk, those occurring beyond our control, led authorities in search of a number that represented zero risk, or at least a probability so low that statistical sampling would be unable to distinguish the value from zero. The etiology of this number representing essentially zero risk was traced in a delightful paper (Kelly 1991) that details how the arbitrary value of one in a million—that is, 10^{-6}—became a *de facto* safe threshold in human health risk assessments.

Formal constructs for dealing with human health risk assessment were in place before ecological risk assessment (EcoRA) was developed, and they had great influence on the procedures used to evaluate ecological risks. In many ways this was unfortunate, because throughout the development of ecological approaches, there were pressures to establish a structure that paralleled the human health risk assessment process. There are many fundamental similarities, but also many differences. Perhaps the greatest shortcoming arises from considering the target receptors (human or otherwise) at the organism level, which often divorces the subjects from the dynamic interactions they really experience in their ecological setting. Separation of human and ecological risk procedures, in addition to creating unnecessary compartmentation, led to two lexicons—often having identical terms, but with different definitions or usage. This leads to communication challenges that become especially difficult for the end users of the risk information. When compartmentalized, separate risk assessments for humans and for all ecological receptors can cause much frustration. Against this backdrop, there is a challenge to develop a holistic approach that merges these two parallel approaches into one integrated realm that could improve the usefulness of the entire effort (Kapustka et al. 2008; also see Chapter 8). A holistic assessment, by extension of the multiple interactions, demands an assessment with a landscape perspective.

Near-term and long-term environmental and socioeconomic effects of environmental planning options must be characterized and evaluated to understand both the benefits and consequences of each option. A formal Environmental Risk Assessment (EnRA) can minimize unwanted or unexpected consequences by identifying those actions most likely to be harmful to humans or ecological resources and ranking the acceptable and unacceptable effects of the various options. Consideration of only short-term profits or benefits from the use of environmental resources will likely produce harmful long-term environmental consequences. Proper use of the risk assessment process will identify threats to environmental resources and allow decision-makers to select management options that have the least negative impact.

A formal risk assessment process has several advantages as applied to environmental planning and management of hazards (Suter 1993):

- It can provide the quantitative basis for comparing and prioritizing risks.
- It can provide informed, science-based input for benefit–cost analyses.
- By expressing results as probabilities, it acknowledges the inherent uncertainty in predicting future environmental states, thereby making the assessment more credible.
- It separates the scientific process of estimating the magnitude and probability of effects from the process of choosing among alternatives and determining the acceptability of risks.

The last point is clearly the most important argument in favor of the risk assessment process. Values, biases, and societal influences define the questions asked in the risk assessment process and influence the management decisions. However, the processes for determining the threat from a proposed action and magnitude of the exposure to stressors are grounded in the sciences. Characterization of the probability that an adverse outcome will occur as a result of the exposure to the stressor in question also is based on interpretation of science-based facts. The results of a risk characterization should be stated as a probability that an event will occur, because the occurrence of future effects in ecological systems cannot be predicted with certainty. This is not merely a limitation of the science. Rather, it is an acknowledgment of the variable and chaotic patterns of ecological systems. A risk assessment can also provide a series of probabilities that an adverse effect will occur under different scenarios.

ASSESSMENT TIERS

The assessment process generally is staged with the initial stages (i.e., scoping and screening) designed to be quick and relatively inexpensive. The early assessment stages

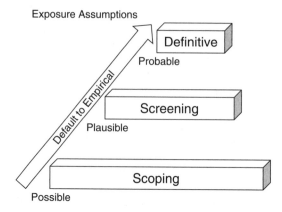

Figure 2.1. Progression from default assumptions toward empirical evidence in risk assessment.

use broad-brush assumptions that skew the outcome in a manner that minimizes the chances of making Type II errors—that is, declaring a situation to be safe when it is not. These early stages are intended to narrow the focus of definitive assessments, if needed, onto the important questions. Basically, these different stages move from highly protective assumptions (e.g., ones that assume toxicity responses at the lowest concentrations and realization of maximum exposures) to ones that more closely align with the real-world setting for the situations in question (Fig. 2.1, Table 2.1).

A most important aspect of these scoping/screening efforts is that the triggers or "bright lines" relating to environmental concentration (or equivalent measurement of magnitude for other stressors) are to be used in a one-way test. That is, if the concentration of a chemical of interest is below the threshold, then the risks are deemed to be in the *de minimis* zone and are dropped from further consideration (Fig. 2.2). However, if concentrations are above the threshold, then further evaluation *may* be warranted—that is, site-specific factors affecting bioavailability or other moderating

Table 2.1. Degree of Rigor and Content in Risk Assessment Tiers

Tiers	Content
Scoping (Tier 1)	Coarse
	Minimum data acquired
	Highly protective default assumptions
Screening (Tier 2)	Some refinement
	More data acquired
	Still relying on protective default assumptions
Definitive (Tier 3)	Finer detail
	Considerable data acquired
	Greater realism replaces default assumptions

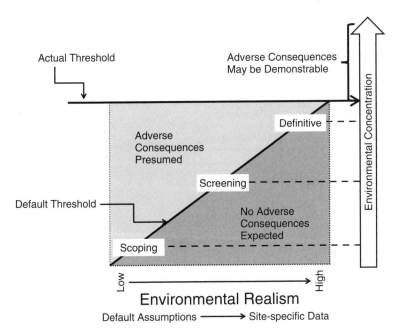

Figure 2.2. The iterative nature of environmental risk assessments.

aspects affecting exposure can be evaluated. Protective (default) thresholds are intentionally set substantially below the expected "true" threshold to minimize Type II errors.

Thus, if the scoping or screening assessments provide insufficient evidence to dismiss stressor–receptor combinations, then refinements using more realistic assumptions are incorporated in the next tier. The intent is to reduce uncertainty as greater site- or case-specific information is incorporated into the risk assessment.[1] The test used to decide whether to continue toward a definitive risk assessment is whether there is sufficient information to make the (informed) environmental management decisions called for (Fig. 2.3).

Allard et al. (2010) have noted that hazard quotients (HQ) should be restricted to use in screening or hazard assessments of chemicals. The assumptions used are such that when a chemical exposure is compared to a toxicity threshold (e.g., TRV), that an HQ < 1 becomes a default *de minimis* value. The assumptions are intentionally skewed toward protectiveness to avoid Type II error (i.e., to conclude that there is not significant or unacceptable hazard when in fact there is). However, given the high probability of a Type I error (i.e., to conclude unacceptable risks when in fact

---

[1]Assessors and end-users of assessment conclusions should be cognizant of the possibility for greater uncertainty as refinements are introduced. This may occur particularly if the refinements impose new assumptions, as often is the case with computational models, or if new data gaps pertaining to the additional parameters reduce the number of significant figures that can be justified in the outcome.

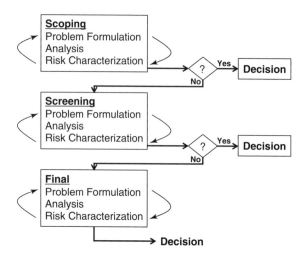

Figure 2.3. Relationship of different degrees of rigor in EcoRA and actual environmental concentrations.

there are none), HQs $\geq$ 1 typically require further studies to refine uncertainties before making risk management decisions. Hope (2008) and Allard et al. (2010) have drawn a distinction between the early stages of scoping or screening as merely being hazard assessments and not risk assessments because the hazard assessments lack both probability and magnitude terms.

But what happens if chemical stressors are not the most important factors determining the continuation of a valued ecological resource or a human community? The procedures that work quite well for known chemical stressors may fail to forewarn about companion stressors that have little or nothing to do with the chemicals: the haul roads severing the continuity of migration routes of caribou; housing developments in the winter range for elk; day-use parks that bisect the foraging range of cougar; the blockage of streams used for spawning salmon; the influx of workers who may be insensitive to the cultural ethos of an aboriginal community. The structure of risk assessments to address the narrow constructs of chemical impacts seem adequate in many instances. However, consideration of other stressors requires that we step back to see how we might improve the process—to include habitat requirements for the species of interest and to capture to proper scales of space and time so that cultural dynamics are reflected in the resulting characterizations of risk that are relevant to the affected stakeholders.

THE RISK ASSESSMENT PROCESS

Risk assessment, whether in finance, engineering, human health, ecological, or a holistic construct, is structured to evaluate one or more scenarios to predict qualitatively or quantitatively the likelihood of some event occurring. The focus may be viewed from either side of a risk–reward divide: What is the chance I may win in blackjack if I'm holding two eights? If I'm the dealer, what is the chance I will have to payout if I hit 17? A regulator mandated with protecting humans and ecological resources, but without frivolously burdening industry with sets of interventions, referees a very challenging situation of "keeping the house honest." Among stakeholders, there are some who must see both sides of the risk–reward conundrum. If more stakeholders could gain a deep-seated appreciation of the process, the uncertainties, and the certitudes, we might be able to proceed with less contentious encounters. In the descriptions of the process that follow, we attempt to illustrate the challenges faced in getting the proper focus for a risk assessment. The analysis will lead to the clear conclusion that a landscape perspective viewed through the lens of systems interactions is crucial to the success of risk assessment and environmental management.

Environmental management evokes tradeoffs; structuring the relevant scenarios quickly becomes complex as relevant scales and relationships are considered in combinatorial fashion (Table 2.2). In building the appropriate scenarios, there must be clarity of the context in terms of space and time relevant to the project. The target receptors—that is, the values to be protected—further establish the context for the assessment. Once the temporal and spatial relationships are defined and the endpoints are agreed, the pathways that affect exposures can be described.

Multiple interactions among receptors, as well as their relationships to the physical environment, make the task of choosing scales complex and complicated. All introductory biology courses before 1975 (and too many still) introduced the concept of ecological hierarchy as a linear progression. Students learned by rote that atoms make up molecules, which are organized in structures called organelles, which make up cells, tissues, organs, organisms, populations, species, communities, biomes, and finally the biosphere. Embedded in this lesson was an implicit (and sometimes explicit) assumption that information at one hierarchical level would transfer to knowledge about the next—a relationship that seldom prevails. The part that was missed in this historical litany is the realization of emerging properties—a population has characteristics that are different than the sum of individual organisms that make up that population; communities function differently than the sum of species assembled in a particular time and space. A hierarchical view directs us to examine one level above the assessment target for context and one level below for mechanisms (McCormick et al. 2004).

The reality of emerging properties along the hierarchical schema has substantive consequences in the conduct of a risk assessment and the subsequent application to environmental management. Information pertaining to the toxicity response of

Table 2.2. Choices to Be Made in Framing Scenarios to Be Evaluated

Spatial Scale	Temporal Scale	Pathways	Consequences
Site	Acute	Biotic	Organisms
Reach	Episodic	Abiotic	Populations
Watershed	Generational	Combined	Species
Region	Eral		Communities
Global			Ecological systems

individual organisms does not necessarily translate into a population-level effect. If one wishes to assess population effects, one better ask questions that are relevant to population dynamics for the species in question, and such questions must be tailored to the proper spatial and temporal scales. The general failure to structure these considerations into risk assessments could rightly be considered to be errors of omission—a situation that plagues most contemporary risk assessments.

Problem Formulation Phase

We can begin to address the concerns of affected stakeholders in an holistic fashion if we expand the focus of assessments from the very start, namely in the Problem Formulation phase, which clearly is the most critical phase in conducting an environmental risk assessment. Several "products" are generated in Problem Formulation that are achieved through iterative explorations of issues and options by affected stakeholders focused on achieving consensus throughout the process. Problem Formulation provides the initial description of the ecological system to be evaluated, along with the overall project goals, objectives, and general extent (scope or "level-of-effort") of work anticipated (Suter 1989, 1993; US EPA 1992, 2003).

There often seems great urgency to launch into the disciplinary work of a risk assessment—get into the field, get into the literature, run an exposure model. Such work tends to be done by junior staff in consulting and contract firms, whether engaged by proponents of a project or regulators. Protocols for sampling media are in place, a general budget was agreed before work began, and staff need to be kept busy. This rush to begin may well be the main reason that so many risk assessments are built on sketchy, unrelated, often irrelevant data sets. The budgets are blown before the management goals or the important questions have been formulated.

Effective risk assessments proceed through a series of actions in a deliberative, iterative fashion striving to achieve consensus (Table 2.3). On high-profile projects, there are likely to be many contentious issues to address at several of the points. Any issue that is skirted likely will emerge later in the process as a deal-breaker. As the various issues are debated, it becomes clear that senior experienced practitioners, ones empowered to make decisions on behalf of their organization, must be engaged in the discourse; and they must at least be sympathetic to holistic, integrated assessments with a landscape perspective.

Even projects that appear to be routine or repetitious of recent risk assessments should be examined anew to revisit each of the actions, if for no other reason than the realization of another project in an area may elevate affected stakeholders' concerns about cumulative effects. Whether a novel project or a familiar one, there are guiding principles to consider in problem formulation (Table 2.4).

As a general rule, both the regulators and the regulated community are wary of stakeholders. Bad experiences have had lasting influence. Yet, in the modern age with sustainability evoked as often as climate change, there has perhaps never been a more compelling situation where engagement is beneficial.

The US Presidential/Congressional Commission on Risk Assessment and Risk Management (1997) identified several benefits stemming from engaging stakeholders,

Table 2.3. Basic Actions and Points of Agreement in Problem Formulation

1. Clear articulation of specific management goals and objectives.
2. Delineation of the landscape of interest to affected stakeholders.
3. Development of site- or project-specific conceptual models that depict the
   - ecological context of the project both in terms of biotic and abiotic components
   - receptors of interest
   - primary stressors as well as ancillary stressors that are part of the landscape of interest—the driving forces within the ecological setting of the project
   - pathways from points of origin or release of stressors to point of impingement or exposure to the receptors of interest (including acknowledgement of secondary effects)
   - interactions among stressors
   - interactions among receptors
4. Explicitly state the assessment endpoints (the values to be protected).
5. Define data quality objectives—the levels of precision and accuracy needed to evaluate relationships among stressors and receptor effects.
6. Describe analytical methods and measurements to be used.
7. Produce a project-specific sampling and analysis plan.
8. Produce a project-specific quality assurance plan.

Table 2.4. Degree of Rigor and Content in Risk Assessment Tiers

- Engage disciplines in genuine dialogue at the earliest opportunity.
- Think beyond the permit—the stakeholders do!
- Anticipate closure objectives.
- Manage operations to minimize clean-up/rehabilitation activities.
- Manage landscape to convert liabilities into assets.
- Complete problem formulation activities before drafting discipline workplans.
- Get the right questions right!
- Identify interactions/synergies.
- Match data needs with decisions to be made.
- Achieve efficiency in sampling and analysis.
- Outcome determines
  - how the remainder of the assessment is conducted.
  - the robustness of the analysis.
- Should constitute at least 50% of the cost of an assessment.

including that it is consistent with democratic decision-making and ensures that public values are considered in the process. Affected stakeholders tend to improve the knowledge base upon which decisions may be made. Selfishly, good experiences from stakeholder involvement can improve the credibility of agencies managing the risks and tends to generate better acceptance of the risk management decisions. Perhaps most importantly, but also typically overlooked in our short-term budget horizon, is that effective stakeholder engagement can reduce the overall time and expense associated with managing the project to address risks. The US EPA (2001a, 2001b) promoted complementary views.

Examining a hypothetical situation (Fig. 2.4), we can perhaps understand the wariness to engage stakeholders. The line in the figures labeled "Acceptable Progress" reflects the linear thinking that is deeply embedded in the execution of engineering/construction projects so typical in environmental management. Software for managing project budgets equates time and money. And there is an expectation that for every unit of time and money spent, there should be an incremental gain in the percentage of the project completed. The contracts may even have incentives for earlier-than-expected completion, the "Idealized Progress" line with the contractor garnering a larger profit for the efficient completion of the project. But in those projects that eventually stand for public comment, new concerns unveiled and new questions raised can send the project back to the drawing board. This "common path" can lead to complete shutdown of the project or at best a serious setback leading to considerable time extensions and escalating budgets.[2] The messy job of working with

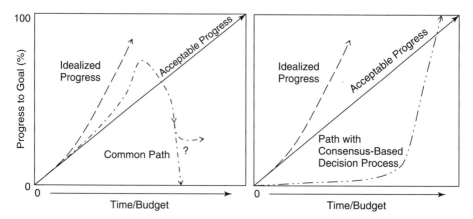

Figure 2.4. Stages in a project with and without effective stakeholder engagement.

[2]A good example is the Mackenzie Valley gas pipeline project that was proposed in the mid-1970s. The first pause in the plans to construct the 1220-km pipeline from the Beaufort Sea to Alberta's northern border came in 1977. It sat dormant until 2000, when those initially opposed to the pipeline seemed to be on board. Five years later, the proponents were threatening to withdraw the project if the negotiations

the affected stakeholder from the outset makes project managers accustomed to steady progress quite nervous. As the clock ticks and the budget is expended, there is no overt indication of progress toward completion being made.

Gordon MacKenzie (1996) portrayed the nervousness of a farm manager who observes dairy cattle between the hours of 9 A.M. and 5 P.M.—they eat and loll around, but produce nothing; he misses the milking times, when of course all of the benefits of the "off" time are realized. So it is with the stakeholder engagement. The benefits are realized when consensus is reached, when obstacles are removed, and when the stakeholders become proponents of the project. This typically occurs away from the main venues of action. It occurs quietly in communities, at family tables or diners, at council meetings, or at church gatherings. Acceptance in some cases comes simply because the stakeholders were afforded the respect of being heard; in other cases, serious concessions must be negotiated to achieve equitable tradeoffs. Regardless of the reasons for acceptance occurring, the time and money used to gain the consensus often brings it to fruition earlier than would have occurred without meaningful stakeholder involvement.

Getting started on stakeholder engagement can be challenging. Fortunately, there are proven ways of doing so. The Consensus-Based Environmental Decision-making (CBED) process is a stakeholder-empowered, community-specific process established to assess, prioritize, and select actions. The process is designed with the goal of optimizing environmental decision-making with respect to human health, ecological, sociocultural, and economic impacts (ASTM-I 2006). The most significant feature of CBED is that it begins with the selection of participants and the establishment of a Stakeholder Committee. The Stakeholder Committee then considers environmental issues from many perspectives: science, culture, economics, environmental justice, and so on. The CBED process is iterative across five main steps: (1) affected stakeholder identification and formation of the Stakeholder Committee; (2) information gathering; (3) forecasting; (4) establishment of informed consent; and (5) implementation and evaluation of initiatives.

The CBED process facilitates decision-making through negotiations among affected stakeholders with agreed (consensual) decision rules. An important part of the process is to determine and clearly communicate to all participants the rules to be followed about fairness, transparency, procedures that will be used to reach consensus, and the delineation of the ultimate decision-making authority. The CBED process is especially appropriate for situations in which there are significant public concerns about specific health, environmental, cultural, social, or economic issues **and** where there are not settled procedures for addressing controversial issues such as those surrounding the "cumulative effects" debate.

The CBED process allows the impact of any project-related or issue-related decision to be assessed. The process provides all affected stakeholders with scientific

didn't move forward quickly. As of July 30, 2009 conflicts still have the project on hold. For additional details see the following websites (accessed in July 2009): http://www.mackenziegasproject.com/ http://www.cbc.ca/news/background/mackenzievalley_pipeline/index.html http://business.theglobeandmail.com/servlet/story/RTGAM.20090330.wrenbridge30/BNStory/Business/home

and legal analyses and decision criteria that are prepared and interpreted by scientific, technical, and legal experts, as well as relevant qualitative experiential knowledge and values-based decision criteria. The Stakeholder Committee decides the relevance and importance of the criteria to the decision under consideration. In the case of issues involving regulator or proponent decisions, it is made clear from the outset that final decisions rest with the regulator or the proponent. In such cases, the other stakeholders play a strong advisory role and are not directly tasked with making the decisions. Involving affected stakeholders actively in the decision-making process reorients that process from one dominated by regulators and proponents to one that includes those who live with the consequences of the decision. This not only increases the successful implementation of decisions, but also can promote greater trust in government, industry, and other institutions.

Broad stakeholder involvement may be especially important for risk assessments concerned with issues that range over large areas and for long periods. But just as there generally has been limited effective stakeholder involvement on projects, there are several other aspects of risk assessment that also have had relatively little attention in the typical risk assessment. Due to the added importance for the landscape perspective, some elaboration is warranted here. In the following subsections, I describe aspects of conceptual models, assessment endpoints, measurement endpoints, data quality objectives, sampling and analyses plans, and quality assurance plans.

Conceptual Models. The Conceptual Model is the focal point for effective communication for all aspects of a risk assessment. As a pictorial and narrative description of the ecological setting for the project, it in effect becomes the blueprint that guides the technical work to be done in characterizing risks and connecting management options to the relevant spatial and temporal scales of the issues being addressed. The process of developing a conceptual model for contaminated sites was standardized by ASTM-I (2008). The basic steps can be translated to work for any risk assessment, including ones with a landscape perspective.

During construction of a project-specific conceptual model, affected stakeholders gain an appreciation for the scope of the project and often gain deeper understanding of the connectivity among different objectives they are seeking to achieve. As schematic summaries of the environmental resources and processes under investigation, the conceptual model illustrates the major relationships between the proposed activities and the values to be protected. Arriving at an agreed depiction of the ecological setting and the linkages of the actions and the values typically requires extensive dialogue among affected stakeholders. Common points of tension that arise relate to setting the relevant spatial and temporal boundaries for the assessment. A proposed activity that occurs over a few years may have long-lasting social and ecological consequences. Changes over a 20-year period, while generational from a human viewpoint, are instantaneous for many ecological processes. Deciding responsibility over an extended period can be quite challenging. Seemingly simple questions: "Who gains? Who pays? How much? For how long?" often become major points of contention (this series of questions is developed further in Chapter 16). Much goodwill is needed to address these questions fairly. For projects involving substantive cross-cultural differences, there are likely to be differences with respect to risk acceptance or aversion. Behaviors leading to possible exposures may require development of a conceptual model that addresses unique cultural features. Even cities with the same ethnic and cultural heritage in relatively close proximity to each other may differ widely in terms of economic base and social context in their acceptance of risk.

As the blueprint takes shape, depicting the critical dynamics occurring in the ecological setting, the management goals and the values to be protected can be honed and refined around working hypotheses and alternative scenarios. Thought experiments can be used effectively to explore *what if* questions. Settling on the specific values to be protected becomes the next major challenge.

Assessment Endpoints. The assessment endpoints explicitly detail the values to be protected (US EPA 2003). Agreeing on assessment endpoints is an interactive process that incorporates stakeholder views and technical feasibility to refine the questions. Questions about appropriateness, reasonableness, and achievability are posed and addressed. Regulatory requirements, relationships to ongoing activities, intended land uses (both current and projected), cultural setting, timeframe for completion, the environmental setting, monetary resources, natural variability of the ecological resources, and scientific or technological feasibility to assess risks are likely to be debated to some degree in the process. Operationally, this stage often begins with a broad range of diverse considerations, which are winnowed to a discrete set of concise questions to be addressed. Though many aspects of this stage are scientific in nature, non-science considerations greatly influence the character of the assessment endpoints.

As the process moves from general discussions to specifics, the assessment endpoints take shape. Characteristically, an assessment endpoint is a statement that identifies an entity, an attribute, location, and duration.[3] It should also be judged in terms of ecological relevance, susceptibility to known or potential stressors, linkage to management goals, and amenability to prediction and measurement, and it must reflect societal values (Table 2.5).

Often risk assessments are criticized for being limited to a few charismatic species and arbitrarily omitting too many others, for focusing on species unlikely to occur on the site because of habitat or range limitations, and for ignoring habitat quality in the exposure assessment. Risk assessments can be made more relevant to stakeholders, including risk managers, if greater attention is given to characterization of landscape features that influence ecological resources (Kapustka 2003, 2005; Kapustka et al. 2001, 2004; Linkov et al. 2004).

---

[3]Note that the US EPA guidance does not include location or time as requisite components of an assessment endpoint. Perhaps this is for those generic hazard assessments that are intended to be applicable anywhere. For EcoRAs and for holistic integrated risk assessments, there must be careful delineation of the relevant locale and the particular time span of interest. As part of the stakeholder dialogue, the temporal and spatial boundaries of the assessment can become critical negotiating points. The direct impact of a short-lived herbicide (e.g., chlosulfuron, half-life 6 weeks) applied to a crop could warrant a very different time span for assessment than, say, a tailings pond at a uranium mine site.

Table 2.5. Definition and examples of Assessment Endpoints

Assessment endpoints include an entity, an attribute, a location, and a time span:
- An entity (e.g., a species or population of interest)
- An attribute (e.g., number, size, rate, condition)
- A location (e.g., a specific reach of a stream)
- A time span (e.g., when does the project/activity begin and how long do we consider the consequences?)

Examples—Ecological receptors: {<u>entity</u>} {**attribute**} {*location*} {time}

- The **growth** of <u>trout</u> in *Fish Creek* over the next five years
  [An organism attribute associated with the individuals in an assessment population]
- The **productivity** of the <u>trout population</u> in *Fish Creek* over the next decade
  [A population attribute associated with an individual assessment population]
- The **average productivity** of <u>trout populations</u> in *Region Y*
  [A population attribute associated with a set of populations]

After the assessment endpoints have been drafted, the next task is to determine what information will be needed to evaluate the effect the project may have relative to the assessment endpoints. This in turn leads to questions as to how that information will be obtained, that is, what are the measurement endpoints?

Measurement Endpoints. A measurement endpoint is the categorical or quantitative expression of an observed or measured parameter and is linked directly to the assessment endpoint. For example, a school of fish may be observed directly or the effect of a substance may be evaluated through inference using toxicity tests from indicator/surrogate species and predicted exposure for the members of the population. If a suitable measurement is not available or feasible for the project, the choice of assessment endpoint should be reconsidered.

One of the largest disconnects between assessment and measurement endpoints, especially as we consider landscape perspectives, comes from failure to comprehend the consequences of changing scale. As will be detailed later (Chapters 4, 5, and 17), grain and extent and therefore resolution changes with scale. If information pertaining to type is not incorporated in a sampling effort, or if it is not discernible in *post hoc* classification of data, then there is almost certainly an unknowable uncertainty introduced. Huge costs can mount on a project, simply due to the failure to incorporate data collection efforts within the construct of iterative problem formulation. Detailed vetting of measurement endpoints is essential if the goal is to produce a technically sound risk assessment. The iterative process for selecting measurement endpoints moves forward to look at data quality objectives (DQOs) and backward to reflect on the assessment endpoints.

Data Quality Objectives. The DQOs explicitly state requirements of data, vis-à-vis measurement endpoints, in terms of measurement resolution and accuracy.

This ensures that measurements taken for a project will result in useful information to evaluate the assessment endpoints. For example, to determine whether a chemical agent is likely to cause harm, toxicity threshold values can be used to set the minimum acceptable detection limits for chemical analyses. The lowest toxicity threshold values for humans, animals, and plants may be identified from published values and compared directly to the concentrations expected or encountered during the risk assessment. The DQOs are generally established as a value lower than the respective toxicity threshold for target receptors. Where analytical or chemistry procedures allow, the DQOs are typically set at one-tenth the toxicity threshold concentration. Defining DQOs in the planning stages of a project maximizes the effectiveness and usefulness of data collected and minimizes the project cost by reducing unproductive efforts.

Once consensus is achieved for the management goals, conceptual model, assessment endpoints, measurement endpoints, and DQOs, the study plan and quality assurance plan can be drafted. As the technical details are fleshed out, it is common to revisit endpoints, precision, and accuracy. Unanticipated complexity and variability over the spatial extent of the project can make an initial plan cost-prohibitive. Recognizing the limitations of budgets may force renegotiation of some endpoints, but such renegotiation occurs from common agreements among stakeholders, not some arbitrary decision made in isolation by the party paying for the work—deciding among tradeoffs is part of the process.

Sampling and Analysis Plan. The companion activities critical to the risk assessment process (literature review, data acquisition, verification, data analysis, and monitoring studies) are incorporated into a project-specific sampling and analysis plan (SAP). New data are frequently required to conduct analyses that are performed during the risk assessment. Verification studies can corroborate the predictions of a specific risk assessment and evaluate the usefulness of various data. Epidemiological studies can provide direct evidence of effects on people or some of the more prominent wildlife species. Similarly, ecological effects or exposure monitoring can aid in the verification process and suggest ways to improve future assessment of risks to plants and animals.

A SAP includes a description of the project area of interest and a summary of key information needed to complete an environmental assessment. It contains information about the management goals and objectives, assessment endpoints, DQOs, measurement endpoints, methods, and sampling strategy. The SAP is a planning document or blueprint of the project used to explain how existing information and new data are to be obtained and analyzed. It contains all the details of what, where, when, how, and why a specified number of samples will be taken; it describes the ways that published information will be accessed; and finally it lays out the strategies and methods for modeling and statistical analyses that will be used in the project. Finally, the SAP should detail the decision criteria that will be used to select management options.

Quality Assurance Project Plans. Quality Assurance Project Plans (QAPPs) are detailed and concise narratives describing the planned efforts to ensure the quality, integrity, and accuracy of the project data (US EPA 2002). All measurements required for a project must be completed in a manner that ensures that data are representative of the samples, media, and conditions measured. Quality assurance objectives for measurement data are expressed in terms of accuracy, precision, completeness, representativeness, and comparability. The purpose of the QAPP is to establish procedures that minimize errors in sampling, measuring, and analyzing environmental situations. The document becomes a backstop for discussion involving uncertainties because the verified quality of data is useful in controlling and accounting for different types of error.

If we remain cognizant that the purpose of a risk assessment is to assist decision-makers in constructing informed choices and that the quality of the decisions ultimately will be judged by affected stakeholders, we might make a better effort to engage the stakeholders from the outset. The risk assessment process contains a mix of natural science and social science (NRC 1994). Kapustka (2001) portrayed the role of the natural sciences as less than half as important as the social sciences in problem formulation and decision-making (Fig. 2.5)—perhaps that explains why so many risk assessments, run largely by engineers and scientists, rush to the analysis and characterization phases without giving proper attention to setting the stage in problem formulation.

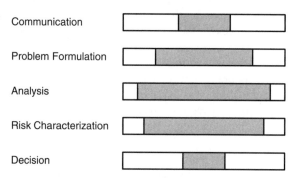

Figure 2.5. Relative importance of social and natural sciences in the components of EcoRAs.

Analysis of Exposure and Effects

By the time an assessment has entered the analysis phase, most of the critical decisions should have been made and articulated in the SAP. In analyzing either exposures or effects, the assessor proceeds through the workplan.

The analysis phase develops profiles of environmental exposure and the effects of the agent. The exposure profile characterizes the ecological systems in which the agent (a.k.a. stressor) may occur as well as the biota that may be exposed. It also describes the magnitude and spatial and temporal patterns of exposure. The effects profile summarizes data on the toxicological and ecological effects of the agents.

The analysis phase of an EcoRA consists of an evaluation of exposure to agents and effects based on the conceptual model developed during problem formulation. Both toxicological and ecological effects are considered with exposure in an iterative and integrated fashion.

Exposure. Exposure analysis describes the conditions and processes under which organisms come into direct contact with an agent. The amount of a chemical that reaches a target tissue is known as the target dose. Many physical, chemical, and biological events influence the target dose an organism receives. In soils, particle size, parent material (mineral content and structure), organic matter, ion exchange capacity, pH, complexing agents, and temperature are a few physical–chemical factors that influence dose. In aquatic systems, exposure and dose are influenced by similar analogous factors. Biological factors that influence the target dose include growth rates, metabolic rates, nutritional status, behavior, and reproductive condition.

Under conservative (protective) assumptions, exposure is often taken to be the amount of contaminant in the environment. However, the true exposure concentration or dose generally is substantially less. Toxicity tests of environmental samples have been used to estimate the effective concentration that results in an adverse response. Under field conditions, behavior and other dynamic activities can greatly influence exposure.

For chemical agents, fate and transport models are often required to estimate changes in concentration in various media (e.g., air, water, soil, or biota). These models consider chemical reactions such as photo-oxidation, biodegradation, solubility, and absorption. Whereas some chemicals degrade quickly, others may persist and concentrate in certain media, increasing the potential for exposure.

The transport and eventual fate of a chemical in the soil, water, air, vegetation, and other environmental components are estimated for different routes of exposure. Several complex media interactions can be combined in this procedure so that availability of the chemical to humans, other animals, and plants can be evaluated in a single model or as a chain of models in which the output from one model becomes the input to another. As the complexity of such models increases, there may be an expanding magnitude of error as well as expanding uncertainty. Assumptions used in constructing such models can be very difficult to trace and unwittingly can contribute to larger uncertainty.

Care also needs to be exercised with respect to significant figures when developing cascading models. Recalling that the number of significant figures is dictated by the parameter with the fewest significant figures (a lesson many scientists educated after the demise of the slide rule seem to have missed) might cause many modelers to opt for simpler, more decipherable models. If the ultimate output will be truncated at one or two significant figures, how much effort is warranted in refining some terms to four, five, or six figures?

Effects. Chemicals and other stress agents act on individual plants, microorganisms, humans, or other animals. An EcoRA usually is concerned with the aggregate of individual organism within the assessment area, but alternatively may focus on effects to population dynamics or community composition and structure as agreed in the assessment endpoints. The effects assessment can incorporate both toxicological and ecological effects.

Toxicological effects assessments for plants and animals are similar to those conducted for non-cancer health effects in human health risk assessments. Environmental concentrations or dose are related to mortality, growth, or reproduction rates. Typically, data are available for a few standard test species, which serve as surrogates for the vast array of species considered in risk assessment.

Ecological effects assessments depend upon the question being asked. If the concern is maintenance of an animal population, then population-level effects need to be measured, modeled, or otherwise estimated. This may be done by measuring demographic information such as age-specific mortality and reproduction and modeling the population growth rates. Alternatively, population density indices can be developed from field observations. Many approaches exist to assess population viability. Plant effects generally are directed at community composition and structure. Ecosystem processes and functions such as primary productivity, decomposition, or nitrogen fixation rates may be assessed.

In addition to estimating the direct effects that agents have on individuals, populations, or communities, indirect effects should also be considered. Indirect effects result from changes in a critical support component of the species' environment. For example, an animal population decline may be due to habitat loss following application of an herbicide; a fish population may decline due to siltation of spawning beds.

Each level of ecological complexity requires different approaches for assessing potential direct and indirect effects of agents. Measuring effects at one level (e.g., organism) allows for some predictive ability to the next (e.g., population), but uncertainty increases rapidly when predictions are made across many levels. Therefore, whenever feasible, effects assessments should use data that are directly applicable to the level of ecological organization of concern.

Characterization of Risk

Risk characterization integrates the exposure and effects profiles. Risk can be expressed as a qualitative or quantitative estimate, depending on available data and the data quality objectives. In this step, the risk assessor also:

- describes risk in terms of assessment endpoints;
- discusses the ecological significance or the human health consequences of the effects; and
- summarizes overall confidence in the assessment.

The risk estimates are evaluated in context of different aspects of uncertainty. Ecological risk can be expressed in qualitative terms or true probabilistic estimates of both the adverse effect and exposure elements.

The information used in assessing risk also may be qualitative or quantitative. Descriptive or narrative information is qualitative. Data in the form of a metric, with some associated relative scale, are quantitative. In the early tiers of an EnRA, qualitative data are often sufficient to answer simple, but important questions. For example, in the scoping tier such questions as "Present" or "Not Present" help focus on the important assessment endpoints. Conversely, in the definitive tier the emphasis often shifts toward quantitative data, and answering such questions as "How many more individuals" or "What is the spatial extent of the stressors" are needed to make decisions.

The risk characterization stage provides the final input to the decision-makers. Whether qualitative or quantitative, the input needs to be responsive to the overall management objectives that were articulated at the outset of problem formulation. The details that were captured in the conceptual model become part of the narrative in describing how the risks might be mitigated. What if stress is modulated for one receptor group? Will that translate into attainment of the ultimate management objectives? Could the proposed remedial action result in greater stress to the target receptors? Will the affected stakeholders benefit equitably with a certain remedial action? If the risk characterization fails to address these types of questions, it will not have done the job—an indictment all too common regarding contemporary risk assessments (NRC, 2005).

CONCLUSIONS

Risk assessment has evolved considerably over the past three decades. Yet, as we move into an era marked by growing awareness by the public of their sense of place and the impact they may have on environmental resources, it is clear that more improvements are needed. In this chapter, I have focused on the details of setting up an assessment in the problem formulation phase. The key refinements to the process that are needed are available now, but are seldom incorporated into projects. These refinements include greater dialogue with affected stakeholders, not only as a means of information exchange, but primarily to vet critical issues embedded in a landscape perspective. In particular, it is the affected stakeholders who are best positioned to identify the valued resources to be protected, though in some situations society may have imposed additional sets of values to be protected (e.g., endangered species). Iterative dialogue about valued resources leads to definitions of spatial extent and time span of interest. How different an assessment in the Northwest Territories of Canada becomes if one looks at the migration corridors of caribou and the travel distances of a workforce employed at a site; how different the assessment becomes when community well-being becomes part of the conceptual model for the assessment. How different it becomes if nonchemical stressors are included in the assessment. The iterative dialogue should also explore benefits, costs, feasibility, and effectiveness of different methods that may be used to gather, analyze, and interpret data before settling on a course of action. These and related topics are examined in subsequent chapters, especially Chapter 8 (on integrated holistic assessments) and Chapters 16, 17, and 18 (on economical ecology).

REFERENCES

Allard P, Fairbrother A, Hope BK, Hull RN, Johnson MS, Kapustka L, Mann G, McDonald B, Sample BE. 2010. Recommendations for the development and application of wildlife toxicity reference values. *Integr Environ Assess Manage* **6**:28–37.

ASTM-I (American Society for Testing and Materials—International). 2006. E2248-06 standard guide for framework for a consensus-based environmental decision-making process. In *The ASTM Annual Book of Standards*. ASTM-I, West Conshohocken, PA.

ASTM-I (American Society for Testing and Materials—International). 2008. E1689-08 standard guide for developing conceptual models for contaminated sites. In *The ASTM Annual Book of Standards*. ASTM-I, West Conshohocken, PA.

Froese K, Ornstein M. 2010. Integrating health in environmental risk assessments. Chapter 8, this volume.

Gladwell M. 2005. *Blink: The Power of Thinking Without Thinking*. Back Bay Books, Little Brown and Company, New York.

Hope BK. 2009. Will there ever be a role for risk assessments? *Human Environ Risk Assess* **15**:1–6.

Kapustka L. 2003. Rationale for use of wildlife habitat characterization to improve relevance of ecological risk assessments. *Human Ecol Risk Assess* **9**:1425–1430.

Kapustka LA. 2005. Assessing ecological risks at the landscape scale: Opportunities and technical limitations. *Ecol Society* **10**:11. [Online] URL: http://www.ecologyandsociety.org/vol10/iss2/art11/

Kapustka LA. 2006. Current developments in ecotoxicology and ecological Risk assessment. In Arapis G, Goncharova N (Eds.), *Ecotoxicology, Ecological Risk Assessment, and Multiple Stressors*. Kluwer Press, The Netherlands, pp. 3–24.

Kapustka LA, Galbraith H, Luxon M, Yocum J. 2001. Using landscape ecology to focus ecological risk assessment and guide risk management decision-making. *Toxicol Indust Health* **17**:236–246.

Kapustka LA, Galbraith H, Luxon M, Yocum J, Adams B. 2004. Application of habitat suitability index values to modify exposure estimates in characterizing ecological risk. In Kapustka LA, Galbraith H, Luxon M, Biddinger GR (Eds.), *Landscape Ecology and Wildlife Habitat Evaluation: Critical Information for Ecological Risk Assessment, Land-Use Management Activities, and Biodiversity Enhancement Practices*. ASTM-I STP 1458, American Society for Testing and Materials-International, West Conshohocken, PA, pp. 169–194.

Kapustka L, McCormick R, Froese K. 2008. Social and ecological challenges within the realm of environmental security. In Linkov I, Ferguson E, Magar VS (Eds.), *Real-Time and Deliberative Decision Making*. Springer, The Netherlands, pp. 203–211.

Kelly KE. 1991. The myth of 10^{-6} as a definition of "acceptable risk." Presented at the *84th Air & Waste Management Association Meeting*, Vancouver, BC. June 16–21, 1991, 91–175.4, 10 pp.

Linkov I, Kapustka LA, Grebenkov A, Andrizhievski A, Loukashevich A, Trifono A. 2004. Incorporating habitat characterization into risk-trace: Software for spatially explicit exposure assessment. In Linkov I, Ramadan A (Eds.), *Comparative Risk Assessment and Environmental Decision Making*. Kluwer Press, The Netherlands, pp. 253–265.

MacKenzie G. 1996. *Orbiting the Giant Hairball: A Corporate Fool's Guide to Surviving with Grace*. Penguin Putnam, New York.

McCormick RJ, Zellmer AJ, Allen TFH. 2004. Type, scale, and adaptive narrative: Keeping models of salmon, toxicology, and risk alive to the world. In Kapustka LA, Galbraith H, Luxon M, Biddinger GR (Eds.), *Landscape Ecology and Wildlife Habitat Evaluation: Critical Information for Ecological Risk Assessment, Land-Use Management Activities, and*

Biodiversity Enhancement Practices. ASTM-I STP 1458, American Society for Testing and Materials—International, West Conshohocken, PA, pp. 69–83.

NRC (National Research Council). 1983. *Committee on the Institution of Means for Assessment of Risks to Public Health. Risk Assessment in the Federal Government: Managing the Process*. National Academy Press, Washington, DC.

NRC (National Research Council). 1994. *Science and Judgment in Risk Assessment*. National Academy of Sciences, Washington, DC.

NRC (National Research Council). 2005. *Superfund and Mining Megasites: Lessons from the Coeur D'Alene River Basin*. National Research Council of the National Academies, The National Academies Press, Washington, DC.

Schwabe CW, Riemann HP, Franti E. 1977. *Epidemiology in Veterinary Practice*. Lee and Febiger, Philadelphia.

Suter GW. 1989. Ecological endpoints. In Warren-Hicks W, Parkhurst BR, Baker SS (Eds.). *Ecological Assessments of Hazardous Waste Sites: A Field and Laboratory Reference*. EPA/600/3-89/013. United States Environmental Protection Agency, Washington, DC, pp. 2–1 to 2–28.

Suter GW. 1993. *Ecological Risk Assessment*. Lewis Publishers, Boca Raton, FL.

Suter GW. 2008. Ecological risk assessment in the United States Environmental Protection Agency: A historical overview. *Integr Environ Assess Manage* **4**:285–289.

US EPA (United States Environmental Protection Agency). 1992. *Framework for Ecological Risk Assessment*. EPA/630/R-92/001, Washington, DC.

US EPA (United States Environmental Protection Agency). 1998. *Guidelines for Ecological Risk Assessment*. EPA/630/R-95/002F, Washington, DC.

US EPA (United States Environmental Protection Agency). 2001a. *Improved Science-Based Environmental Stakeholder Processes*. EPA-SAB-EC-COM-01-006.

US EPA (United States Environmental Protection Agency). 2001b. *Understanding Public Values and Attitudes Related to Ecological Risk Management: An SAB Workshop Report of an EPA/SAB Workshop*. EPA-SAB-EC-WKSP-01-001.

US EPA (United States Environmental Protection Agency). 2002. *Guidance for Quality Assurance Project Plans*. EPA/40/R-02/009, Washington, DC. Also see http://epa.gov/quality/qs-docs/g4-final.pdf.

US EPA (United States Environmental Protection Agency). 2003. *Generic Ecological Risk Assessment Endpoints (GEAEs) for Ecological Risk Assessment*. EPA/630/P-02/004F, Washington, DC.

US EPA (United States Environmental Protection Agency). 2006. *Guidance for Systematic Planning Using the Data Quality Objective Process*. EPA/240/B-06/001, Washington, DC. Also see http://epa.gov/quality/dqos.html.

US Presidential/Congressional Commission on Risk Assessment and Risk Management. 1997. Final Report, Vols. 1 and 2, Washington, DC.

3

POPULATIONS, HABITAT, AND ECOLOGICAL SYSTEMS: ELUSIVE BUT ESSENTIAL CONSIDERATIONS FOR A LANDSCAPE PERSPECTIVE

Lawrence A. Kapustka

In this chapter I explore three elusive concepts that are central to our ability to evaluate and manage ecological risks. I begin with the challenges of population approaches, move on to habitat characterization, and conclude with some limitations in our knowledge of ecological systems.

POPULATIONS

The term *population* has been used in many ways. In a nonexhaustive listing of different definitions Menzie et al. (2008) cited a score of definitions of the term, albeit with much overlap. If consideration of *meta-populations* were included, the list would grow even longer. With many different nuances regarding population, misunderstandings are sure to rise, especially when the issues are transported into legal and regulatory usage. To finesse the regulatory conundrum, the US EPA (2003) introduced a variant, the *assessment population*, for use in risk assessments, namely "those members of a species residing, foraging, or otherwise using the specific area of interest in the assessment."

Choosing to focus on the "assessment population" can change the spatial extent of the assessment landscape. Traditionally, at least within the Superfund Program in the United States, the boundary of a site was considered to be the practical maximum

Environmental Risk and Management from a Landscape Perspective, edited by Kapustka and Landis
Copyright © 2010 John Wiley & Sons, Inc.

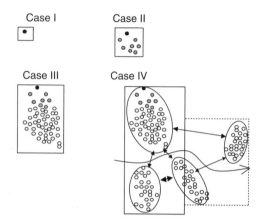

Figure 3.1. Representation of the assessment population concept relative to meta-populations. The solid lines represent assessment boundaries defined by the extent of the chemical of concern. Proceeding from Case I through IV, the relative positions of individual organisms of the population are identifiable by position. The dashed outline in Case IV illustrates the assessment boundary expanded to circumscribe the subpopulations that are connected to the core area [After Fig. 3.3 in Barnthouse et al. (2008).]

extent of the chemicals of concern.[1] This policy directed that an assessment could, and often did, extend beyond the property boundaries of the probable responsible party. Defining the site as the extent of contamination provides administratively manageable units. However, such operational units seldom have obvious connection to ecologically relevant structures that correspond to vegetation type or wildlife use patterns. Let's look at four cases representing assessment population boundaries (Fig. 3.1):

Case I. The assessment population theoretically could be limited to a single organism of the species in question, a situation that could occur if the species of interest has a large home range relative to the contaminated area.

Case II. More typical, may circumscribe of a number of organisms, but is not so large as to encompass a biological population.

Case III. Contains an entire biological population of the species of interest, a situation that may occur if the site is large and the species of interest has a relatively small home range.

Case IV. If the assessment population is recognized to exist as a collection of subpopulations (i.e., a meta-population), then the area requiring assessment extends beyond the zone of contamination, representing an increase in the area of interest over the common practice.

[1]The interpretation of practical in some cases was related to concentrations of the chemical of concern that could be detected above background concentrations or above toxicity threshold concentrations.

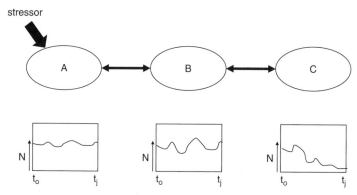

Figure 3.2. Relationship between stressor impingement point and meta-population responses. Note the trend lines of subpopulation numbers (N) depicted below the respective subpopulations (ovals). [Adapted from Spromberg et al. (1998) and Macovsky (1999).]

Work by Landis and his students (Spromberg et al. 1998, Macovsky 1999) examined the rationale for expanding the area of evaluation to encompass the meta-population and further argued that the population is the appropriate unit for species-specific risk assessments (Landis 2002). Their work demonstrated that the subpopulation farthest from the point of impingement of a stressor is the first to exhibit a decline in number and is the most likely to be extirpated (Fig. 3.2). To appreciate the importance of this finding, consider a typical hazardous waste site investigation in which data on population size were collected at the site (for example, Case III, as illustrated in Fig. 3.1) and for comparative purposes, comparable data were collected at a "reference site" (for example, any of the outlying areas supporting a subpopulation, as illustrated in Case IV in Fig. 3.1). A likely, though unwittingly erroneous, conclusion could be as follows[2]:

> Obviously the chemicals of concern are not harmful as there are more animals at the contaminated site than the uncontaminated reference site.

The way one defines population has other consequences in several aspects of risk assessment. For example, if one chooses to use a collection of individual organisms as the target for assessment, then traditional toxicity data may be used to describe likely effects of a statistical population. However, the statistical population does not possess the emergent properties of a biological population. Therefore, an unknowable degree of uncertainty is incorporated into the risk that is characterized.

---

[2]The dynamics of meta-populations is one strong reason, though not the only one, that the common practice of using reference sites should be abandoned. [See Landis and Yu (2004) for a cogent critique of the reference site concept and the recommendation to use gradient analyses or to define reference conditions that can be evaluated relative to the site under investigation.] Unfortunately, some regulatory jurisdictions have codified requirements to establish a reference site as part of the site assessment process. When confronted with a regulatory requirement for a reference site, risk assessors should strive to negotiate the removal of the requirement for the project; better yet seek amendment of the requirement.

Ignorance of the dynamics of the system breeds false confidence regarding likely consequences. What we don't know, or choose not to acknowledge, can have significance. If we miss the underlying population dynamics in crafting the risk management responses, we will not be able to evaluate the influence of the management decisions, because we will not know what to measure or how to interpret the data that are obtained. Ignoring population dynamics also constrains any ability to evaluate cumulative effects that multiple stressors may have in a region.

There are tradeoffs as well in defining the target population for assessment as a biological population. For starters, it is no easy task to define the relevant limits of a population, especially across landscapes devoid of sharp boundaries.[3] Operationally, where would a probable responsible party's liability end if the species of interest were a neotropical migrant? Even if reasonable spatial boundaries could be agreed, obtaining good measurements of population sizes and rates of change can be quite challenging and costly. For cryptic species, getting such data can be impossible or essentially so.

Another confounding aspect in assessing biological populations is that much of the information from toxicity tests has limited relevance to populations. This is a direct result of the assumptions and restrictions incorporated into toxicity tests. The toxicity information can be irrelevant because the conditions of the tests, generally focused on individual organisms, do not account for dynamic relationships that occur in populations. One example: Death of some members of a population is often compensated by increased fitness of the surviving members. This may occur because more food may be available and so fecundity may increase, birth or hatch weights of offspring may be larger, and survival of offspring may increase. Though we are conditioned to think of death as a more substantive endpoint than say a slight impairment of neurological function, the response at the population level can be just the reverse. The explanation for this seeming contradiction is that the slightly impaired individuals drawing upon limited resources may lower the collective fecundity of the population, whereas the population that had been reduced in number by the death of some members may experience increased reproductive success.

Obviously, there are some tough questions remaining to be asked and answered to overcome some of the limitations of population-level risk assessments. Nevertheless, some progress is being made. The rationale and approaches for performing population-level risk assessments is established in several chapters of Barnthouse et al. (2008).

---

[3]History of science buffs will recall the entertaining though intense debates of the 1960s between the Wisconsin (Grant Cottom and John Curtis) and Washington State (Rexford Daubenmire) schools regarding the distribution of plant communities. The arguments hinged on the assertion of gradients of environmental variables (Wisconsin) that may be discontinuous along the physical landscape versus sharp boundaries (Washington) observable in mountainous terrain and other locales that may have abrupt changes in soil type for example. Barbour (1996) captures the essence of this debate that began quietly in the 1950s. A series of commentaries in *Science* following the publication by Daubenmire (1966) revealed the intensely caustic disagreements that prevailed between the two schools that really began with Frederick Clements and Henry Gleason in the first two decades of the 20th century. Also see Nicolson and McIntosh (2002) for a historical analysis of Gleason's Individualistic concept and refer back to Chapter 1 for additional references on this topic.

Making this transition from the more common organism-level risk assessments will not be easy or without a new set of limitations. However, population-level risk assessments should be more defensible than strictly organism-level assessment because practitioners and risk managers alike would become more aware of the dynamics underlying their conclusions and management actions. The process and the risk management arena would reflect ecological realities to a greater extent than typical of organism-level assessments.

HABITAT

Habitat, like population, can be a fuzzy concept relating to broad categories such as "forest habitat" or "coastal habitat," but it can also refer to narrower constructs such as "elk habitat" or "black-capped chickadee habitat." In these examples, the term may be referring to a biome, an ecoregion, or the specific environmental conditions favorable to a single species. In its loosest usage, habitat is merely a place where organisms may reside.

Historically, political interest in biogeography was spurred by desires to catalogue resources of the New World (including the infamous trek of Lewis and Clark and their entourage of 1803–1804 along well-established aboriginal routes) and to attract settlers into the potentially arable lands of the North American continent. Predicting the potential of the landscape in various regions took on both practical and academic import.

The earlier work of Carl Ludwig Willdenow at the close of the 18th century, along with the work of his protégé Alexander von Humboldt, laid the foundation for correlating patterns of plant adaptations and vegetation types to a particular climate. Advancements in phytogeography by Clinton Hart Merriam at the close of the 19th century refined the concept as he described the patterns of vegetation defined by elevation, latitude, and aspect in his Lower Sonoran, Upper Sonoran, Transition, Canadian, Hudsonian, and Arctic-Alpine zones. Vegetation type could be predicted in broad terms from a few environmental variables. Forests, grasslands, and other vegetation types were mapped in part to characterize resources for timber sales or agricultural potential (see Barbour et al 1987, Comer et al. 2003).

As with all mapping, the final product depends on many choices related to objectives, scale (grain and extent), resolution, and definitions of type. If the map is intended to portray resources of a continent or smaller subdivisions for planning over several decades, does one illustrate what is there now or what one presumes will be there in the future? Does one illustrate the cleared land after a harvest or, alternatively, illustrate the forest that will surely re-grow? Kuchler (1964) opted to map the "potential natural vegetation" of the conterminous United States. Here the choices were to exclude all urban areas, agricultural lands, and the like, substituting in their stead the vegetation type that had previously occupied the space. But the emphasis on "potential" transformed the map from a historical document to one of supposition based on prevailing views of successional trajectory. Kuchler's successional trajectories were heavily influenced by mono- and poly-climax concepts of Clements (1916) and Daubenmire

(1952). Thus with respect to plant cover types, habitat became an admixture of climate (temperature, rainfall, evaporative potential), soil, elevation, and aspect—all tempered with ones assumptions of succession. Similar approaches to that of Kuchler evolved in the development of ecoregion maps intended to inform land management activities (Omernik 1987, 1995, Bailey 1998) and biophysical depictions in Canada (see http://www.env.gov.bc.ca/wld/catalogue/class_interp.html accessed July 16, 2009, http://www.geology.gov.yk.ca/biophysical_mapping.html accessed July 16, 2009).

Currently, mapping of resources has taken on the role of identifying spatial distributions of species, communities, and ecosystems to guide efforts in the conservation of biodiversity. Here the greatest challenge seems to be in finding the middle ground between fine-grained depictions of ecological communities and coarse-grained ecoregion units. It is this mid-range that is most critical to land management activities (see Comer et al. 2003).

As emphasized in the discussion on risk assessment (Chapter 2), defining the objectives of the assessment has paramount importance. Deciding the appropriate spatial and temporal scale requires careful consideration of the purpose and level of effort to be undertaken. But perhaps most important to the whole effort, there needs to be common understanding that information from one scale cannot readily be translated to substantially finer or coarser grain with any great utility. The plants and animals occupying a small hazardous waste site may bear little resemblance to the assemblages that define the prevailing conditions of the ecoregion in which the site is found. Thus as is emphasized in several places in this book, the local conditions, bounded by the focus of the assessment and determined in large part by the species of interest, dictate which data will be useful as well as how the information should be evaluated.

A general definition of habitat is "the collection of biological, chemical, and physical features of a landscape that provide conditions for an organism to live and reproduce" (ASTM-I 2004). This definition relates to "n-dimensional niche space," an observation that organisms have acclimated (physiological coping mechanisms) and adapted (evolved) to particular combinations of environmental features (Whittaker 1975). Consequently, a given plant species grows better in certain soils with particular pH range, in a particular micro-climate, and so on. Animals are drawn to suitable physical structure and food availability, while avoiding areas of lower quality. In addition, the preferred habitat characteristics are likely to shift with life stages and with seasons.

A given area may be excellent for one species and uninhabitable for another. Consider the different preferences/requirements of ruffed grouse and burrowing owl; yak and gazelle; large-mouth bass and rainbow trout; chickadee and meadow lark; black spruce and red oak. A location that is ideal for a species during one season may be inhospitable during different seasons—for example, the high arctic for many waterfowl in summer with the long days and abundant food versus the brutal weather of winter; or the bulging mountain streams of the Pacific Northwest during the spring melt for Chinook salmon versus the shrunken and warm waters of late summer.

There are differential area use rates by different species. Animals are drawn to particular features of the landscape for foraging, loafing, nesting/birthing, and so on. Some species are attracted to disturbance zones and edges, but others avoid such areas. Resting and feeding areas are often located apart from each other. In such cases, the ideal habitat for a species must have heterogeneous landscape features in close proximity [e.g., northern bobwhite quail (Schroeder 1985a) eastern wild turkey (Schroeder 1985b), moose (Allen et al. 1987)].

Vegetation on a site is determined by the soil type, nutrient status, pH, organic carbon content, drainage, slope, elevation, aspect, and history of biotic interactions (e.g., grazing, fire, or other determinants of successional trajectory). In turn, the type and condition of the vegetation along with other landscape features such as the availability of food items, combined with various physical components, determine the quality of habitat for a particular wildlife species in terrestrial settings; similar parameters determine habitat quality for aquatic animals.

The concept of habitat assumes an underlying scale of space and time. For these reasons, clarity in communication about habitat condition, status, or successional trajectory requires the descriptions of habitat to be coupled with the species of interest, the location, and the period of time. Consequently, when referring to different quality of habitat at a particular location (*good, poor, degraded*, or *improved*), the term begs for clarification: Good for which species? In which season?

Efforts to bring habitat considerations into ecological risk assessment have been promoted since the mid-1990s (Freshman and Menzie 1996, Rand and Newman 1998). Reasons to consider habitat requirements of the species of interest were to take into account spatial differences in contaminant concentrations that typically occur across foraging areas and to relate the proportion of a local population likely to be exposed to the contaminants. The importance of this refinement is linked directly to exposure assessment. For example, if landscape characteristics deter animals from foraging in the areas where toxicants occur, then the exposure pathway may be nonexistent. Alternatively, if the landscape features attract wildlife to where the toxicants occur, then one has the prototypical "attractive nuisance."

Several exploratory studies using hypothetical placeholder values for habitat quality were made to illustrate the relative importance of habitat as a modifier of exposure calculations (Hope 2000, 2001, 2004; Wickwire et al. 2004, Linkov et al. 2001, 2002, 2004). Kapustka (2003, 2005), Kapustka et al. (2001, 2004), and Linkov et al., (2004) advance the idea of using habitat suitability index (HSI) values as the habitat quality parameter to modify exposure calculations.

Kapustka et al. (2001) identified five of 12 possible scenarios in which an ecological risk assessment could benefit by use of habitat characteristics or preferences of the species of interest (Fig. 3.3). The scenarios that qualify for habitat considerations are situations where the landscape is heterogeneous with respect habitat parameters or where the distribution of stressors (chemicals) is patchy. The size of the site relative to home range or foraging range of individuals of a species is an important consideration as to whether habitat characteristics should be incorporated into exposure assessments (Carlson et al., 2004). Though precise determination of foraging range for most wildlife is difficult if not elusive, intuitively we know that there is some minimum area required before habitat characterization is warranted.

There are three different ways of characterizing habitat quality that range in degrees of sophistication: (1) qualitative categories, (2) semiquantitative indices, and

spatial relationship				
habitat	homogeneous	homogeneous	heterogeneous	heterogeneous
agent	homogeneous	heterogeneous	heterogeneous	homogeneous
	O	O	+	O
	O	O	+	O
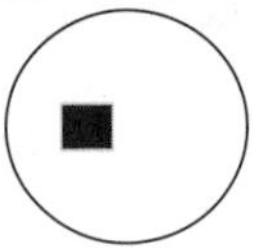	O	O	+	+

Figure 3.3. Contingency table illustration relationships of home range (white circle) relative to site size (black square)—cases where habitat characterization may be useful in reducing uncertainty of exposure estimates (+) and cases where habitat considerations may be moot (O). [Adapted from Kapustka et al. (2001).]

(3) quantitative models. Which avenue to take in selecting the different habitat characterization options (if any) should be explored rigorously in problem formulation: first, determining if habitat considerations would be useful and cost-effective; second, iteration of data quality objectives, measurement endpoints, and management goals.

- If categorical classes of habitat quality are sufficient for the problems being addressed, then two simple categories of "suitable = 1" and "unsuitable = 0" may be used.
- If there is more knowledge of the system, more classes could be devised with operational definitions for "good" (2), "okay" (1), and "poor" (0) could be used.
- If greater rigor in characterizing habitat is warranted, then structured semiquantitative approaches such as HSI models could be used.

Characterization of habitat for certain species was formalized by the US Fish and Wildlife Service in the 1990s (Schroeder and Haire 1993, US FWS 1981).[4] Currently, there are more than 160 HSI models published, though usage is limited for quantitative

---

[4]The HSI has a minimum value of 0.0, which represents unsuitable habitat, and a maximum value of 1.0, which represents optimal habitat. An HSI model produces an index value between 0.0 and 1.0, with the assumption that there is a relationship between the HSI value and carrying capacity.

predictions of population densities (Terrell and Carpenter 1997). Kapustka et al. (2001, 2004) compiled data from the terrestrial HSI models. They described how the data from the various models could be used in selecting the species to be assessed either directly or as surrogates, how the data requirements for calculating indices could feed into the data quality objectives process, and finally how the HSI could be used to modify exposure estimates [see discussion below and also ASTM-I (2004)]. Alternatively, detailed site characterization of population density and demographic structure may be incorporated into population matrix models or various multiple regression models to establish greater ecological realism.

Landscape relationships have been considered in efforts to incorporate ecological dynamics into risk assessments by another group within the US EPA. The Program to Assist in Tracking Critical Habitat (PATCH) model uses a GIS platform that allows user input in defining polygons and their characteristics (Shumaker 1998; www.epa.gov/wed/pages/models.htm).

Proof of concept work regarding spatially explicit exposure models such as SEEM (Wickwire et al. 2004), which is a component of the U.S. Army Risk Assessment Modeling System (ARAMS) (www.wes.army.mil/el/arams/arams.html) and Risk-Trace (Linkov et al. 2004a, 2004b), have been developed. These models simulate the stochastic movement of one or more receptors across the landscape in relation to the quality of the habitat for the species. Predicted dietary exposure concentrations are compiled using assumptions of feeding rates and contaminant concentrations across the foraging range of the individuals. These models currently are restricted to calculating risk quotient (RQ) or hazard quotient (HQ) based on point estimates of toxicity (e.g., toxicity threshold values) and as such are limited to screening-level assessments (following arguments by Hope 2009, Allard et al. 2010). However, it should be possible to modify the models so that full toxicity response relationships could replace the point estimate comparisons.

Johnson et al. (2007) conducted field tests to evaluate the predictions of SEEM. They compared outputs of traditional deterministic exposure models with a spatial model and compared the results of both with blood-lead concentrations from songbird species at two small arms range complexes. Lead concentration in soils, food items, and feces were compared with lead in blood and feathers of adults and nestlings. The spatially explicit model SEEM more closely predicted the observed concentrations of lead in birds, whereas the conventional deterministic methods estimated 10-fold higher concentrations of lead in receptors.

Minor modifications of the ecological risk assessment process allow incorporation of habitat quality as a modifier of exposure estimates. In general, this modification entails substituting a measure of habitat quality for the commonly used area use factor (AUF). Two situations dictate the manner in which habitat is used in the exposure estimates [from Kapustka et al. (2001), incorporated into ASTM-I E2385 (2004)]:

- **Situation a**—for species with relatively small home ranges; estimating the numbers of animals in each habitat subdivision or polygon delineated for the project.

The number of animals likely to use an area is a function of their social organization, their territory or home range sizes, and the quality of the habitat. For any species, the density of individual organisms will vary across a site, depending on the spatial variation in habitat quality, with higher densities (with smaller home ranges) inhabiting areas of higher habitat quality. The number of organisms of a given species that are likely to occupy any habitat subdivision can be approximated using the area of the subdivision, the HSI score for that subdivision, and information on either the range of the animals' density or home range sizes, as illustrated in the following equations:

$$\text{For use with home range data}: \quad N_s = \frac{A_s}{HR_s} \tag{1}$$

$$\text{For use with density data}: \quad N_s = A_s \times CC_s \tag{2}$$

where N_s is the number of individual organisms likely to inhabit the subdivision, A_s is the area of the subdivision, HR_s is the approximate home range size of the animals within the subdivision, and CC_s is the approximate carrying capacity of the subdivision (carrying capacity is an estimate of expected density).

HR_s in this equation is a function of the HSI score for the polygon and the approximate home ranges used by the species. There is an assumption that animals inhabiting zones with moderate quality habitat (scoring about 0.5 on the HSI scale) will have home ranges and densities that approximate the central tendency. Conversely, areas that score closer to the two extremes (0.0 and 1.0) will have home ranges and densities that are closer to the minimum and maximum, respectively.

- **Situation b**—for species with relatively large home ranges; estimating the proportion of time animals would spend in each area of contamination.

The proportion of its time that a wide-ranging organism is likely to spend on a site will be a function of the size of the site relative to its home range requirements, the quality of the habitat on the site relative to its surroundings, and the rate at which habitat quality may change through time.

If the contaminated site's habitat quality is approximately equal to that of the site surroundings, the proportion of time that an animal will spend on the site can be estimated using a simple proportion calculation [if the site is roughly 25% of the animals home range, then it will spend approximately 25% of the time there and obtain 25% of its diet (or incidental ingestion) there].

If the habitat on the site is of lower or higher quality than the surroundings, the animal is likely to spend proportionally less or more of its time there respectively. However, the relationship between habitat quality and use in a wide-ranging organism is unlikely to be linear. No matter how good the habitat on the site is, it is unlikely that the organism will be able to obtain all its life-history requirements there—it will be predisposed to spend some of its time in the surroundings. Also, an efficient organism living in an area of temporally varying habitat will probably track shifts in habitat quality and thus visit all parts of its home range, at least intermittently.

Once again, this time allocation may be estimated using HSI scores, except that here an HSI score that represents the habitat quality in the areas that the organism may use off-site is required. By comparing the size of the site relative to the animals' home range and the habitat quality on- and off-site or within any polygon of the site, the approximate time allocation can be estimated, as illustrated in Eq. (3):

$$P_s = \frac{A_s/HR_s}{\sum_{s=1}^{n} (A_s/HR_s)} \tag{3}$$

where: P_s is the proportion of time spent foraging in sub-area s, A_s is the area of sub-area s, and HR_s is the home range size associated with habitat quality in sub-area s.

As in Eq. (1), HR_s in this equation is a function of the HSI score for the polygon and the approximate home ranges of the species. The assumption is that animals inhabiting zones of medium-quality habitat (scoring about 0.5 on the HSI scale) will have home ranges that approximate the central tendency of home range of the species. Conversely, areas that score closer to the two extremes (0.0 and 1.0) will have home ranges that are closer to the minimum and maximum, respectively.

As in the analysis phase, risk characterization calculations also differ according the relative size of home ranges of the assessment species and contaminated areas. The two alternative approaches follow.

Determining Risk for Species Relatively Small Home Ranges

Once the density of animals in each polygon is determined, the proportion of the site population exposed at contaminant concentrations higher than acceptable levels can be easily determined. The appropriate contaminant concentration in each habitat subdivision is input into individual based wildlife exposure models to characterize exposure in each habitat subdivision. The proportion of the population at risk is then determined by summing the number of individuals in polygons where exposure is above an acceptable threshold, then dividing this number by the total number of individuals in all sub-areas.

Estimating Exposure to Organisms with Relatively Large Home Ranges

An HSI is a numerical index that represents the capacity of a given habitat to support a selected fish or wildlife species. For habitat evaluation, the value of interest is an estimate or measure of habitat condition in the study area, and the standard of comparison is the optimum habitat condition for the same evaluation species. Therefore,

HSI = (Study area habitat conditions)/(Optimum habitat conditions)

HSI models can be used in cases where the required output is a measure of the probability of use of an area either by individual organisms or by a population. In applying organism HSIs to exposure assessment for animals with exclusive or non-exclusive home ranges larger than the contaminated site, HSI output should be viewed as a measure of the probability of an individual using a given subsection of its home range. For this type of application, output from the HSI model can be used to estimate the proportion of its time that an individual will spend exploiting a given area within its home range. For organisms with home ranges smaller than the contaminated area, the HSI of a section of the site may be treated as a surrogate for carrying capacity or population density. For organisms with large home ranges the proportion of time that the organism spends on the site is incorporated as an area use factor (AUF) in the dietary exposure equation as follows:

$$\text{ADD}_{\text{pot}} = \sum_{s=1}^{m} P_s \left[\sum_{j=1}^{n} (C_{js} \times \text{FR}_{js} \times \text{NIR}_j) + (D_s \times \text{FS} \times \text{FIR}_{\text{total}}) \right] \quad (4)$$

where ADD_{pot} is the Potential average daily dose, P_s is the AUF; the proportion of time spent foraging in sub-area s [Eq. (2)], C_{js} is the average concentration of contaminant in food type j in sub-area s, FR_{js} is the fraction of food type j contaminated in sub-area s, NIR_j is the normalized ingestion rate of food type j, D_s is the average contaminant concentration in soils in sub-area s, $\text{NIR}_{\text{total}}$ is the normalized ingestion rate summed over all foods, and FS is the Fraction of soil in diet.

The AUF is incorporated as

$$\text{ADD}_{\text{pot}} = \sum_{s=1}^{m} P_s \left[\sum_{j=1}^{n} (C_{js} \times \text{FR}_{js} \times \text{NIR}_j) + (D_s \times \text{FS} \times \text{FIR}_{\text{total}}) \right] \quad (5)$$

where P_s is the AUF; the proportion of time spent foraging in sub-area s [Eq. (3)], C_{js} is the Average concentration of contaminant in food type j in sub-area s, FR_{js} is the Fraction of food type j contaminated in sub-area s, and D_s is the average contaminant concentration in soils in sub-area s.

ECOLOGICAL SYSTEMS

If populations and habitat are challenging concepts, then beware of the dynamics of ecological systems. Here, I highlight three inherent limitations faced in our efforts to evaluate and manage risks of ecological systems, namely stochasticity, elusive baselines, and unknowns. These are drawn from earlier papers by Kapustka (2006, 2008).

Stochasticity

May's (1976) examination of patterns in population dynamics moved ecological analysis into a new era. May's observations echoed contemporary discoveries in meteorology and fluid dynamics and launched a new way of thinking about nature (Gleick

1987).[5] This new science of chaos demanded a different realm of predictions. Previously, scientists assumed that models could predict outcomes accurately through a series of equations. Errors in predictions were thought to be due only to imperfect knowledge, correctable by adding one more coefficient, one more term, or one more equation. The sea change in analytical thought caused by chaos theory is that we can no longer expect to predict the precise outcome of a population, a community, or an ecological system's functional processes; rather we need to express possible outcomes as conditional probabilities.

Ecological systems are stochastic. Slight variations in the initial conditions of a population, along with the magnitude of stressors (such as toxicants that alter survival or fecundity) acting on that population, will result in several possible outcomes—predictions about each possible outcome are thus expressed as a probability.[6] When evaluating predictions developed in the context of an ecological risk assessment, we might now conclude the following:

- **Case 1**. Population A exposed to stress X is predicted as having an 80% chance of extirpation over 10 years.
- **Case 2**. Population B exposed to stress Y is predicted as having a 10% chance of extirpation over 10 years.
- **Case 3**. Stress Z is predicted to have no adverse effect on population C.

As risk assessors, we are now faced with the reality that certain uncertainties are certain—we cannot know precisely how a particular population will respond. This introduces several challenges regarding communication of risk predictions, as well as being prepared for the wrath that is sure to come when the population responds against the odds.

For example, what if in

- Case 1, after 20 years, population A is thriving;
- Case 2, within 1 year, population B is extirpated; and
- Case 3, population C declines 20% per year over 4 years?

Just as meteorologists express the likelihood of rain occurring in a specified area and a specified time, risk assessors can only express the likelihood that a particular outcome will occur. Similarly, in ecology, we need to communicate the boundaries of our predictions. This requires both a degree of humility and a need to educate non-science audiences about the predictions. Unless the probability of an outcome is 0 or 1.0, any of the possible outcomes may occur at any given time. It is the occurrence of the rare, even mostly unexpected, event that can have profound influence on a system as so eloquently portrayed in *The Black Swan* (Taleb 2007).

---

[5]Barbour (1996) attributed the roots of a major paradigm shift toward Gleason's Individualistic view (from the organismal or super-organismal view of Clements) during the post–World War II period in the 1950s. It is possible that without the quiet transition to individualism, May's work may have been less likely.

[6]See Chapter 13 in Landis and Yu (2004) for a discussion of the Community Conditioning Hypothesis.

The Elusive Baseline

A common concern of environmental management and the focus of ecological risk assessments is the establishment of a reference baseline condition that can be used to evaluate pre- and post-conditions for specified endpoints. The rationale for establishing a reference baseline carries some intriguing philosophical baggage. Implicit in the pursuit of the baseline is an assumption that, except for the actions of humans, there exists a stable ecological condition. In North America this is often thought of, either implicitly or explicitly, as a time prior to major European settlement. How frequently these discussions begin with some reference to pristine conditions. But as Krech (1999) established, Native American activities very measurably altered the North American landscape prior to European settlement. More fundamentally, with or without humans acting on the landscape, climate-driven ecological succession was occurring since the last glacial epoch waned some 10,000 to 15,000 years ago.[7]

In large part, the search for a reference baseline reflects our collective desire to define the environmental conditions we wish to manage toward (Landis and McLaughlin 2000). However, the search for this elusive ecological baseline will be difficult. At best, we can describe a snapshot view, a fixed point in time, in which we characterize static conditions. Chaos theory shows the near-impossibility of establishing a clear trajectory of the vegetation and wildlife of interest at any historical moment. The one thing we can be certain about is that change was occurring. If some prior landscape condition is desired, then ecologists, it would seem, have an obligation to clearly describe those conditions that are possible and those that are unattainable. Even that which is possible may be prohibitively costly in terms of energy or other resources that would have to be traded to achieve the ends.

Even if we agree on the description of desired baseline conditions, there remain many challenges in monitoring the changing status of those conditions. Referring to the earlier discussion of temporal scale, one challenge is to match the frequency of observation to the relevant scale at which ecological processes occur—that is, over decades, centuries, millennia, and beyond. The problem with only making observations over short time periods (seasons or a few years as is typical of research projects and risk assessments) is that short-term trajectories may give false indication of long-term trends. Observing a "fortuitous change" that coincides with, but is not fully consistent with, one hypothesis can be misleading (Fig. 3.4). Consider how remarkably different the conclusions might be among the four different sampling periods illustrated, and how such conclusions would alter environmental management decisions, if not viewed in the context of a longer trajectory.

The challenge of tracking changes in ecological resources is addressed in part through adaptive management, a tool used by most resource agencies such as fish and game management departments, forestry, and water resource management units. However, adaptive management still is subject to uncertainties about baseline trajectory.

---

[7]Of course, climate has been changing since the beginning of time, but the Holocene Epoch is the most relevant for considering human influences across landscapes.

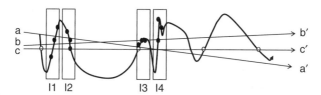

Figure 3.4. Stylized trajectories (lines a–a′, b–b′, or c–c′) superimposed on an ecological condition (a population or a process such as productivity) over time with four arbitrary sampling periods I1 to I4 (rectangles and closed circles). [Line c–c′ (open circles) would imply constancy (stability), a condition that does not exist]. (Adapted from Kapustka 2008.)

Systems-Level Unknowns

Two areas of study relevant to modeling exposures for terrestrial EcoRAs that are particularly lacking are plant uptake processes and food web dynamics. These two perspectives of exposure assessment have had relatively little attention during the development of risk assessment methods and contribute much to uncertainty in risk assessments.

The foundation of plant uptake kinetics was derived mostly from research that used seedlings of herbaceous plants. Some research was done on a very refined scale by applying substances to selected regions of root hairs, the specialized epidermal cells attributed as the site for uptake of water and nutrients by terrestrial vascular plants. Indeed most descriptions of plant uptake processes found in physiology texts focus on the root hair as the critical site of uptake. Unfortunately, these descriptions have limited relevance to plants in environmental settings. The limitation arises from the fact that most plants in terrestrial environments are mycorrhizal within a few days or weeks after germination; and with the onset of colonization by mycorrhizal fungi, root hair formation typically is suppressed. Thus, most plants in the field have few or no root hairs. For the most part, plant uptake models have been developed to represent non-mycorrhizal herbaceous dicotyledonous plants. Can such models reasonably predict plant uptake for older plants or woody species?

The relationships among terrestrial plants and rhizosphere flora are dynamic and complex. But in the field, several other layers of complexity must be considered to depict the true situations governing wildlife exposures to chemicals. These additional complexities require analyses of food webs. Plants encounter a wide variety of bacteria and fungi ranging across a continuum from lethal pathogens to obligate symbionts (Kapustka 1987). In a manner analogous to the establishment and function of mammalian intestinal flora, plants harbor bacteria and fungi on their roots. These rhizosphere microbes alter patterns of root growth, affect nutrient relationships of the plant, affect water uptake by plants, change metabolic processes in plant cells, protect plants from pathogens, and differentially influence phytotoxic responses (Fitter 1985, Harley and Smith 1983, Kapustka et al. 1985, Smith and Read 1997). The interaction between soil, mycorrhizal fungi, bacteria, organic matter, and plant roots creates a mutually supportive ecological system (Fig. 3.5). Mycorrhizal fungi form the most widespread associations between microorganisms and higher plants. The fungi

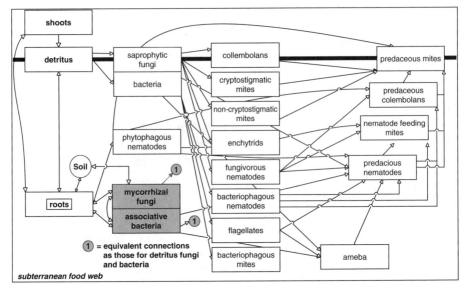

Figure 3.5. Stylized depiction of the below-ground food web in a terrestrial system. (Adapted from Kapustka 2006.)

are obligatorily dependent on the plant for carbon sources, but the plant may or not benefit from the fungus by obtaining important nutrients such as nitrogen and phosphorous. On a global basis, mycorrhizae occur in 83% of dicotyledonous and 79% of monocotyledonous plants and all gymnosperms are mycorrhizal (Marschner 1995).

Cohen et al. (1990) published data on 113 food webs; only 24% described terrestrial systems. Of these, only one-third of the webs considered subterranean species. The terrestrial webs spanned numerous ecotypes from prairie to forest to desert. Even for the above-ground terrestrial food webs, there has been little recent work.

The importance of understanding the true nature of food web interactions is illustrated in potential exposure pathways. The routes of exposure may depend upon the functional presence of mycorrhizal associations. If mycorrhizae do not influence the uptake of a substance from soil into plant shoots, the maximally exposed animal groups would be herbivores (Fig. 3.6, Case 1). However, if mycorrhizae function as intermediaries in the uptake of the substance into plants, then the maximally exposed animal groups above ground would be insectivores (Fig. 3.6, Case 2). Currently, there are no tools to predict which pathway will be most prominent; site-specific data would be required to make that determination.

CONCLUSIONS

There are great challenges posed in adapting risk assessments to focus on populations, habitat characteristics, and ecological systems relevant to the species of interest. Whether constrained by policy or precedent, there appears to be more excuses to

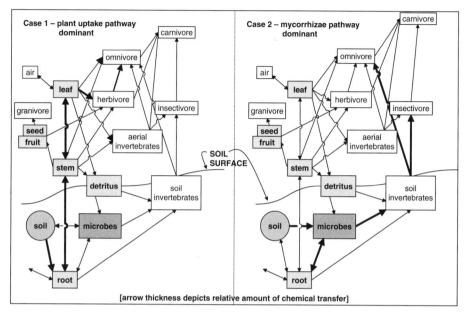

Figure 3.6. Conceptual model of exposure pathways for two cases that differ by the functional condition of mycorrhizae. (Adapted from Kapustka 2006.)

avoid these challenges than there are compelling a better effort. Yet, if the goal is to manage the risks posed to humans and the ecological resources in our surroundings, then these challenges must be accepted. A landscape perspective empowers risk assessors and managers to address real-world issues in the most relevant manner. Risk assessments that ignore these challenges are deficient, in that from the outset we know they have little relevance to what will occur in nature. Management of risk based on such assessments cannot be evaluated properly due to the many inherently unknowable uncertainties.

REFERENCES

Allard P, Fairbrother A, Hope BK, Hull RN, Johnson MS, Kapustka L, Mann G, McDonald B, Sample B. 2010. Recommendations for the development and application of wildlife toxicity reference values. *Integr Environ Assess Manage* **6**:28–37.

Allen AW, Jordan PA, Terrell JW. 1987. *Habitat Suitability Index Models: Moose, Lake Superior Region*. US Fish and Wildlife Service Biological Report 82(10.155). Fort Collins, CO.

ASTM-I. 2004. Standard Guide for Estimating Wildlife Exposure Using Measures of Habitat Quality. *Annual Book of Standards*, Vol. **11**.05. American Society for Testing and Materials, Conshohocken, PA.

Bailey RG. 1998. *Ecoregions: The Ecosystem Geography of the Oceans and Continents*. Springer-Verlag, New York.

Barbour MG. 1996. Ecological fragmentation in the fifties. In Cronon W (Ed.), *Uncommon Ground: Toward Reinventing Nature*. Norton Press, New York, pp. 233–255.

Barbour MG, Burk JH, Pitts WD. 1987. *Terrestrial Plant Ecology*, second edition. Benjamin/Cummings, Menlo Park, CA.

Barnthouse LW, Munns WR, Sorensen MT. 2008. *Population-Level Ecological Risk Assessment*. Taylor & Francis, Boca Raton, FL.

Carlsen TM, Coty JD, Kercher JR. 2004. The spatial extent of contaminants and the landscape scale: an analysis of the wildlife, conservation biology, and population modeling literature. *Environ Toxicol Chem* **23**:798–811.

Clements FE. 1916. *Plant Succession: An Analysis of the Development of Vegetation*. Carnegie Institution of Washington Publication 242. Washington, DC.

Cohen JE, Briand F, Newman CM. 1990. *Community Food Webs: Data and Theory*. Springer-Verlag, New York.

Comer P, Faber-Langendoen D, Evans R, Gawler S, Josse C, Kittel G, Menard S, Pyne M, Reid M, Schulz K, Snow K, Teague J. 2003. *Ecological Systems of the United States: A Working Classification of U.S. Terrestrial Systems*. NatureServe, Arlington, VA.

Daubenmire RF. 1952. Forest vegetation of northern Idaho and adjacent Washington and its bearing on concepts of vegetation classification. *Ecol Monogr* **22**:301–330.

Daubenmire RF. 1966. Identification of typal communities. *Science* **151**:291–298.

Fitter AH. 1985. Functional significance of root morphology and root system architecture. In Fitter AH, Atkinson D, Read DJ, Usher MB (Eds.), *Ecological Interactions in Soils: Plants, Microbes, and Animals*. British Ecological Society Special Publication No. 4, Blackwell Press, London, pp. 87–106.

Freshman JS, Menzie CA. 1996. Two wildlife exposure models to assess impacts at the individual and population levels and the efficacy of removal actions. *Human Ecol Risk Assess* **2**:481–498.

Gleick J. 1987. *Chaos, Making a New Science*. Penguin Books, New York.

Harley JL, Smith SE. 1983. *Mycorrhizal Symbiosis*. Academic Press, New York.

Hope B. 2000. Generating probabilistic spatially-explicit individual and population exposure estimates for ecological risk assessments. *Risk Anal* **20**:573–589.

Hope B. 2001. Consideration of bioenergetic factors in spatially-explicit assessments of ecological receptor exposure to contaminants. *Toxicol Ind Health* **17**:322–332.

Hope BK. 2004. Approaches to spatially-explicit, multi-stressor ecological exposure estimation. In Kapustka LA, Galbraith H, Luxon M, Biddinger GR (Eds.), *Landscape Ecology and Wildlife Habitat Evaluation: Critical Information for Ecological Risk Assessment, Land-Use Management Activities, and Biodiversity Enhancement Practices*. ASTM STP 1458, ASTM International, West Conshohocken, PA, pp. 311–323.

Hope BK. 2009. Will there ever be a role for risk assessments? *Human Environ Risk Assess* **15**:1–6.

Johnson MS, Wickwire WT, Quinn MJ, Ziolkowski DJ, Burmistrov D, Menzie CA, Geraghty C, Minnich M, Parsons PJ. 2007. Are songbirds at risk from lead at small arms ranges? An application of the spatially explicit exposure model. *Environ Toxicol Chem* **26**:2215–2225.

Kapustka LA. 1987. Interactions of plants and non-pathogenic soil microorganisms. In Newman DW, Wilson KG (Eds.), *Model Building in Plant Physiology/Biochemistry, Volume 3*. CRC Press, Boca Raton, FL, pp. 49–56.

Kapustka L. 2003. Rationale for use of wildlife habitat characterization to improve relevance of ecological risk assessments. *Human Ecol Risk Assess* **9**:1425–1430.

Kapustka LA. 2005. Assessing ecological risks at the landscape scale: Opportunities and technical limitations. *Ecol Soc*. **10**: 11. [Online] URL: http://www.ecologyandsociety.org/vol10/iss2/art11/.

Kapustka LA. 2006. Current developments in ecotoxicology and ecological risk assessment. In Arapis G, Goncharova N (Eds.), *Ecotoxicology, Ecological Risk Assessment, and Multiple Stressors*. Kluwer Press, The Netherlands, pp. 3–24.

Kapustka L. 2008. Limitations of the current practices used to perform ecological risk assessment. *Integr Environ Assess Manage* **4**:290–298.

Kapustka LA, Arnold PT, Lattimore PT. 1985. Interactive responses of associative diazotrophs from a Nebraska Sand Hills grassland. In Fitter AH, Atkinson D, Read DJ, Usher MB (Eds.), *Ecological Interactions in Soils: Plants, Microbes, and Animals*. British Ecological Society Special Publication No. 4, Blackwell Press, London, pp. 149–158.

Kapustka LA, Galbraith H, Luxon M. 2001. Using landscape ecology to focus ecological risk assessment and guide risk management decision-making. *Toxicol Industr Health* **17**:236–246.

Kapustka LA, Galbraith H, Luxon M, Yocum J, Adams B. 2004. Application of habitat suitability index values to modify exposure estimates in characterizing ecological risk. In Kapustka LA, Galbraith H, Luxon M, Biddinger GR (Eds.), *Landscape Ecology and Wildlife Habitat Evaluation: Critical Information for Ecological Risk Assessment, Land-Use Management Activities, and Biodiversity Enhancement Practices*. ASTM STP 1458, American Society for Testing and Materials International, West Conshohocken, PA, pp. 169–194.

Krech S. 1999. *The Ecological Indian: Myth and History*. WW Norton & Company, New York.

Kuchler AW. 1964. *Potential Natural Vegetation of the Conterminous United States*. American Geographical Society, Special Publication No. 36, Washington, DC.

Landis WG, McLaughlin JF. 2000. Design criteria and derivation of indicators for ecological position, direction and risk. *Environ Toxicol Chem* **19**:1059–1065.

Landis WG, Yu M-H. 2004. *Introduction to Environmental Toxicology*, third edition. Lewis Press, Boca Raton FL.

Landis WG. 2002. Population is the appropriate unit of interest for a species-specific risk assessment. (Learned Discourse) *SETAC Globe* **3**:31–32.

Linkov I, Grebenkov A, Baitchorov VM, et al. 2001. Spatially explicit exposure models: Application to military sites. *Toxicol Ind Health* **17**:230–235.

Linkov I, Burmistrov D, Cura J, et al. 2002. Risk-based management of contaminated sediments: Consideration of spatial and temporal patterns of exposure modeling. *Environ Sci Technol* **36**:238–246.

Linkov I, Grebenkov A, Andrizhievski A., Loukashevich A., Trifonov A. 2004a. Risk-trace: Software for spatially explicit exposure assessment. In Kapustka LA, Galbraith H, Luxon M, Biddinger G (Eds.), *Landscape Ecology and Wildlife Habitat Evaluation: Critical Information for Ecological Risk Assessment, Land-Use Management Activities, and Biodiversity Enhancement Practices*. ASTM STP 1458, American Society for Testing and Materials International, West Conshohocken, PA, pp. 286–296.

Linkov I, Kapustka LA, Grebenkov A, Andrizhievski A, Loukashevich A, and Trifono A. 2004b. Incorporating habitat characterization into risk-trace: Software for spatially explicit exposure assessment. In Linkov I, Ramadan A (Eds.), *Comparative Risk Assessment and Environmental Decision Making*. Kluwer Press, The Netherlands, pp. 253–265.

Macovsky L. 1999. *A Test of the Action at a Distance Hypothesis Using Insect Metapopulations*. M.S. Thesis, Huxley College, University of Western Washington, Bellingham, WA.

Marschner H. 1995. *Mineral Nutrition of Higher Plants*, second edition. Academic Press, San Diego.

May RM. 1976. Simple mathematical models with very complicated dynamics. *Nature* **261**:459–467.

Menzie C, Bettinger N, Fritz A, Kapustka L, Regan H, Moller V, Noel H. 2008. Population protection goals. In Barnthouse LW, Munns WR, Sorensen MT (Eds.), *Population-Level Ecological Risk Assessment*. Taylor & Francis, Boca Raton, FL, Chapter 3, pp. 41–68.

Nicolson M, McIntosh RP. 2002. H.A. Gleason and the individualistic hypothesis revisited. *Bull Ecol Soc Am* **April**: 133–142.

Omernik JM. 1987. Ecoregions of the conterminous United States. Map (scale 1:7,500,000). *Ann Assoc Am Geogr* **77**:118–125.

Omernik JM. 1995. Ecoregions: A spatial framework for environmental management. In Davis WS, Simon TP (Eds.), *Biological Assessment and Criteria: Tools for Water Resource Planning and Decision Making*. Lewis Publishers, Boca Raton, FL, pp. 49–62.

Rand GM, Newman JR. 1998. The applicability of habitat evaluation methodologies in ecological risk assessment. *Human Ecol Risk Assess* **4**:905–929.

Schroeder RL, Haire SL. 1993. *Guidelines for the Development of Community-Level Habitat Evaluation Models*. Biological Report 8, US Department of Interior, US Fish and Wildlife Service, Washington, DC.

Schroeder RL. 1985a. *Habitat Suitability Index Models: Northern Bobwhite*. US Fish and Wildlife Service Biological Report 82(10.104), Fort Collins, CO.

Schroeder RL. 1985b. *Habitat Suitability Index Models: Eastern Wild Turkey*. US Fish and Wildlife Service Biological Report 82(10.106), Fort Collins, CO.

Shumaker NH. 1998. *A Users Guide to the PATCH Model*. EPA/600/R-98/135. Environmental Research Laboratory, US Environmental Protection Agency, Corvallis, OR, also see www.epa.gov/wed/pages/models.htm.

Smith SE, Read DJ. 1997. *Mycorrhizal Symbiosis*, second edition. Academic Press, Cambridge.

Spromberg JA, John BM, Landis WG. 1998. Metapopulation dynamics: indirect effects and multiple distinct outcomes in ecological risk assessment. *Environ Toxicol Chem* **17**:1640–1649.

Taleb NN. 2007. *The Black Swan: The Impact of the Highly Improbable*. Random House, New York.

Terrell JW, Carpenter J. 1997. *Selected Habitat Suitability Index Model Evaluations*. USGS/BRD/ITR 1997-0005, US Department of Interior, US Geological Survey, Washington, DC.

US EPA (United States Environmental Protection Agency). 2003. *Generic Ecological Risk Assessment Endpoints (GEAEs) for Ecological Risk Assessment*. U.S. Environmental Protection Agency, Risk Assessment Forum. EPA/630/P-02/004F, Washington, DC.

US FWS (US Fish and Wildlife Service). 1981. *Standards for the Development of Habitat Suitability Index Models*. ESM103, Washington, DC.

Whittaker RH. 1975. *Communities and Ecosystems*, second edition. Macmillan, New York.

Wickwire WT, Menzie CA, Burmistrov D, Hope BK. 2004. Incorporating spatial data into ecological risk assessments: Spatially explicit exposure module SEEM for ARAMS. In Kapustka LA, Galbraith H, Luxon M, Biddinger G (Eds.), *Landscape Ecology and Wildlife Habitat Evaluation: Critical Information for Ecological Risk Assessment, Land-Use Management Activities, and Biodiversity Enhancement Practices*. ASTM STP 1458, American Society for Testing and Materials International, West Conshohocken, PA, pp. 297–310.

4

RELEVANCE OF SPATIAL AND TEMPORAL SCALES TO ECOLOGICAL RISK ASSESSMENT

Alan R. Johnson and Sandra J. Turner

> Monk: All is reduced to One. What is this One reduced to?
> Bassui: One inch long, one hundred meters short.
> Monk: I can't understand.
> Bassui: Go and take some tea.
>
> *Stryk and Ikemoto (1981, p. 38)*

As this dialogue between Bassui and an unnamed monk illustrates, discussions of scale can be confusing. Is one inch *long*? Is 100 meters *short*? It's all relative, of course. Perhaps the best course *is* simply to retreat and enjoy a cup of tea. However, scientists in many disciplines grapple with issues of scale in both space and time. We summarize some of the resulting literature, with an emphasis on ideas from landscape ecology, and discuss their relevance to ecological risk assessment.

The world can look very different depending upon the scale of observation. The processes relevant to our description are often scale-dependent. In the macroscopic world, describing the dynamics of planetary motion requires only one physical force: gravity. But among the forces recognized in modern physics, gravity is the weakest, and its effects are only apparent over relatively large distances. Descriptions of phenomena at the increasingly microscopic scales of molecules, atoms, and subatomic particles requires stronger, but shorter-range, forces to be invoked, and gravity can essentially be ignored.

Environmental Risk and Management from a Landscape Perspective, edited by Kapustka and Landis
Copyright © 2010 John Wiley & Sons, Inc.

On the other hand, sometimes things look rather similar across a range of scales. As Benoit Mandelbrot, the inventor of fractal geometry, has observed, when one examines a cloud or a coastline closely, it doesn't become smooth, but reveals irregularities at many scales. Understanding what changes, and what stays the same, across varying scales of observation is one of the foundations of clear thinking about scale.

The concept that natural phenomena and anthropogenic change have characteristic temporal and spatial scales is critically important if we are to narrow any risk assessment to those things that specifically and coherently address the focal question of the assessment. We should incorporate these concepts explicitly into the design and implementation of ecological risk assessments. Identifying the appropriate observation and measurement criteria and the appropriate temporal and spatial scales for measurements is crucial. Investigations must consider interactions across multiple scales to quantify probable effects of environmental stressors on the chosen risk assessment endpoints. But before we can examine that issue, we must introduce some terminology that will allow for a more precise discussion.

A TYPOLOGY OF SCALE

Wu (2007) has systematized concepts of scale by distinguishing between dimensions, kinds of scale, and components of scale. Dungan et al. (2002) also discuss the terminology used to describe various aspects of scale in ecology. We present our own treatment of the subject (Table 4.1) that borrows heavily from these two sources.

Dimensions of Scale

The dimensions that scale can refer to are either temporal or spatial. Time is inherently one-dimensional. Space, at least in our ordinary experience, is three-dimensional, although for the sake of analysis we sometimes restrict our attention to two-, one-, or zero-dimensional subspaces. That is, we live a world with volume, but sometimes talk about surfaces, lines, or points (embedded in three-dimensional space). So the first issue in discussing scale is to state the dimensions under consideration. Are we referring to temporal scale, spatial scale, or both? And, if space is involved, are we measuring scale in terms of length, area, volume, or some combination?

Wu (2007) also identifies levels of hierarchical organization as a dimension of scale. We do not follow this part of his scheme, because we view hierarchical structure as being conceptually different from dimension or scale. The tendency, prevalent in the literature, to equate hierarchical levels with scale has caused much confusion. However, as has been persuasively argued elsewhere, scales and levels are conceptually different things and should not be confused (Ahl and Allen 1996, Allen and Hoekstra 1992, King 2005).

Kinds of Scale

The concept of scale can be applied to different aspects of reality (or our understanding of it). The kinds of scale considered here reflect those with particular relevance to

Table 4.1. A Typology of Scale, Adapted from Wu (2007) and Dungan et al. (2002), with Some Modifications

Dimensions of Scale	
Space	Dimensions that allow for characterizing the size and location of objects and phenomena
Time	Dimensions that allow for characterizing the sequence, duration, and frequency of events
Kinds Scale	
Intrinsic	The spatial or temporal scale at which a pattern or process occurs
Observational	The spatial or temporal scale of measurements or other data used to study a pattern or process
Experimental	The spatial or temporal scale of manipulations or treatments used in the scientific investigation of a pattern or process
Analysis or modeling	The spatial or temporal scale imposed or assumed in data analysis or model construction
Policy	The spatial or temporal scale at which policies or management activities affect the system
Components of Scale	
Length, period, or frequency	An estimate of the intrinsic scale of a pattern or process
Grain	The finest level of spatial or temporal resolution allowed by the observational, analysis or modeling scale
Extent	The greatest spatial or temporal span captured by the observational, analysis, or modeling scale
Coverage	Proportion of the spatial or temporal extent actually sampled; related to sampling density or intensity

ecology and environmental risk assessment. For any phenomenon in the material world, we can imagine an intrinsic scale. What is the intrinsic scale of a hurricane? For a single event, we can discuss spatial scale in terms of the instantaneous size of the storm or the cumulative area affected by its passage. The temporal scale is set by the storm's duration. For hurricanes as a class of disturbance events, the frequency of recurrence is a component of temporal scale.

The concept of intrinsic scale (Table 4.1) can be applied widely to physical, chemical, biological, ecological, and social phenomena. The combined effect of settling rates and turbulent mixing rates determine how long particulates emitted from a stack will remain in the atmosphere. When the advection rate is included, the spatial scale of the particle fallout is determined. The dispersal rate of organisms places a constraint on the rate of spread of an invasive species. The intrinsic population growth rate constrains the time scale required for recovery of a population subsequent to intense harvesting or massive mortality from an oil spill. Human settlements grow, and sometimes decline, at various rates. New technologies, with concomitant benefits and risks, diffuse through society at rates affected by economic and cultural factors.

Although it is often convenient to talk about intrinsic rates of phenomena, we in fact have no direct knowledge of any "intrinsic" properties of nature. What we, as scientists, can obtain are observations, which we analyze in various ways and from which we build models or construct theories to account for the observed phenomena. Any set of observations can be characterized by its spatial and temporal scale. If the observations come from a manipulative experiment, our experimental treatments occur at a particular spatial and temporal scale. As we analyze the data, we may subset it, or pool observations, or compute averages, or do other manipulations that have the effect of changing the spatial or temporal scale. If we build a model, or elaborate a quantitative theory, this too has an associated scale, although sometimes the scale is implicit (and perhaps obscure) rather than explicitly recognized. We may use our models or theories to make inferences at spatial or temporal scales different than the experimental or observational scales associated with our data. For example, responses of plants in the laboratory or small-scale field plots to ozone may be used to model changes in regional forest productivity (Laurence et al. 2000). This approach requires extrapolation of effects across scales, which is difficult and risky (Woodbury 2003), but is often demanded to generate policy. Policies are a product of human institutions and operate at particular temporal and spatial scales. Discussions of scale must be careful to distinguish the kind of scale under consideration (Table 4.1).

Components of Scale

So far our typology of scale has focused on qualitative features. We turn now to components of scale, which address quantitative aspects of scale (Table 4.1).

When we seek to quantify the intrinsic scale of a phenomenon in space, we usually do so in terms of a characteristic size or range of sizes. Size in space is often represented in terms of a length scale (e.g., a radius or diameter), but area, volume, or mass units may also be used. Thus, nanoparticles can be characterized by a distribution of particle sizes, usually log-normal, which may be related to the physical processes of their formation (Kiss et al. 1999). Similarly, in aquatic ecosystems, the biomass spectrum has frequently been studied, with bioenergetic and ecological interpretations given to the positions and temporal changes in the modes (peaks) in the observed distribution of organism sizes (e.g., Kerr and Dickie 2001). In landscape ecology, metrics such as area-weighted average patch size are used to characterize the scale of heterogeneity in patchy landscapes (e.g., Gardner et al. 2008).

Along the temporal dimension, the most straightforward analog of size is duration. However, time scales are also often characterized according to frequency. The motivation for this comes from statistical time-series analysis, in which a given time series, representing the fluctuations of some ecologically relevant quantity, can be converted into a sum of sine and cosine functions via the Fourier transform (Platt and Denman 1975, Turner et al. 1991). For many environmental quantities, it is observed that variation at certain frequencies contribute more heavily to the observed dynamics, as indicated by peaks in the frequency spectrum. These peaks may be said to represent dominant or characteristic frequencies for the phenomenon. For instance, in reconstructed regional dynamics spanning centuries, forest disturbance processes such as fire and defoliating insect outbreaks exhibit quasi-periodic behavior that may be related to climatic variability or other factors (Swetnam and Betancourt 1990, Swetnam and Lynch 1993).

Ecological phenomena can be said to have an intrinsic scale, expressed in terms of characteristic size and frequency along spatial and temporal dimensions, but our knowledge of these scales ultimately rests on observational data. Two key components of observational scale are referred to by landscape ecologists as grain and extent. The term *grain* can be understood by analogy to grain in photographic film. In film, images are recorded by photochemical changes of a silver halide into actual grains of metallic silver. In some situations, the grains become perceptible to the human eye, yielding a coarse texture to the image. Clearly, the photograph is unable to record details finer than the size of the individual grains. Thus, landscape ecologists use the term *grain* to mean "the finest level of spatial or temporal resolution of a pattern or dataset" (Wu 2007). In a digital image, the grain would be equivalent to the pixel size. The term *resolution* is often used to express a concept that is either equivalent to or closely related to grain. For instance, it is often stated that Landsat-TM imagery has a spatial resolution of 30 m, which is the approximate linear distance represented by a single pixel, a measure of grain as we have defined it.

Extent refers to the spatial or temporal span of an observation set. In space, this would be equivalent to the size of the area represented on a map or in a digital image, or the size of a study area from which environmental samples or data were collected. This may be expressed in units of area, or by a representative length scale (e.g., the length of one side of a square image). In time, extent would be equivalent to the duration of the study or time span covered by the observations.

Consider a digital satellite image of a portion of the Earth's surface. For convenience, assume a 1000- $\times$ 1000-pixel image taken from a Landsat-TM scene. As noted above, we can take the grain to be 30 m, the spatial resolution of an individual pixel. The extent can be calculated as 1000×30 m $= 30,000$ m $= 30$ km. Now, assume that we apply some procedure to classify each pixel into a land cover category (e.g., water, urban, cropland, forest). We may then want to "ground-truth" our classification by visiting locations on the ground and determining whether the actual land cover corresponds with the prediction of our classified image. Visiting the location represented by each individual pixel is an unrealistic proposition: There are 10^6 pixels in the image, not to mention the difficulty of accurately locating a particular 30- $\times$30-m area on the ground. Instead, we will visit selected locations scattered throughout the area represented by the image, and gather land cover data at some spatial resolution. Thus, our ground-based data will have essentially the same extent as the image, but the grain may be different. More noticeably, the classified pixels provide 100% coverage of the area in terms of predicted land cover (assuming no cloud cover in the original image), but our field data will provide substantially less coverage. The coverage will depend upon how many field sites we visit. A related issue is the spacing of the field sites. The locations visited may be selected in a uniform grid pattern, or randomly placed, or selected by some other procedure. Generally, the more sites selected, the closer they will be to each other, at least on average.

The notions of grain, extent, and sample spacing are relevant to the temporal dimension as well. In time-series analysis, sampling frequency is a key feature of a dataset. Sampling frequency is the number of observations per unit time and, thus, is inversely related to sample spacing (i.e., the time interval between observations). Observations are often treated as instantaneous (grain = zero), but sometimes the time interval associated with individual measurements or observations is important. The extent (duration of the time series) sets limits on the low-frequency phenomena that can be extracted by the analysis. A rule of thumb is that the data must encompass three to four complete cycles for a phenomenon to be detected. On the other hand, high-frequency phenomena with a periodicity less than the sampling interval will not be detected either. However, a data artifact termed *aliasing* can lead to false inference of variation at a periodicity longer than the true periodicity (lower than the true frequency) if the sampling interval is set too wide (Turner et al. 1991).

SPACE–TIME RELATIONSHIPS

Thus far we have addressed space and time as separate dimensions of scale. However, in ecological phenomena, spatial and temporal scales are often correlated, where processes that operate at broad spatial extents often display slow temporal dynamics, compared with more rapid, but spatially localized, processes. This trend of "small-fast" versus "big-slow" has been repeatedly noted. We will consider its expression in graphical terms and in ecological theory.

A useful place to begin is with the description of space–time relationships in physical oceanography, as presented by Stommel (1963) in his classic paper "Varieties of Oceanographic Experience." Stommel presents a graphical depiction of variability in sea level at various spatial and temporal scales, ranging from centimeters to the globe and from seconds to millennia. Obviously, the diagram is a schematic, since no existing dataset spans such grain and extent in both space and time. Yet, it clearly depicts the emergent understanding of oceanographic processes as inferred from multiple empirical studies at various scales. Stommel's diagram shows peaks at certain space–time scales, corresponding to particular processes that contribute substantially to sea level variation. There is a general ridge with local maxima, trending from gravity waves (~ 10–100 m, 10 s) to ice age variations (global, $\sim 10^{12}$ s). In general, phenomena at a broader spatial scale are also characterized by a longer time scale. Tides are an exception; even though they vary on a short time scale, they are manifest at local *and* global scales.

Many other authors have used simple graphic depictions to indicate space and time scales associated with phenomena. We can take the diagrams offered by Suter (2007, p. 99) as exemplary for our purposes. Suter's diagrams show ovals placed in a two-dimensional graph in order to array phenomena according to their spatial and temporal scales. One graph contains ovals labeled with various ecological phenonmena, such as macroorganism physiology (~ 1 m, ~ 1 day) up through population (~ 100 m, ~ 10 years) and ecosystem dynamics ($\sim 10^4$ m, ~ 100 years). The other graph depicts various anthropogenic stressors, ranging from individual pesticide applications

(~ 100 m, ~ 1 hr) and chemical spills (~ 100 m, ~ 1 days) up to pesticide use ($\sim 10^6$ m, ~ 10 years) and climate change (global, ~ 100 years) An implication of these diagrams is that responses to anthropogenic stressors are most likely to be manifest in ecological phenomena of a similar scale. For instance, a single pesticide application or small chemical spill may well affect individual organisms but changes in the population or ecosystem dynamics are likely to be associated with stressors of broader spatial extent and longer duration, such as (a) repeated exposures of routine pesticide use or (b) climate change.

Most space–time diagrams array phenomena roughly along a diagonal line, with broader spatial scales being associated with longer temporal scales. Hierarchy theory, as applied to ecological systems by O'Neill et al. (1986), offers an explanation for this pattern. Johnson (1996) summarized the argument in terms of three properties assumed to apply to ecological systems. The first is near-decomposability, as posited by Simon (1969), in which ecological systems can, to a good approximation, be regarded as hierarchical structures with groups of strongly interacting components, which in turn form the components for interactions at the next hierarchical level. Secondly, the hierarchical structure is nested [*sensu* O'Neill et al. (1986)], such that higher levels physically contain lower levels, and therefore higher levels are necessarily larger. Finally, the system is rate structured, such that higher levels operate at a slower rate than lower levels. Taken together, this implies that ecological phenomena associated with various hierarchical levels will be arrayed along a diagonal in a space–time diagram.

Holling (1992) puts forth a more detailed theory of ecosystem structure and dynamics based on space–time relationships. A key element of the theory is the extended keystone hypothesis, which suggests that all terrestrial ecosystems are controlled and organized by a small number of biotic and abiotic processes. As a corollary, the entrainment hypothesis predicts that ecological time-series data should have periodicities clustered in a few sets, reflecting the characteristic frequencies of the structuring variables. However, adequate time series (long extent, short sampling interval) to directly test this hypothesis are hard to come by. As an alternative, Holling uses adult body mass as a surrogate for the spatial and temporal scales at which terrestrial animals operate. Empirically examining the distribution of body masses of birds and mammals in boreal forest and short-grass prairie ecosystems, he finds that they are clumped into a few relatively distinct clusters. Several hypotheses are examined to account for this discontinuous distribution, and the only one not rejected by the data is the textural discontinuity hypothesis, which assumes that the discontinuities in body mass distribution reflects a hierarchical landscape structure to which animals respond in terms of foraging and other behaviors that are correlated with body size.

PROBLEMS OF INCOMMENSURATE SCALES

When designing an observation or measurement protocol, the grain, sampling interval, and extent of the data to be collected should be chosen to capture the characteristic scales of the phenomena of interest. If there is too coarse a grain or too wide a sampling interval, then the measurements may fail to capture small- (spatial) or high-frequency (temporal) processes. If there is too narrow an extent, then large- (spatial) or low-frequency (temporal) phenomena may be missed. This requirement is easy to justify theoretically, but sometimes hard to meet in practice (Turner and Johnson 2001).

Stommel (1963) gives an example of oceanographic measurements that failed to capture the desired phenomenon. The Pacific equatorial undercurrent is long and narrow, with a relatively stable position, so it can be accurately mapped based on short-duration current meter readings along north–south sections at widely spaced east–west intervals. Oceanographers had assumed that the equatorial Indian Ocean would be similar, with the exception of an annual reversal due to the monsoon winds. A cruise was designed based on this assumption, but the data revealed that the expected large-scale semi-annual pattern did not materialize, and the observations were too broadly spaced in space and time to accurately map the current at its actual characteristic spatiotemporal scale.

Mismatched time and space scales also plague efforts at ecological risk assessment. Cramer and Hobbs (2005) discuss the problem of assessing ecological risk, using vegetation composition as the endpoint, due to salinity and water-logged soils arising from human-caused alterations in the hydrology of arid Australian landscapes. Human impacts on the water table are typically predicted on the basis of hydrologic modeling at the watershed (catchment) scale. However, ecological studies of the impacts of altered water tables and salinity are usually conducted at a much smaller patch scale. How does one bridge the gap between the stressor as represented in a broad extent, coarse grain and long characteristic time scale of catchment hydrology model and the receptors as characterized by relatively short studies of relatively small patches? Extrapolating from one scale to another is complicated by the fact that ecological variables often display nonlinear, threshold responses, such that small changes in a driving variable may cause large changes in ecological state. Additionally, these changes may not be readily reversible: They may display hysteresis, where returning the driving variables (e.g., water table depth) to what it was before the change of state may not return the ecological variables (e.g., vegetation composition) to their previous values. Cramer and Hobbs do not solve the problem of extrapolation across scales for the system under consideration, but they do provide a conceptual framework to guide further research.

Ecotoxicologists frequently face the problem of incommensurate scales, although the issue is not always explicitly framed as such, when it comes to using laboratory data to infer ecological effects of toxicants in the field (Johnson and Rodgers 2005). Laboratory toxicity experiments typically focus on relatively short-term, single-species bioassays that are used to predict long-term responses of populations in the field. Even when multispecies tests or microcosms are used, the laboratory systems are usually much smaller than the ecosystems they are assumed to represent. Exigencies of cost and practicality push scientists toward experimental systems that are relatively small in spatial extent and short in temporal duration. But policymakers are concerned with systems that are much larger and with consequences that may arise relatively slowly. Effective application of ecotoxicological data to policy decisions requires some method of bridging the disparity in scales.

Perhaps one of the clearest examples of disparate scales imposed by policy demands is in the risk assessment of long-lived radioactive materials. To regulate disposal of high-level radioactive waste, the United States Congress passed the Nuclear Waste Policy Act of 1982. In response, the EPA attempted to promulgate quantitative standards governing release of radiation from a high-level waste repository (1992, 40 CFR § 191). These included quantitative exposure limits for individuals (whole body and critical organ) for the first 1000 years of disposal, as well as the requirement that predicted premature cancer deaths should not exceed 1000 over a 10,000-year period (Edwards 1993). The challenges of estimating individual exposures or excess cancer deaths centuries or millennia in the future is daunting and is inevitably fraught with large uncertainties. In the words of Chapman (2002), the task becomes one of evaluating "hypothetical doses to hypothetical people" in a time as distant in the future as the ancient Egypt is in the past. Little wonder the public is skeptical about the ability of experts to make reliable predictions over such long time scales. In fact, the United States First Circuit court vacated the EPA's original individual exposure limits, citing an inadequate basis (Edwards 1993), although the EPA's final standards contained a modified individual standard (2001, 40 CFR § 197).

PITFALLS IN EXTRAPOLATION ACROSS SCALES

Ecologists have a wealth of data and models representing processes over relatively small, homogeneous areas. However, these models cannot be used to represent the same processes in the broader landscape without careful attention to the effects of scale and heterogeneity. King et al. (1991) explored the problem of "spatial transmutation," in which the mathematical function relating a process to parameters or other input variables changes with scale due to the effects of nonlinearity and heterogeneity. One manifestation of this is known as the "fallacy of averages" (Welsh et al. 1988). Assume that at a fine-scale, a process y can be expressed as some function of a state variable, x, and parameter, p:

$$y = f(x, p)$$

Now x and p can generally be assumed to vary in space. We wish to compute Y, the expected value of the process over a larger landscape:

$$Y = E(y) = E(f(x, p))$$

It is tempting to compute the expected value (i.e., the average) of $f(x, p)$ by computing its value when x and p take on their expected (average) values. Thus, we might assume

$$E(f(x, p)) = f(E(x), E(p)) = f(\overline{x}, \overline{p})$$

Unfortunately, this is generally *not* true, unless f is a linear function, which it seldom is in ecological systems. Assuming that this equation holds when it doesn't is the fallacy of averages.

As an example, consider temperature dependence in the degradation rate of a contaminant in an aquatic environment. We assume first-order kinetics with temperature dependence that can be modeled by the Arrhenius equation:

$$k = Ae^{-E_a/RT}$$

where k is the rate constant, A is a constant, E_a is the activation energy, R is the ideal gas law constant, and T is temperature on an absolute scale (e.g., Kelvin). Assume that a contaminant is uniformly distributed in a lake that is strongly stratified, with (a) 50% of the lake volume in the epilimnion at a temperature of 25°C and (b) the remainder in the hypolimnion at a temperature of 15°C. Further assume $A = 9.74 \times 10^{25}$ day^{-1} and $E_a = 35.5$ kcal mol^{-1}. This example is not intended as a model of any actual situation, because it ignores factors such as seasonal mixing, but environmentally realistic parameters were chosen by approximating the temperature-dependent degradation kinetics of the pesticide carbaryl (Lartiges and Garrigues 1995). At the fine-scale, the local concentration of the contaminant declines exponentially with time, at a rate set by the local temperature. For the lake as a whole, however, the contaminant degradation has two phases: a rapid initial decline as the epilimnion is depleted, followed by a slower decline as the hypolimnion dominates the dynamics (Fig. 4.1). This change from a simple exponential decay at the local scale to a more complex dynamic at the lake scale is an example of transmutation. If one were to commit the fallacy of averages and try to predict the lake-scale dynamics as an exponential decay at the average lake temperature (i.e., 20°C), one would underpredict the amount of degradation in the early days and overpredict the degradation in the longer-term (Fig. 4.1).

A related problem often encountered in extrapolations across scale has been called "the ecological fallacy." This terminology arises from the early recognition of the problem in the social sciences, and it refers to improper inferences made from data where individual responses are aggregated into groups (Robinson 1950). Thus, the term relates to an aggregation problem, and not to the discipline of ecology, although the relevance to ecological data analysis has been noted (e.g., Wu 2007).

Robinson (1950) distinguishes between "individual correlations," in which data represent the properties of individuals (persons, in his examples), versus "ecological correlations," in which each data point represents the properties of a group of individuals. Robinson demonstrates that ecological correlations are often misleading if used as indicators of statistical relationships among individuals. Although this is not a scale issue *per se*, it is true that as scientists deal with data over broad spatial extents, individuals are often aggregated into spatially defined groups, such as census tracts, counties, or states. Changing the spatial grain of the data, by aggregating individuals or small groups into larger groups, is a form of extrapolation across scale that may affect computed correlations.

Robinson (1950) demonstrates the difference in individual versus ecological correlations by examining the relationship between race and illiteracy in the United States in 1930. For a sample size of nearly 100,000 individuals, a 2×2 contingency table is presented in which individuals are categorized as "White" or "Negro" and as "Literate" or "Illiterate." The Pearson correlation coefficient computed on the individual data

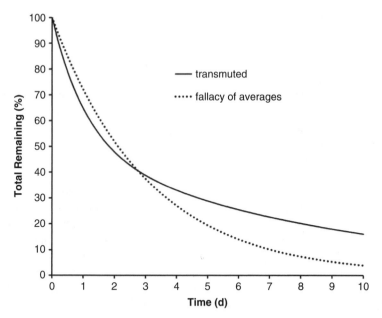

Figure 4.1. Predicted degradation of a hypothetical contaminant in a thermally stratified lake. Using the assumptions fully described in the text: Locally, degradation kinetics are first-order, predicting an exponential decline, but for the lake as whole, this is "transmuted" into a biphasic response, with a relatively rapid initial decline and slower decline after a few days (solid line). The average parameters in the local first-order kinetic model would incorrectly predict the dynamics at the scale of the lake, an example of the "fallacy of averages" (dotted line).

is +0.203, indicating a statistically significant, but relatively modest, association. If the data are grouped into states or broader geographic regions, the corresponding eco-logical correlations are +0.773 and +0.946, respectively, indicating a much stronger association between the aggregate variables of "% Negro" and "%Illiterate." To exam-ine these high correlations and assume that they indicate the strength of association between race and illiteracy for individuals is the ecological fallacy. It is an error that Robinson found his colleagues too commonly making, and which researchers in many disciplines are still prone to make.

In a study of selenium (Se) and mercury (Hg) concentrations in aquatic organ-isms in lakes surrounding the Sudbury metal smelters, Belzile et al. (2006) conclude that their data "suggest that Se plays an important role in limiting the whole-body assimilation of Hg at lower levels of the aquatic food chain." This sounds like a claim about individual organisms. However, the data presented to support this conclusion are tissue concentrations of MeHg or Total Hg versus tissue or aqueous concentrations of Se, computed as average values for each lake. The conclusion that can be validly derived from the data presented is that in those lakes where organisms *on average* have higher tissue concentrations of Se, they tend *on average* to also have lower tissue concentrations of total Hg. To extend this to infer that individuals with higher Se concentrations tend to have lower total Hg concentrations is to commit the ecological fallacy. It may or may not be true, depending upon the individual correlations within groups (i.e., within lakes).

The ecological fallacy may not always be a significant impediment to ecological risk assessment. Cohen has examined epidemiological data relating average radon concentrations (exposure) to lung cancer mortality (response) in a large number of counties in the United States [Cohen (1990) and subsequent publications, summarized in Seiler and Alvarez (2000)]. Cohen (1990) correctly states that, as the data are spatially aggregated, the observed negative correlation between exposure and response at the county level cannot, in general, be used to infer a negative relationship between low levels of radiation exposure and cancer among individuals. However, if the frequently used linear no-threshold model is assumed to be valid for individual responses, then the mortality rate *is* assumed to be proportional to the average exposure. Seiler and Alvarez (2000) argue that the ecological fallacy is irrelevant for risk assessment, as the goal is to predict population-level risks rather than to infer the true dose-response relationship for individuals. In general, risk assessors need to be aware of the phenomena described as the ecological fallacy, but whether or not it presents a problem depends upon the nature of the data and the conclusions to be drawn.

VALID APPROACHES TO EXTRAPOLATION ACROSS SCALES

Extrapolations across spatial or temporal scales are crucial in ecotoxicology and environmental risk assessment (Munns 2004). Johnson and Rodgers (2005) provide an overview of several promising approaches, both theoretical and empirical, for developing scaling relationships that allow for extrapolation across scales. A variety of models may be employed, including statistical models, mathematical models, computer simulations, and physical models (such as microcosms and mesocosms). Specific approaches include dimensional analysis, allometric scaling, fractal geometry, and microcosm/mesocosm experiments in which scale is appropriately manipulated as a treatment variable.

In thermodynamics, a distinction is made between extensive and intensive properties (DeVoe 2001, p. 9). Extensive properties are directly related to the size of the system, whereas intensive properties are independent of system size. For instance, if the system under consideration is a gas at thermodynamic equilibrium in a closed container at constant temperature and pressure, then the mass of the contained gas is an extensive property, whereas the density (mass/volume) is an intensive property. There are probably very few truly intensive properties of relevance to ecological systems. For instance, by analogy to molecules of gas in a container, one might assume that population density (individuals/area) would be independent of area. However, empirical data generally exhibit a negative correlation between population density and the area censused (Gaston et al. 1999). Various explanations have been offered to account for observed density–area relationships, including ecological factors (e.g., Buckley

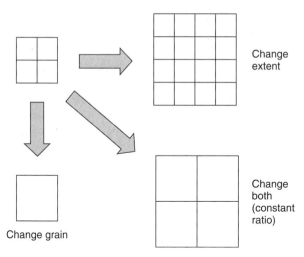

Figure 4.2. Illustration of changes in spatial scale for a grid-based (pixel or raster) dataset. Changes in scale can involve a change in the area spanned (extent), or a change in the spatial resolution of the data (grain), or both.

and Roughgarden 2006) and artifacts due to biases in sampling (e.g., Pautasso and Weisberg 2008). In either case, it is clear that scale-dependence of ecological properties is a pervasive consideration when comparing or extrapolating between various ecological systems.

Extrapolations across scale typically involve changes in grain, or extent, or both. For example, consider changes in scale for raster-based spatial data, in which grain is determined by the size of a grid cell, and extent is determined by the size of the array. Extrapolation from a fine-scale to a coarser-scale grain would involve aggregating or averaging data to the larger grid cell. Extrapolation to a broader extent would involve expanding the area represented by the entire array of grid cells. Of course, both processes may occur simultaneously, perhaps keeping a constant ratio between grain and extent (Fig. 4.2).

Similarity-Based Scaling

If the property of interest is intensive (independent of scale), then special extrapolation techniques are not needed. For extensive properties, one strategy is to normalize or re-scale the property of interest to a quantity that is independent of scale. This approach relies on the concept of scale-invariance and is the basis of similarity-based scaling [sometimes abbreviated SBS (Wu 2007)]. For instance, consider the collective motions of a large number of random walkers (which could be used to simulate molecular diffusion or the movements of organisms). The average distance of the walkers from their place of origin increases nonlinearly but predictably with time. In one dimension, the distribution of the walkers approximates a Gaussian (normal) distribution, with the mean of the distribution, μ, at the common origin point, and a standard deviation,

σ, that increases in proportion to the square-root of time, t. Thus, the properties μ and $\sigma(t)^{-1/2}$ are scale-invariant (i.e., independent of time duration). This model (Fickian diffusion) has been widely applied to physical processes such as Brownian motion or molecular diffusion, and it can be applied to model organismal movements in some situations. A refinement is to assume that organisms move in a correlated random walk, in which they have a greater probability of persisting in an established direction of movement. This model predicts that the dispersion of walkers from their origin increases more quickly than the Fickian model, with a scaling exponent that depends upon the degree of correlation between steps. However, this behavior only persists up to a point: In the long-run (and at broad spatial scales), a crossover to Fickian diffusion occurs (Johnson et al. 1992).

At times, a theoretical basis for renormalizing a property of interest into a scale-invariant quantity is lacking. We are then left with the phenomenological approach of empirically quantifying the scale dependence of the property, perhaps in terms of a regression or other statistical model. An important example of this approach is the exploration of scale dependence in the correlation between human populations and species richness reported by Pautasso (2007). Synthesizing multiple studies of human presence (population size, housing density, etc.) and species richness of various taxa (birds, plants, etc.), Pautasso found that the correlation between human presence and biodiversity varies systematically with both the grain and the extent of the study. Over at least four orders of magnitude, the correlation varies linearly with the logarithm of scale (grain or extent). For studies with grain >1 km, the correlation is generally positive. Thus, as coarse resolution, humans and other species co-occur in similar environments, probably reflecting patterns of productivity and resource availability. However, at the finer-scale, human presence has negative effects on the diversity of other species. Recognition of this scale dependence goes a long way toward elucidating some of the contradictory claims that have been made regarding the impact of urbanization on biodiversity.

Model-Based Scaling

In cases where no theoretical or empirical scaling relation exists, the alternative is to use ecological models (analytical or simulation) to make the extrapolation. There are a variety of approaches to model-based scaling (MBS), and Wu (2007) provides a useful summary. We will briefly describe some of the more common techniques, with a focus on the problem of upscaling (i.e., extrapolation from fine-grained, narrow extents up to coarse-grained, broad extents). The reverse process of downscaling has also received some attention in the ecological literature, often in the context remote sensing and discussion of "subpixel" properties. While a given fine-scale pattern, in principle, has a unique expression at a broader, coarser scale, the reverse is not true. A given broad-scale ecosystem property (e.g., total annual nutrient export from a watershed) may be compatible with a large number of different fine-scale patterns (e.g., proportions and spatial arrangement of various land cover types). Thus, the downscaling problem may not have a well-defined solution, unless additional constraints can be imposed (correlation with fine-scale data on soils, topography, etc.).

Upscaling an ecological process involves computing the broad-scale value of the process, Y, given information about the corresponding fine-scale process at a set of locations, $\{y_i\}$. In general, we do not know the y_i's directly, but we compute them according to some model. Using notation similar to that in our discussion of transmutation (above), we obtain

$$y_i = f(x_i, p_i)$$

In this context, f may not represent some analytically tractable mathematical function, but rather a computer simulation model. The problem is to calculate Y given f and some knowledge about $\{x_i\}$ and $\{p_i\}$.

The simplest method, called extrapolation by lumping, is to assume

$$Y = f(\overline{x}, \overline{p})$$

As was noted in our discussion of spatial transmutation, this can only be assumed to be true if f is linear. Otherwise, we commit the fallacy of averages. Accordingly, this approach has limited applicability.

A somewhat more general approach, extrapolation by effective parameters, assumes

$$Y = f(x_{\text{eff}}, p_{\text{eff}})$$

where x_{eff} and p_{eff} are effective state and parameter values. This approach is commonly used in modeling flow in porous media, such as groundwater hydrology. It can be shown that, for saturated flow through multiple blocks with differing hydrologic conductivities, the effective conductivity is the arithmetic mean if the blocks are arranged in parallel, and the harmonic mean if the blocks are in series. For unsaturated flow, there is no general solution, but the geometric mean is found to serve as an effective parameter in some situations (Wu 2007).

Both of the approaches considered so far assume that the same function or model, f, serves to describe the process at local and broader scales. As demonstrated in our discussion of spatial transmutation, this is not necessarily the case. A more general approach, called direct extrapolation, is to assume one of the following:

$$Y = \langle y_i \rangle = \langle f(x_i, p_i) \rangle \qquad \text{or} \qquad Y = \sum_{i=1}^{N} y_i = \sum_{i=1}^{N} f(x_i, p_i)$$

such that the expression of the process in the broader landscape is either the average or the sum of the local processes. In direct extrapolation, one runs the model at each local site (e.g., grid cell or patch) with appropriate state and parameter values, and then he/she averages or sums the outputs from these individual simulations. Inputs (state variables and parameters) for each site may be derived, for example, from GIS coverages. In some cases, certain inputs may not be available for every local site; for example, some variables may only be available from field measurements at a number of locations. In this case, one may fit a statistical distribution to the variable (e.g., lognormal) and then use Monte Carlo techniques to sample from the assumed distribution when running the model at a local scale.

Even in direct extrapolation, it is assumed that the value of y_i depends only upon the value of local inputs (x_i and p_i). In other words, the process y in the ith patch (or grid cell) does not depend upon what is happening in other patches, so each local simulation can be run independently. This is probably a fairly good assumption if one is modeling net primary production or CO_2 flux for various grid cells representing a terrestrial landscape. However, if one is modeling metapopulation dynamics, the dynamics of a local population in one patch may very much depend upon what is happening in other patches to which it is linked via dispersal. In that case, extrapolation can only be done by spatially interactive modeling, in which both (a) spatial heterogeneity in input variables and (b) dynamic interactions between patches are included in the simulation.

INCORPORATING SCALE IN ECOLOGICAL RISK ASSESSMENTS

To this point, we have reviewed a number of concepts, primarily arising in the context of landscape ecology, which appear relevant to the conduct of ecological risk assessment. We turn now to a brief consideration of the extent to which issues of scale are currently incorporated in the practice of ecological risk assessment.

A recent paper, entitled "An Examination of Ecological Risk Assessment and Management Practices" (Hope 2006), provides an illustrative starting point. The paper contains a section on "Temporal and spatial scales of assessments." However, the temporal issues addressed focus on the distinction between retrospective versus prospective assessment, and spatial issues aren't discussed. Despite the section heading, there is no explicit discussion of scale (temporal or spatial). Spatial issues are raised elsewhere in the paper, in a section on "Spatially explicit risk assessments," which summarizes the use of spatial simulation models and other spatial analysis tools. Again, scale is not explicitly considered, but issues of grain and extent are implicit in the development and application of these methods. Finally, in considering future directions for the field, Hope (2006) foresees "increasing focus on assessments that evaluate risk over larger, more ecologically relevant spatial scales."

Most ecological risk assessments conducted today do not explicitly address issues of spatial and temporal scale, although scale issues are implicit in the data used and many of the assumptions made during analysis. There is often a desire to expand the analysis to a "larger spatial scale," although the criteria for choosing the most relevant scale or scales are often left unspecified. The term "scale," in either a spatial or temporal sense, may be employed, but explicit scale considerations or use of scaling relationships or explicit extrapolations across scale are either lacking or relatively superficial. Discussions of "scale" in the current ecological risk assessment literature generally demonstrate a high level of interest, but not much in the way of practical applications of existing theory.

There are, of course, examples that provide a more substantive consideration of scale in ecological risk assessment. Hope (2005) presents a spatially explicit modeling approach for assessing exposure via ingestion of contaminated food by foraging individuals. The simulations represent hypothetical situations, but the potential for adapting the approach to model real-world situations is evident. Scale issues are explicitly discussed in defining the extent of the assessment area (encompassing locations where the stressor occurs), the forage area (associated with an individual organism), and the assessment population area (encompassing the foraging areas of all individuals that might encounter the stressor). Other scale issues that might be important, but are not treated in depth, are (a) the choice of appropriate spatial and temporal grain in the model and (b) the coverage required to characterize contaminant spatial heterogeneity for the model.

Landis (2003) provides an explicit discussion of scale issues for ecological risk assessment, framing his remarks in the context of a "hierarchical patch dynamics paradigm." In this paradigm, ecological systems are conceived of as nested hierarchical systems, with the temporal dynamics and the emergence of spatial patterns at multiple scales. Colnar and Landis (2007) provide an example of the paradigm applied to risk assessment for an invasive species, the European green crab (*Carcinus maenas*) in the Pacific Northwest region of the United States (also see Chapter 12). The conceptual model developed for the assessment includes three distinct spatial scales: a local scale that provides the mechanistic basis for the overall model, a focal scale that includes the Cherry Point, Washington (USA) area, and a regional scale that set a broad context for the model.

FINAL THOUGHTS AND CONCLUSIONS

We began this paper with a typology of scale adapted from Wu (2007) and Dungan et al. (2002). This included an examination of various aspects of scale, including its dimensions (temporal and spatial), kinds (intrinsic, observational, experimental, etc.) and components (grain, extent, etc.). We examined the relationship between temporal and spatial scales, noting that ecological phenomena are often observed to cluster on a continuum ranging from "small-fast" to "big-slow," and briefly summarized two theoretical frameworks offered to account for this pattern (O'Neill et al. 1986, Holling 1992).

We are seldom in possession of data adequately characterizing ecological phenomena or potential stressors across a broad range of scales in both space and time. Rather, we face the challenge of piecemeal data (and models) addressing various aspects of the ecological risk assessment at incommensurate spatiotemporal scales. This requires that we find ways to extrapolate across scales, in order to apply the information we have to make predictions at the scales relevant for risk management.

There are many potential pitfalls that must be avoided in making extrapolations across scale. Two that have received particular attention are the fallacy of averages and the ecological fallacy (Welsh et al. 1988, Robinson 1950). Avoiding such pitfalls, there are a number of conceptually valid approaches to extrapolation across scales. The most straightforward is similarity-based scaling, which uses quantitative scaling relationships (or, equivalently, normalizing factors to achieve scale-invariance). These scaling relationships may be empirical; or they may be motivated by, or completely derived from, theoretical considerations. When simple similarity-based scaling is not possible, the alternative is usually model-based scaling, often employing computer simulation models.

Our review of scale considerations in the scientific literature on ecological risk assessment found a general interest in the phenomenon, but only a handful of instances where substantive use of scale or scaling concepts had been applied. Undoubtedly, additional examples of careful scale considerations or use of spatiotemporal scaling techniques in risk assessment exist. Yet, the few examples summarized here suffice to illustrate the tenor of current discussions. There is much more that can be, and needs to be, done. Progress will not be easy, because issues of scale have a way of being deceptively simple at first glance, and stubbornly intractable when examined in depth. However, like the monk advised to take a cup of tea, perhaps we can regain our mental clarity and try to tackle the difficult problem once again.

REFERENCES

Ahl V, Allen TFH. 1996. *Hierarchy Theory. A Vision, Vocabulary, and Epistemology*. Columbia University Press, New York.

Allen TFH, Hoekstra TW. 1992. *Toward a Unified Ecology*. Columbia University Press, New York.

Belzile N, Chen Y-W, Gunn JM, Tong J, Alarie Y, Delonchamp T, Lang C-Y. 2006. The effect of selenium on mercury assimilation by freshwater organisms. *Can J Fisheries Aquatic Sci* **63**:1–10.

Buckley LB, Roughgarden J. 2006. A hump-shaped density–area relationship for island lizards. *Oikos* **113**:243–250.

Chapman N. 2002. Long timescales, low risks: Rational containment objectives that account for ethics, resources, feasibility and public expectations—some thoughts to provoke discussion. In: *The Handling of Timescales in Assessing Post-closure Safety of Deep Geological Repositories, Workshop Proceedings*. Nuclear Energy Agency, OECD, pp. 145–153.

Cohen BL. 1990. A test of the linear no-threshold theory of radiation carcinogenesis. *Environ Res* **53**:193–220.

Colnar AM, Landis WG. 2007. Conceptual model development for invasive species and a regional risk assessment case study: The European green crab, *Carcinus maenas*, at Cherry Point, Washington, USA. *Human Ecol Risk Assess* **13**:120–155.

Cramer VA, Hobbs RJ. 2005. Assessing the ecological risk from secondary salinity: A framework addressing questions of scale and threshold responses. *Aus Ecol* **30**:537–545

DeVoe H. 2001. *Thermodynamics and Chemistry*. Prentice Hall, Upper Saddle River, NJ.

Dungan JL, Perry JN, Dale MRT, Legendre P, Citron-Pousty S, Fortin MJ, Jakomulska A, Miriti M, Rosenberg MS. 2002. A balanced view of scale in spatial statistical analysis. *Ecography* **25**:626–640.

Edwards N. 1993. Yucca Mountain or: How we learn to stop worrying and love the Department of Energy's high-level waste disposal guidelines. *Virginia Environ Law J* **12**:271–297.

Gardner RH, Lookingbill TR, Townsend PA, Ferrari J. 2008. A new approach for rescaling land cover data. *Landscape Ecol* **23**:513–526.

Gaston KJ, Blackburn TM, Gregory RD. 1999. Does variation in census area confound density comparisons? *J Appl Ecol* **36**:191–204.

Holling CS. 1992. Cross-scale morphology, geometry, and dynamics of ecosystems. *Ecol Monogr* **62**:447–502.

Hope BK. 2005. Performing spatially and temporally explicit ecological exposure assessments involving multiple stressors. *Human Ecol Risk Assess* **11**:539–565.

Hope BK. 2006. An examination of ecological risk assessment and management practices. *Environ Int* **32**:983–995.

Johnson AR. 1996. Spatiotemporal hierarchies in ecological theory and modeling. In Goodchild MF, Steyaert LT, Parks BO, Crane MP, Johnston CA, Maidment DR, Glendinning S (Eds.), *GIS and Environmental Modeling: Progress and Research Issues*. GIS World, Ft. Collins, CO, pp. 451–456.

Johnson AR, Rodgers JH. Jr. 2005. Scaling in ecotoxicology: Theory, evidence and research needs. *Aquat Ecosyst Health Manage* **8**:353–362.

Johnson AR, Milne BT, Wiens JA. 1992. Diffusion in fractal landscapes: Simulations and experimental studies of tenebrionid beetle movements. *Ecology* **73**:1968–1983.

Kerr SR, Dickie LM. 2001. *The Biomass Spectrum: A Predator–Prey Theory of Aquatic Production*. Columbia University Press, New York.

King AW. 2005. Hierarchy theory and the landscape...level? or, Words do matter. In Wiens JA, Moss MR (Eds.), *Issues and Perspectives in Landscape Ecology*. Cambridge University Press, Cambridge, UK, pp. 29–35.

King AW, Johnson AR, O'Neill RV 1991. Transmutation and functional representation of heterogeneous landscapes. *Landscape Ecol* **5**:239–253.

Kiss LB, Söderlund J, Niklasson GA, Granqvist GC. 1999. New approach to the lognormal size distributions of nanoparticles. *Nanotech* **10**:25–28.

Landis WG, 2003. The frontiers in ecological risk assessment at expanding spatial and temporal scales. *Human Ecol Risk Assess* **9**:1415–1424.

Lartiges SB, Garrigues PP. 1995. Degradation kinetics of organophosphorus and organonitrogen pesticides in different waters under various environmental conditions. *Environ Sci Technol* **29**:1246–1254.

Laurence JA, Ollinger SV, Woodbury PB. 2000. Regional impacts of ozone on forest productivity. In Mickler RA, Birdsey RA, Hom J (Eds.), *Responses of Northern U.S. Forests to Environmental Change*. Springer, New York, pp. 425–453.

Munns WR. 2004. Axes of extrapolation in risk assessment. *Human Ecol Risk Assess* **8**:19–29.

O'Neill RV, DeAngelis DL, Waide JB, Allen TFH. 1986. *A Hierarchical Concept of Ecosystems*. Princeton University Press, Princeton, NJ.

Pautasso M. 2007. Scale dependence of the correlation between human population presence and vertebrate and plant species richness. *Ecol Lett* **10**:16–24.

Pautasso M, Weisberg PJ. 2008. Negative density–area relationships: The importance of zeros. *Global Ecol Biogeogr* **17**:203–210.

Platt T, Denman KL. 1975. Spectral analysis in ecology. *Annu Rev Ecol Syst* **6**:189–210.

Robinson WS. 1950. Ecological correlations and the behavior of individuals. *Am Sociol Rev* **15**:351–357.

Seiler FA, Alvarez JL. 2000. Is the "ecological fallacy" a fallacy? *Human Ecol Risk Assess* **6**:921–941.

Simon HA. 1969. The architecture of complexity. In Simon HA (Ed.), *The Sciences of the Artificial*. MIT Press, Cambridge, MA, pp. 192–229.

Stommel H. 1963. Varieties of oceanographic experience. *Science* **139**:572–576

Stryk L, Ikemoto T. 1981. Zen: *Poems, Prayers, Sermons, Anecdotes, Interviews*, second edition. Swallow Press, Chicago, IL

Suter GW II. 2007. *Ecological Risk Assessment*, second edition. CRC Press, Boca Raton, FL.

Swetnam TW, Betancourt JL. 1990. Fire–southern oscillation relations in the southwestern United States. *Science* **249**:1017–1020.

Swetnam TW, Lynch AM. 1993. Multicentury, regional-scale patterns of western spruce budworm outbreaks. *Ecol Monogr* **63**:399–424.

Turner SJ, Johnson AR. 2001. A theoretical framework for ecological assessment. In Jensen ME, Bourgeron PS (Eds.), *A Guidebook for Integrated Ecological Assessments*. Springer, New York, pp. 29–39.

Turner SJ, O'Neill RV, Conley W, Conley MR, Humphries H. 1991. Pattern and scale: statistics for landscape ecology. In Turner MG, Gardner RH (Eds.), *Quantitative Methods in Landscape Ecology*. Springer-Verlag, New York, pp. 17–49.

Welsh AH, Peterson AT, Altmann SA. 1988. The fallacy of averages. *Am Natur* **132**:277–288.

Woodbury PB. 2003. Do's and don'ts of spatially explicit ecological risk assessments. *Environ Toxicol Chem* **22**:977–982.

Wu J. 2007. Scale and scaling: A cross-disciplinary perspective. In Wu J, Hobbs RJ (Eds.), *Key Topics in Landscape Ecology*. Cambridge University Press, Cambridge, UK, pp. 115–142.

5

QUANTITATIVE MEASURES AND ECOLOGICAL HIERARCHY

G. Darrel Jenerette and Jianguo Wu

Societies are becoming increasingly vulnerable to the multiple negative impacts associated with environmental changes. Societal vulnerability includes both (a) the risks associated with disturbances and (b) coping measures available to mitigate the consequences of disturbances (Turner et al. 2003). Because risks and availability of coping measures vary spatially and temporally, understanding the spatiotemporal pattern of vulnerability is central to managing environmental changes. Quantifying vulnerability is challenging because the social and ecological processes that respond to and cause vulnerability vary at scales from individuals to the entire earth. Meeting this challenge requires an interdisciplinary approach reconciling pattern, process, and scale between social and ecological sciences. Landscape ecology provides both a guiding theory and methodology for conducting such interdisciplinary research. In this chapter we provide a conceptual framework for the quantification of ecological hierarchies and then discuss the uses of different quantitative approaches to understand these hierarchies. We conclude by describing how these approaches can be used and identifying important problem domains where quantitative analyses of ecological hierarchies are needed to better understand vulnerability.

Environmental Risk and Management from a Landscape Perspective, edited by Kapustka and Landis
Copyright © 2010 John Wiley & Sons, Inc.

SCALE AND HIERARCHY—THE OPERATIONAL CONTEXT
FOR UNDERSTANDING ECOLOGICAL DYNAMICS

Ecological landscapes are multiple-scaled heterogeneous areas where living and non-living components interact. Many ecological processes affect and are influenced by spatial heterogeneity, creating a reciprocal relationship between ecological processes and landscape pattern (Pickett and Cadenasso 1995, Wu and Loucks 1995). Landscape heterogeneity results from patches and gradients occurring over multiple scales. Within a given scale window, spatial pattern describes the relative distribution and shape of landscape elements. Scaling relations describe the sensitivity of spatial patterns to changes across multiple scale windows (Wu 2004). Scaling relations commonly exhibit substantial and idiosyncratic behaviors for a broad range of pattern descriptions (Turner et al. 1989, Wickham and Riitters 1995, Wu and Loucks 1995, Wu et al. 2002). Such scale dependencies have important functional consequences for ecosystem organization because these scale variations often reflect the limited scale domain of individual processes and the hierarchical organization of multiple processes (O'Neill et al. 1986, Wu and Loucks 1995, Simon 1996).

Many definitions are commonly used by ecologists and researchers in other disciplines when describing scale, thus necessitating a clear articulation of terms (Jenerette and Wu 2000). A major distinction in scale descriptions differentiates between cartographic and ecologic definitions. Cartographic scale describes the ratio between distance on a map and distance on the earth. Increasing the cartographic scale of a map reduces the amount of real-world area displayed. Appropriately quantifying the cartographic scale is essential for building many useful representations of the real world; computing the actual area of a mapped spatial unit or the actual distance between locations on a map requires precise knowledge of cartographic scale. However, scale influences spatial relationships and the detection of these relationships beyond the cartographic description of scale.

Following Wu and Li (2006a), we articulate three aspects of ecological scale: dimension, kind, and component. Dimensions of scale include space, time, and a hierarchy of organizational levels, which provide the coordinates for measuring and studying scale (compare with alternative perspective regarding hierarchy in Chapter 4). Here we focus on the spatial dimension, considered in depth by landscape ecologists. Space can be described with linear (distance), planar (area), or volumetric (volume) features, and this complexity is quickly encountered when describing the spatial characteristics of the earth's surface. As a spheroid with strong topographic irregularities, linear and planar descriptions of the earth's surface often require abstractions to simplify the three-dimensional complexity.

The dimension of scale provides the context for the kinds and components of scale. The kinds of scale refer to intrinsic, analysis, and policy scales [simplified from Wu and Li (2006a)] (Fig. 5.1). The intrinsic scales are the spatial and temporal dimensions characterizing specific ecological processes. Dispersal distance and home ranges are two common ecological examples of intrinsic scales. The analysis scales are the scales where scientific knowledge of a system is generated. The analysis scales may include observational scales, the scales at which surveys of natural variability or

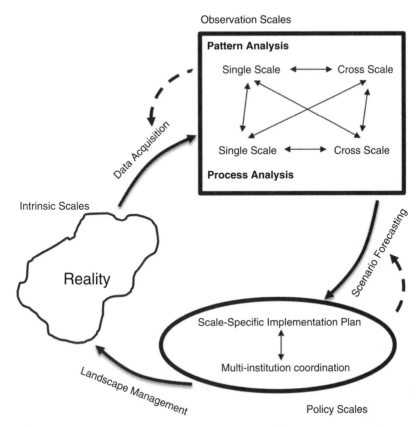

Figure 5.1. Scaling and landscape risk management. The reality, observation, and policy kinds of scale and their interactions are shown. Feedbacks between analysis and data acquisition and policy with scenario forecasting show how processes within the kind of scale affect information flows. Within each kind of scale, interactions within an individual hierarchical level and across multiple levels need explicit recognition. Policy and understanding are adaptive. As the system changes, more robust forecasts can be generated and management alternatives can be refined.

manipulated experiments are implemented. Analysis scales may also include model scales, the scales at which either process or statistical models are constructed and used to interpret and extend observations. Policy scales are the scales where societal decisions can be implemented to affect ecological processes. Aligning the analysis scales with intrinsic scales is necessary for accurate estimation of landscape characteristics, understanding the causes and consequences of landscape variability, and devising policies that meet societal goals. However, identifying intrinsic scales requires estimating scale dependencies, their variabilities, and uncertainties. Furthermore, the kinds of scale are not entirely separable. Policy choices are informed by the scales of analysis; and because specific policies are implemented, they may affect both intrinsic and analysis scales of certain ecological or socioeconomic processes.

The basic components of scale include resolution (spatial, temporal, thematic, etc.), grain, and extent (maximum dimensional range). A common example of grain and extent is depicted by remotely sensed imagery. The size of individual pixels is the grain, and the size of the entire image is the extent. The grain is commonly orders of magnitude smaller than the extent. The grain and extent components apply to all kinds of scale. The observation kind of scale is also composed of a "support" component, which refers to the area over which individual measurements are made or statistically derived. For remotely sensed data, the support is similar to the grain; however, for a field campaign where samples are acquired from individual plots distributed through-out the landscape, the support can be orders of magnitude smaller than the grain. In describing the components of scale, ratios between different components of scale (e.g., grain/extent) often provide useful metrics for describing a dataset (Turner et al. 1989, Schneider 2001, Wu and Li 2006a). By using scale component ratios, comparisons between different landscapes can be facilitated and the magnitude of scale translation from samples to the extent can be quantified (Schneider 2001, Wu and Li 2006a).

When examining patterns of a single system with processes varying in their intrinsic scale, one is quickly confronted by the variation in pattern associated with changes in analysis scale. Changing either the extent or grain of analysis reveals different relationships in spatial variability (Jelinski and Wu 1996, Wu 2004). Over a range of scale, some descriptors of spatial pattern may vary systematically, but most change idiosyncratically (Wu et al. 2002, Wu 2004). The scale dependence of analysis is a consequence of heterogeneity and reflective in part of the hierarchical organization of ecological patterns and processes (O'Neill et al. 1986, Wu and Loucks 1995). Hierarchy theory organizes complex phenomena into levels that correspond to different scales in time and space (O'Neill et al. 1986, Simon 1996, Wu 1999). Modern hierarchy theory suggests that levels are discrete and identifiable (O'Neill et al. 1986, Wu and Loucks 1995, Wu 1999, Holling et al. 2002). Hierarchical levels are identified on the basis of variability, interaction, and organization. Variability refers to the magnitude and configuration in the heterogeneity of state variables or flux rates. Interaction refers to the exchange of material, energy, or information between distinct landscape units. Lower levels occur at a finer scale; they are smaller, faster, and more variable. Higher levels occur at a coarser scale; they are larger, slower, and less variable. Fine subsystems are linked together to form coarser subsystems: these coarser subsystems incorporate additional processes and provide constraints on fine-scale dynamics. Because the separation between levels is incomplete, hierarchy theory considers a complex system to be near-decomposable (Simon 1996, Wu 1999). Much attention of hierarchy theory has been devoted to understanding interactions between levels. Cross-scale exchanges occur primarily through (a) the exertion of constraints from higher levels to lower levels and (b) the propagation of disturbances from lower levels to higher (Holling and Gunderson 2002). Higher levels provide the limits to variation within a level. This constraining of lower levels serves to dampen variability moving up a hierarchy.

The generation of hierarchical levels is a defining characteristic of complex sys-tem dynamics (Levin 1998). In general, within an individual scale domain a consistent suite of processes tend to dominate relationships. This consistency within individual scale domains has been the hallmark of the widely used power-law or fractal-based scaling approaches (Brown et al. 2002, Brose et al. 2004). Within a single scale, the patterns of organization may be independent of initial conditions. However, between scale domains, complex systems exhibit self-organization leading to nested systems that are only loosely connected to the processes governing the dynamics at adjacent domains of scales. Self-organization between scales results in bounded homeostatic regulation of system properties throughout a range of environmental conditions. Within ecological systems, the interplay between fast and slow variables has been cited as the source of self-organization (Carpenter et al. 1999, Carpenter and Turner 2000). These processes are widely found in physical systems of self-organization with interactions between amplification and inhibitory processes (Prigogine and Stengers 1984). As environmental conditions change across boundaries of scale domains, system properties often exhibit abrupt shifts. These shifts between alternate stable states are a key characteristic of hierarchical systems and self-organization (Scheffer et al. 2001, Gunderson and Holling 2002, Scheffer and Carpenter 2003). The presence of alternate stable states also highlights the importance of history: The current state of a system may be more related to previous conditions when self-organization began. The challenges of understanding complex systems are thus twofold, identifying relationships that are consistent within individual levels and identifying shifts in dominant processes across levels.

Hierarchy theory has been applied extensively to understanding ecological characteristics (Pickett et al. 1987, Urban et al. 1987, Kolasa and Pickett 1989, Waltho and Kolasa 1994, Wu and Loucks 1995, Kolasa et al. 1996). Evidence of hierarchical organization in ecological communities and populations is increasing (Waltho and Kolasa 1994, Kolasa et al. 1996, Gillson 2004). Linkages between terrestrial and aquatic components of a landscape are being aided by a hierarchical conceptualization (Fisher et al. 1998a, 1998b; Dent et al. 2001). These studies have shown that for many systems, the separation of processes and patterns into discrete scales is necessary to understand dynamics within and between scale domains.

Hierarchy theory is a viable description of ecosystem organization that provides useful guidelines for understanding how information is translated between scales (Wu 1999, Wu and David 2002). A hierarchical approach has led to the development of multiscaled methods for analysis. A common strategy is to consider simultaneously three levels (or scales) in a study: the level of interest (the focal level), the level above the focal level, and the level below the focal level (Wu and Loucks 1995, Simon 1996, Parker and Pickett 1998). The integration of patch dynamics with hierarchy theory into hierarchical patch dynamics suggests that hierarchical organization is implicit in spatial organization (Wu and Loucks 1995). The application of hierarchical patch dynamics has been shown as a useful framework for identifying risk or disturbances to ecological processes (Gilson 2004, Colnar and Landis 2007). Hierarchical partitioning can often be observed in the spatial structure of a landscape, and the characteristics of the spatial structure can be informative of the processes generating the hierarchy.

Table 5.1. Aspects of Complex Systems and Corresponding Quantitative Measures for Describing Each Aspect

Features of Complex Systems	Quantitative Measures
Spatial heterogeneity	Landscape metrics, geostatistics, spatial statistics
Self-similar patterns	Power laws, allometric scaling
Hierarchical properties and scale breaks	Scalograms, multifractals
Large number of components	Principal components analysis, clustering, discriminant analysis, hierarchical partitioning
Nonlinear interactions	Scale analysis, simulation modeling
Multiple drivers	Multivariate regression, dynamic model simulation
Spatiotemporal cycles	Wavelets, Fourier transforms, ARIMA
Organization	Network analysis, information theory, entropy, simulation
Upscaling	Power laws, multivariate regression, simulation
Downscaling	Unmixing, nonlinear statistics, comparable upscaling approaches

[a]The quantitative measures are not meant as a complete list of approaches but to provide an initial guidance of approaches that have been helpful in the past. Applications and references for each method are described in the text

QUANTITATIVE APPROACHES FOR UNDERSTANDING HIERARCHY AND SCALE VARIATION

Quantitative measures compatible with hierarchical partitioning of environmental patterns are needed to accurately estimate relationships between patterns and processes, understand the causes and consequences of landscape variability, extend knowledge obtained from studies of limited spatial extent, and better forecast ecological dynamics in response to policy implementation (Table 5.1). These needs are met with the dizzying number of quantitative approaches available to landscape ecologists. These approaches offer choices for describing (a) spatial pattern and variability on a single scale and across multiple scales and (b) the system organization associated with these structures (Dale et al. 2002, Dungan et al. 2002, Fortin and Dale 2005, Li and Wu 2007). Of even greater challenge than analyzing landscape patterns is quantifying relationships between patterns and processes over a range of scales, which requires the use of a suite of scaling methods (Wu and Li 2006b).

Quantifying Landscape Heterogeneity

We first describe the quantitative approaches available for describing the scale dependence of landscape patterns and then review approaches for translating relationships across scales. Spatial patterns can be described by single-scale and multiscaled metrics, which include the composition and configuration of a landscape (Li and Reynolds 1994, Gustafson 1998, Li and Wu 2007). Composition metrics describe the diversity and variability in landscape components irrespective of their geometry and location (e.g., various richness and diversity measures for categorical maps and landscape-level means for continuous variables). Configuration metrics describe the spatial arrangement of landscape components (e.g., patch shape and fractal dimension for categorical variables and semi-variance for continuous variables).

Throughout this section, different quantitative approaches will highlight the differences in landscape data. Data describing spatial variability are often in either categorical or continuous type variables. As will be discussed, these different data models use generally distinct analysis tools, landscape metrics and spatial statistics. Landscape data also vary in their spatial representation, with raster and vector data models being common. In both types of data, space is divided into a homogeneous block, which is related to the grain size. Data in a raster format divides space into regularly shaped and consistently sized blocks distributed continuously throughout a landscape. Common sources for these data are remotely sensed imagery. Data in vector format divide space into irregularly shaped and varying-sized units from infinitesimally small points to convoluted polygons. A common source for these data includes governmental unit statistics. These differing data formats require different quantitative tools for their analysis; however, many of the concepts intrinsic to these tools are similar.

Landscape metrics are used commonly to quantifying spatial pattern and variation (McGarigal and Marks 1995, Wu 2004, Li and Wu 2007 for recent reviews). Landscape composition metrics describe the diversity and variability of landscape components. For example, several diversity metrics akin to species diversity indices (e.g., Shannon Diversity) can be computed to describe the relative heterogeneity of landscape composition. Landscape configuration metrics describe the arrangement of patches throughout the landscape. These metrics can be computed for the entire landscape, or individual landscape classes. Both analyses often provide distinct information regarding the variation of landscape heterogeneity. A variety of different metrics are easily computed in several freely available software packages such as Fragstats, Metaland, or RULE (McGarigal and Marks 1995, Cardille et al. 2005, Gardner and Urban 2007). Many metrics are correlated with each other; users should choose metrics to maximize information relevant to a research question (Riitters et al. 1995, Gustafson 1998).

Landscape metrics can also be used to describe the scale dependence of patterns through the construction of scalograms (Wu et al. 2002, Wu 2004). Scalograms of landscape metrics describe the variation in response to successive changes in the scale of a base map. Scale changes of a map can be conducted by both changes to the grain and extent of a base map. To change the grain size, the resolution of a raster map is degraded by aggregating pixels (Turner et al. 1989, Wu 2004). To change the extent, a base map is sampled from an arbitrary initial location and then a series of maps are generated by extending the area included in the sample (Wu et al. 2002). Comprehensive analyses of diverse real and simulated landscapes have shown that landscape pattern metrics exhibit only three kinds of patterns: simple power laws, staircase patterns, and erratic fluctuations (Wu et al. 2002, Shen et al. 2004, Wu 2004).

Landscape data also arrive in the form of variables that are continuous or ordinal and thus not directly amenable to analysis through landscape metrics. These data include both unclassified remotely sensed data such as vegetation indices and field surveys of local species diversity or nutrient contents. Spatial patterns of these variables can be analyzed through spatial statistics, such as autocorrelation indices or auto-variances (Rossi et al. 1992). The correlation in space of point-count data (e.g., infection locations) can be analyzed through Ripley's K (Haase 1995). These approaches compute either the correlation or variance associated with a variable separated by a specified distance or lag. The significance of a given estimate of spatial correlation can be assessed through Monte Carlo methods and neutral landscapes (Fortin and Dale 2005, Li and Wu 2007). Scalograms of these relationships are commonly computed to describe how either the autocorrelations (correlograms) or auto-variances (variograms) change with the lag between units. These analyses are most commonly modeled using one of a few functions to describe the range of spatial dependence, the sill (maximum spatial dependence), and the nugget (variation remaining below the sampling scale). Information from these analyses can further be used to identify trends in mean values, directionality, and local anisotropy.

Regularities in spatial patterns can be detected through approaches that identify spatial scales where patterns are related. Techniques for identifying regularities include wavelets, Fourier transformations, and auto regressive integrated moving averages (ARIMA). Wavelet analyses are increasingly becoming popular in spatial analyses because of the flexibility in specifying the scales where characteristic spatial patterns can be observed and also because they identify the locations where spatial relationships can be identified (Dale and Mah 1998, Grenfell et al. 2001, Sarkar and Kafatos 2004, Keitt and Urban 2005). Fourier transformations and the associated spectral and power analyses have also been widely used (Lobo et al. 1998, Keitt 2000, Lunetta et al. 2006). These approaches identify the scales of cyclical regularities in pattern as derived from Fourier transformations. ARIMA approaches, which have primarily developed for describing time-series data, provide an alternative approach for describing regularities in spatial patterns (Steele et al. 2005, Diez and Pulliam 2007). Combinations of these approaches have also been used to leverage the benefits of each individual technique (Alhamad et al. 2007).

Relationships between multiple variables distributed through space can be described through applications of multivariate regression-based analyses. This family of analyses include techniques such as principal components analysis (PCA), clustering, and discriminant analyses. Principle component analyses are commonly used to reduce a large number of intercorrelated variables into much fewer and orthogonal parameters that describe the variability present in the landscape (Naiman et al. 1994, Deutschewitz et al. 2003). Clustering techniques have been used to identify relatively homogeneous units within spatial data, and the resulting spatial distribution of cluster membership can help describe spatial variability (Hargrove and Hoffman 1999, 2004). Disciminant analysis examines the effectiveness of different partitioning schemes for multivariate data and as such can be used to verify the appropriateness of patch delineations (Herkert 1994, McPherson et al. 2006). Here again, combinations of all multiple methods can allow for robust analyses of landscape heterogeneity (Jenerette et al. 2002).

Occasionally the use of landscape metrics for continuous variables or spatial statistics for categorical variables may be necessary. To use landscape metrics on continuous variables, the data must be discretized for analysis. The discretization procedure is critical to the results because this procedure in part determines the analysis scale. Discretization of continuous patterns should explicitly examine the consequences of chosen break-points for the results of pattern analysis. In contrast, for analysis of categorical data by spatial statistics, data can be separated into a series of binary maps (e.g., habitat suitability). Indicator geostatistics is one approach to describe the patterns of binary classified maps (Boucher and Kyriakidis 2006). For scaling landscape variability, a systematic bootstrapping approach to sampling the landscape at different scales can describe how variability differs between hierarchical levels (Jenerette et al. 2006). While combining methodologies may provide the optimal solution to an analytical problem, these *ad hoc* tools should be thoroughly vetted before interpreting resulting patterns.

Quantifying System Organization

The degree of system organization integrates the relationships between spatial patterns and ecological functioning. Properties of system organization are closely related to the interactions within and between levels and the resulting stability, domains of attraction, and resilience of system (Holling 2001, Scheffer et al. 2001, Carpenter et al. 2005). Many of the approaches for identifying system organization have been described earlier in the context of rescaling analysis—rescaling requires assumptions of some types of system organization. These assumptions can be quantitatively assessed through approaches such as similarity analysis and dynamic model simulations (Urban 2005). An additional approach for describing system organization is becoming available through network analyses (Newman and Watts 1999, Urban and Keitt 2001, Grimm et al. 2005, Kossinets and Watts 2006, Montoya et al. 2006). Network analyses examine the connectivity between distinct system elements and characterize both the overall network topology and the relationships and connectedness of individual or groups of elements.

Scaling Relationships Within and Across Scale Domains

The challenges of scaling become much larger when a researcher moves from describing empirical patterns to using those patterns as tests against alternative hypotheses of functional relationships. Wu and Li (2006a, 2006b) classified a number of scaling methods into two main approaches: similarity-based and dynamic model-based approaches.

Allometric scaling is a common example of the similarity-based approach. The derivation of power-law relationships is grounded in biological allometry, statistical mechanics, and fractal geometry (Mandelbrot 1983, Barenblatt 1996, Schneider 2001, Brown et al. 2002). Fractal analysis, the geometry of self-similarity, addresses the frustration involved with describing real-world objects, such as clouds and mountain ranges, in terms of traditional Euclidean shapes, such as squares and spheres (Mandelbrot 1983, 1998). The classic question asks: How long is a coastline? The answer depends on the length of the ruler used to measure it: With a smaller ruler, one is able to measure more of the small coves and peninsulas, thereby increasing the measured length. For fractal objects the increase in measured length is predicted by a power-law relationship with the ruler length. Such a power-law relationship has been observed for many physical structures including coastlines, topography, and stream networks (Mandelbrot 1983, Phillips 1993, Nikora 1994, Nikora et al. 1999). In ecology, power-law scaling has been commonly used to describe the relationships between organism metabolism and body mass (West et al. 1997, Enquist et al. 1998, West et al. 2003) and between species diversity and habitat area (MacArthur and Wilson 1967). Several other examples abound, including stream discharge frequencies and stream solute concentration distributions, and regional lake water quality can also exhibit power-law relationships (Plotnick and Prestegaard 1993, Jenerette et al. 1998, Kirchner et al. 2000). The fundamental basis of power-law scaling is the hypothesis that the same process is responsible for patterns observed across a broad range of scales. A strong criticism of the allometric approach is the lack of information regarding transitions between scale domains; many seemingly fractal patterns are only observed within constrained scale domains (Avnir et al. 1998, Ludwig et al. 2000, Turcotte and Rundle 2002).

The derivation of allometric functions may take either an analytic or empirical approach (Wu and Hobbs 2002). While the analytic approach often employs such techniques as dimensional analysis, the empirical approach relies heavily on regression analysis of different kinds (Wu and Li 2006b). The choice of regression methods has been much discussed due to misapplication of ordinary least squares (OLS) regression. The OLS method is appropriate when the independent axis is expressly manipulated and the error in this variable is nonexistent. When evaluating scaling relations, this assumption is rarely satisfied; reduced major axis (RMA) regression is more appropriate because it identifies error in both the independent and dependent variables. Only a few examples have applied allometric approaches to understand spatial heterogeneity (Ludwig et al. 2000, Schneider 2001, Miller et al. 2004, Sponseller and Fisher 2006). However, interest in spatial allometry is increasing because the potential benefits of these approaches provide a general method to extrapolate observed patterns over a range of scales with a clear mechanistic explanation (Wu 2004).

A theory of system organization based on power-law distributions in the frequency of events has developed, commonly referred to as self-organized criticality (SOC) (Bak et al. 1988, Bak 1996, Turcotte and Rundle 2002). The theory predicts the frequency of dissipative events within a system scales as a power-law function of the size of the event (Turcotte and Rundle 2002). Large events are rare and small events are common—both are expected. Predictions of individual events is impossible, however, the frequency of any event is predicted by knowing the frequency of a single event size and the scaling relationship between event frequency and size. The SOC theory suggests that these internally driven events are the critical organizing characteristics of self-organizing systems. Examples of seemingly SOC processes important to ecological landscapes include forest fires (Bak et al. 1990, Malamud et al. 1998, Ricotta et al. 1999), co-evolutionary events (Kauffman 1993, Sole et al.

1996, 1999; but see Kirchner and Weil 1998 for contrary evidence), stream structure (Rodriguez-Iturbe et al. 1992, Rinaldo et al. 1993, Rigon et al. 1994, Stolum 1996), earthquakes (Christensen et al. 2002), and landslides (Guzzetti et al. 2002). In these applications of self-organized criticality theories, "input" to a complex system is nearly constant, whereas the "output" is a series of events whose magnitude varies due to internal mechanisms (Turcotte et al. 2002).

Upon initial inspection, hierarchy and power-law scaling seem incompatible (Werner 1999). Hierarchy theory suggests that with changes in scale, controls of the system changes and therefore the behavior of ecosystem attributes should change; there should be scale breaks that are not easily determined *a priori* for a specific system. In contrast, power-law scaling suggests that a single or a universal set of similar processes drives the system at all scales; all sizes of events between the fundamental unit and the maximum system size are expected to occur and are explainable in reference to the same process. However, fractal distributions have only been observed spanning a limited range of scales (Avnir et al. 1998). Finite-size scaling effects are real as dimensions approach the fundamental building blocks or entire extent of the system. These observations are congruent with a hierarchical approach examining scaling relations across multiple domains of scale. In other words, allometric scaling relations may be expected within a single scale domain, but scaling thresholds will result in shifts at the borders between adjacent scaling domains (Wu and Li 2006b).

Incorporating allometric scaling within limited scale domains and the disjunctions in scaling relationships that occur at the boundary of scale domains is a challenge spanning all of science. The more comprehensive scaling approaches allow for key processes to dominate system dynamics, but also allow these processes to vary between hierarchical levels. An integration of these ideas is central to multifractals, mathematical models that allow a scaling exponent to vary with scale (Lavallee et al. 1993, Mandelbrot 1999). Multifractals incorporate ideas of hierarchy theory such as the changes in control with changes in scale, as represented by variation in the scaling exponent. Another integration is suggested by the developing theory of highly optimized tolerance (HOT) (Carlson and Doyle 2002). The HOT theory acknowledges that many systems with high structural complexity have been designed to be robust to much environmental heterogeneity, although they are especially fragile to certain aspects of change. This developing theory provides an alternative to understanding complexity in systems that were designed for specialized functions, and it may be useful for understanding systems managed for the production of ecosystem services.

In contrast to allometric scaling, dynamic model-based scaling involves the extension of mechanistic models from one scale to another. Process-based model extrapolation is being widely used for extending information on biological processes across scales. King (1991) describes four methods general scaling methods: extrapolation by lumping (EL), direct extrapolation (DE), extrapolation by expected value (EEV), and explicit integration (EI). Additional methods include extrapolation by effective parameters (EEP), spatially interactive modeling (SIM), and the scaling ladder method (SL) (Wu and Li 2006b, Wu and Hobbs 2002).

An EL scales to the target scale by parameterizing a local model using the averaged inputs over the landscape. Thus, EL does not consider spatial heterogeneity at the target scale, and consequently it is the simplest and most error-prone upscaling method. It is generally only applicable when the local model is linear and the interactions between patches are weak. A DE scales to the target scale by averaging the outputs of the local model using spatially varying patch specific parameters and inputs. This approach often reduces scaling errors due to nonlinearities in the model response to varying inputs (Bierkens et al. 2000). Both the data and computational intensity required for DE increase nonlinearly with landscape extent and grain size. Similar to DE, EEV scales to the target scale by implementing local models at the patch scale, but in contrast, EEV uses a sampling approach to describe the spatial heterogeneity probabilistically. By using a statistical approach, EEV is amenable to robust uncertainty analysis. An EI scales to the target scale by analytically or numerically integrating the local-scale model. To use EI, spatial heterogeneity must be represented as mathematical functions of space in closed forms, and the indefinite integral of the local model with respect to space must be obtainable. When these requirements are met, EI would be the most efficient and accurate, but meeting these assumptions is usually impossible. The SIM approach differs from the previous approaches by directly modeling the dynamics within and between patches. SIM incorporates feedbacks, time delays, and scale-specific features. All of these methods are primarily short-range scaling procedures that generally are applicable only between adjacent scale domains. The SL method provides a framework for coupling several short-range approaches into a coherent methodology between scales (Wu 1999, Wu and David 2002). It is based on the hierarchical patch dynamics paradigm (HPDP), which synthesizes ideas from hierarchy theory and patch dynamics (Wu and Loucks 1995). The SL approach is implemented by establishing a spatial patch hierarchy consisting of a series of nested scale domains and then using it as a scaling ladder to dynamically transfer information between two adjacent scales, one step a time.

Each of these approaches offers alternative methods for estimating the parameters and inputs to dynamic model at scales larger than those of the original data collection. Because of the central importance of uncertainty analysis in scaling, the EEV approach is often the recommended approach. Advances in uncertainty analysis associated with simulation modeling are rapidly developing and offer powerful alternatives (Li and Wu 2006). Inverse modeling approaches, where model parameters are derived by maximizing the fit between model output and observed patterns, can provide a robust estimate of coupled model and data uncertainty. Monte Carlo Markov chain (MCMC) is a widely used method for parameter estimation and uncertainty assessment (Clark 2005). With the increasing availability of computational power, the dynamic modeling-based approaches will continue providing the greatest flexibility and opportunities for scaling processes operating over a range of scales.

While much of the historical interest in scaling has focused on upscaling, downscaling is also becoming increasingly necessary. The goal of downscaling coarse-grained information is to derive the fine-scale pattern within a given areal unit (e.g., pixel or patch) (Bloschl and Sivapalan 1995, Bierkens et al. 2000, Wu et al. 2006). If only the composition within the coarse-grain information is desired, endmember unmixing approaches may be appropriate. Endmember unmixing is a common approach when a given sample is composed of a mixture of unique elements; the observed pattern at the coarse scale is a combination of several endmembers whose composition to the observation is unique (Sabol et al. 1992, Asner et al. 2005). The most widely used endmember unmixing approaches are linear models, which assume that the observation at a coarse scale is a linear combination of all members at the fine-scale (Buyantuyev and Wu 2007). The relative abundance of each land-cover type within a pixel is obtained by solving a closed system of n linear equations, where n is the number of distinct data sources. Unmixing has been widely used in remote sensing image classification. Remote-sensing scientists have developed a series of subpixel analysis methods to "unmix" individual pixels to estimate the relative areal proportions of different land-cover types within a pixel. In recent years, a number of nonlinear methods have been developed for pixel unmixing, including fuzzy membership functions (Foody 1999), indicator geostatistics (Boucher and Kyriakidis 2006), and neural-network-based methods (Moody et al. 1996). Downscaling has been an important research focus in climatological studies where global climate model output is incorporated into regional scale studies (Wilby and Wigley 1997, Kidson and Thompson 1998, Schmidli et al. 2007). Extensions of downscaling to ecological problems remain an important research challenge.

Problem Domains for Quantitative Scaling

There are many needs to which advances in scaling will help with estimating societal risks and managing landscapes. Previous research has suggested several theoretical challenges for scaling (Wu et al. 2006); here we focus on broad problem domains that require scaling applications. Scaling research is critically needed to downscale the effects of global changes, upscale physiological responses to these changes, and link cross-scale processes between ecological and sociological systems. Many ecologically important variables are changing globally in response to society, such as increasing temperature, drying, disturbances, species migrations, nutrients, and land conversions (Vitousek 1994). How will these global changes be reflected in conditions locally and regionally? The direct impacts of these changes vary in scale, such as the increase in CO_2 concentration in the well-mixed atmosphere and the more constrained deposition of nitrogen particles downwind of an emission site. How landscapes will respond to these altered inputs and new management decisions for producing ecosystem services is unknown (Walker 1994, Bennett et al. 2005). From the bottom-up direction, upscaling physiological processes and responses to environmental conditions is a key research challenge (Harvey 2000). Physiological responses are nonlinear with periods of activity and dormancy coupled with a constrained but dynamic maximization of resource allocation for growth. Both of these processes respond to environmental cues, are interactive, and can be hysteretic. These physiological responses are amenable to scientific analysis and are featured in the plot sizes of most ecological field studies.

Physiological dynamics at the ecosystem and individual organizational levels may be especially relevant to understanding sustainability issues as diverse as freshwater availability, disease spread, disturbance propagation, and natural resource product generation including food, fuel, and fiber. Upscaling knowledge on physiological processes, including responses to global changes, will allow for the feedbacks between local and global changes to be robustly analyzed. The remaining challenge, linking ecological with societal processes, is essential for generating a truly predictive landscape science amenable to assessing risk and providing knowledge useful for managing landscapes. The difficulty in this scaling challenge lies in the mismatch between intrinsic scales of the quasi-independent ecological and sociological systems. The scale mismatch is manifest as seemingly arbitrary relationships between spatial patterns of organization. A common example is shown by river systems. While rivers have often been considered integrative components of the entire watershed ecological functioning, they are often boundaries between political units from individual plots to international borders. Hierarchical approaches will be needed to link processes in both subsystems to understand how they respond to variation in each. These links are strong in many regions: Ecosystems are responding dramatically to societal decisions yet provide many nonsubstitutable services. Will approaches developed from either ecological or sociological traditions be sufficient? Likely a new suite of theories will be required to address this substantial knowledge gap.

CONCLUDING REMARKS

In evaluating environmental risks, the variation of patterns and processes with scale is a core concern. Hierarchical approaches have proven to be useful for describing and predicting scale-dependent variations both within and between domains of scale. Hierarchy theory provides a general framework for implementing a variety of quantitative analyses and understanding the results from diverse sources of knowledge. The spatial structure and scaling relationships in a region often provide insights into the processes shaping landscape patterns and their scale multiplicity. Landscape ecology provides a rich theory and a suite of tools for understanding these process-pattern relationships. Disturbance events, including those with large economic and human health consequences, are expected in complex ecological systems. Resilience and adaptive cycle approaches seem to provide additional tools for blending within and between scale processes (Gunderson and Holling 2002). However, the quantitative tools offered by this approach are still being actively researched. Initial studies of ecological resilience suggest that allowing smaller events to regularly occur can potentially mitigate large disturbance events. This systems-based approach to resilience is in contrast with many sociological concepts of resilience. Even relatively small events at the system level can have catastrophic effects to many individuals. Furthermore, the effect of these events will not be equitably distributed across societal strata because of variation in coping mechanisms. How to simultaneously increase the resilience of individuals without reducing the resilience of the entire system is a fundamental challenge for landscape risk assessments.

ACKNOWLEDGMENTS

This research was supported by the National Science Foundation (NSF), through a Biological Informatics postdoctoral fellowship awarded in 2004 and research grant (CNH-0814692) to GDJ and research grants to JW (BCS-0508002, Biocomplexity/CNH and DEB-0423704, Central Arizona-Phoenix Long-Term Ecological Research). Any opinions, findings and conclusions or recommendation expressed in this material are those of the authors and do not necessarily reflect the views of NSF.

REFERENCES

Alhamad MN, Stuth J, Vannucci M. 2007. Biophysical modelling and NDVI time series to project near-term forage supply: Spectral analysis aided by wavelet denoising and ARIMA modelling. *Int J Remote Sensing* **28**:2513–2548.

Asner GP, Knapp DE, Cooper AN, Bustamante MMC, Olander LP. 2005. Ecosystem structure throughout the Brazilian Amazon from Landsat observations and automated spectral unmixing. *Earth Interactions* **9**:Paper Number 7.

Avnir D, Biham O, Lidar D, Malcai O. 1998. Is the geometry of nature fractal? *Science* **279**:39–40.

Bak P. 1996. *How Nature Works: The Science of Self-Organized Criticality*. Copernicus, New York.

Bak P, Chen K, Tang C. 1990. A forest-fire model and some thoughts on turbulence. *Phys Lett A* **147**:297–300.

Bak P, Tang C, Wiesenfeld K. 1988. Self-organized criticality. *Phys Rev A* **38**:364–374.

Barenblatt G. 1996. *Scaling, Self-Similarity, and Intermediate Asymptotics*. Cambridge University Press, Cambridge, UK.

Bennett EM, Peterson GD, Levitt EA. 2005. Looking to the future of ecosystem services. *Ecosystems* **8**:125–132.

Bierkens MFP, Finke PA, de Willigen P. 2000. *Upscaling and Downscaling Methods for Environmental Research*. Kluwer, Dordrecht, The Netherlands.

Bloschl G, Sivapalan M. 1995. Scale issues in hydrological modeling—A review. *Hydrol Processes* **9**:251–290.

Boucher A, Kyriakidis PC. 2006. Super-resolution land cover mapping with indicator geostatistics. *Remote Sensing Environ* **104**:264–282.

Brose U, Ostling A, Harrison K, Martinez ND. 2004. Unified spatial scaling of species and their trophic interactions. *Nature* **428**:167–171.

Brown JH, Gupta VK, Li BL, Milne BT, Restrepo C, West GB. 2002. The fractal nature of nature: Power laws, ecological complexity and biodiversity. *Philos Trans R Soc London Ser B Biol Sci* **357**:619–626.

Buyantuyev A. Wu J. 2007. Estimating vegetation cover in an urban environment based on Landsat ETM+ imagery: A case study in Phoenix, USA. *Int J Remote Sensing* **28**:269–291.

Cardille J, Turner M, Clayton M, Gergel S, Price S. 2005. METALAND: Characterizing spatial patterns and statistical context of landscape metrics. *Bioscience* **55**:983–988.

Carlson JM, Doyle J. 2002. Complexity and robustness. *Proc Natl Acad Sci USA* **99**:2538–2545.

Carpenter SR, Turner MG. 2000. Hares and tortoises: Interactions of fast and slow variables in ecosystems. *Ecosystems* **3**:495–497.

Carpenter SR, Ludwig D, Brock WA. 1999. Management of eutrophication for lakes subject to potentially irreversible change. *Ecol Appl* **9**:751–771.

Carpenter SR, Westley F, Turner MG. 2005. Surrogates for resilience of social–ecological systems. *Ecosystems* **8**:941–944.

Christensen K, Danon L, Scanlon T, Bak P. 2002. Unified scaling law for earthquakes. *Proc Natl Acad Sci USA* **99**:2509–2513.

Clark JS. 2005. Why environmental scientists are becoming Bayesians. *Ecol Lett* **8**:2–14.

Colnar AM, Landis WG. 2007. Conceptual model development for invasive species and a regional risk assessment case study: The European green crab, *Carcinus maenas*, at Cherry Point, Washington, USA. *Human Ecol Risk Assess* **13**:120–155.

Dale MRT, Mah M. 1998. The use of wavelets for spatial pattern analysis in ecology. *J Veg Sci* **9**:805–814.

Dale MRT, Dixon P, Fortin MJ, Legendre P, Myers DE, Rosenberg MS. 2002. Conceptual and mathematical relationships among methods for spatial analysis. *Ecography* **25**:558–577.

Dent CL, Grimm NB, Fisher SG. 2001. Multiscale effects of surface–subsurface exchange on stream water nutrient concentrations. *J N Am Benthol Soc* **20**:162–181.

Deutschewitz K, Lausch A, Kuhn I, Klotz S. 2003. Native and alien plant species richness in relation to spatial heterogeneity on a regional scale in Germany. *Global Ecol Biogeogr* **12**:299–311.

Diez JM, Pulliam HR. 2007. Hierarchical analysis of species distributions and abundance across environmental gradients. *Ecology* **88**:3144–3152.

Dungan JL, Perry JN, Dale MRT, Legendre P, Citron-Pousty S, Fortin MJ, Jakomulska A, Miriti M, Rosenberg MS. 2002. A balanced view of scale in spatial statistical analysis. *Ecography* **25**:626–640.

Enquist, BJ, Brown JH, West GB. 1998. Allometric scaling of plant energetics and population density. *Nature* **395**:163–165.

Fisher SG, Grimm NB, Marti E, Gomez R. 1998a. Hierarchy, spatial configuration, and nutrient cycling in a desert stream. *Austr J Ecol* **23**:41–52.

Fisher SG, Grimm NB, Marti E, Holmes RM, Jones JB. 1998b. Material spiraling in stream corridors: A telescoping ecosystem model. *Ecosystems* **1**:19–34.

Foody GM. 1999. The continuum of classification fuzziness in thematic mapping. *Photogramm Eng Remote Sens* **65**:443–451.

Fortin MJ, Dale M. 2005. *Spatial Analysis: A Guide for Ecologists*. Cambridge University Press, Cambridge, UK.

Gardner RH, Urban DL. 2007. Neutral models for testing landscape hypotheses. *Landscape Ecol* **22**:15–29.

Gillson L. 2004. Evidence of hierarchical patch dynamics in an East African savanna? *Landscape Ecol* **19**:883–894.

Grenfell BT, Bjornstad ON, Kappey J. 2001. Travelling waves and spatial hierarchies in measles epidemics. *Nature* **414**:716–723.

Grimm V, Revilla E, Berger U, Jeltsch F, Mooij WM, Railsback SF, Thulke H-H, Weiner J, Wiegand T, DeAngelis DL. 2005. Pattern-oriented modeling of agent-based complex systems: Lessons from ecology. *Science* **310**:987–991.

Gunderson LH, Holling CS (Eds.). 2002. *Panarchy: Understanding Transformations in Human and Natural Systems*. Island Press, Washington, DC.

Gustafson EJ. 1998. Quantifying landscape spatial pattern: What is the state of the art? *Ecosystems* **1**:143–156.

Guzzetti F, Malamud BD, Turcotte DL, Reichenbach P. 2002. Power-law correlations of landslide areas in central Italy. *Earth Planet Sci Lett* **195**:169–183.

Haase P. 1995. Spatial pattern-analysis in ecology based on Ripley K-function—Introduction and methods of edge correction. *J Veget Sci* **6**:575–582.

Hargrove WW, Hoffman FM. 1999. Using multivariate clustering to characterize ecoregion borders. *Comp Sci Eng* **1**:18–25.

Hargrove WW, Hoffman FM. 2004. Potential of multivariate quantitative methods for delineation and visualization of ecoregions. *Environ Manage* **34**: S39–S60.

Harvey LDD. 2000. Upscaling in global change research. *Climatic Change* **44**:225–263.

Herkert JR. 1994. The effects of habitat fragmentation on midwestern grassland bird communities. *Ecol Appl* **4**:461–471.

Holling CS. 2001. Understanding the complexity of economic, ecological, and social systems. *Ecosystems* **4**:390–405.

Holling CS, Gunderson L. 2002. Resilience and adaptive cycles. In Gunderson L, Holling CS (Eds.). *Panarchy: Understanding Transformations in Human and Natural Systems*. Island Press, Washington DC, pp. 25–62.

Holling CS, Gunderson L, Peterson GD. 2002. Sustainability and panarchies. pp. 63–102. In Gunderson L Holling CS (Eds.), *Panarchy: Understanding Transformations in Human and Natural Systems*. Island Press, Washington, DC.

Jelinski DE, Wu J. 1996. The modifiable areal unit problem and implications for landscape ecology. *Landscape Ecol* **11**:129–140.

Jenerette GD, Wu J. 2000. On the definitions of scale. *Bull Ecol Soc Am* **8**:104–105.

Jenerette GD, Lee J, Waller D, Carlson RE. 1998. The effect of spatial dimension on regionalization of lake water quality data. In Poiker TK, Chrisman N (Eds.), *The 8th International Symposium of Spatial Data Handling*. I.G.U. G.I.S. Study Group, Burnby, Canada, Vancouver, BC.

Jenerette GD, Lee J, Waller DW, Carlson RE. 2002. Multivariate analysis of the ecoregion delineation for aquatic systems. *Environ Manage* **29**:67–75.

Jenerette GD, Wu J, Grimm NB, Hope D. 2006. Points, patches, and regions: Scaling soil biogeochemical patterns in an urbanizing ecosystem. *Global Change Biol* **12**:1532–1544.

Kauffman SA. 1993. *The Origins of Order: Self-Organization and Selection in Evolution*. Oxford University Press, New York.

Keitt TH. 2000. Spectral representation of neutral landscapes. *Landscape Ecol* **15**:479–493.

Keitt TH, Urban DL. 2005. Scale-specific inference using wavelets. *Ecology* **86**:2497–2504.

Kidson JW, Thompson CS. 1998. A comparison of statistical and model-based downscaling techniques for estimating local climate variations. *J Climate* **11**:735–753.

King A. 1991. Translating models across scale in the landscape. Pages 479–517 *in* Turner MG, Gardner RH (Eds). *Quantitative Methods in Landscape Ecology*. Springer-Verlag, New York, NY.

Kirchner JW, Weil A. 1998. No fractals in fossil extinction statistics. *Nature* **395**:337–338.

Kirchner JW, Feng XH, Neal C. 2000. Fractal stream chemistry and its implications for contaminant transport in catchments. *Nature* **403**:524–527.

Kolasa J, Pickett STA. 1989. Ecological-systems and the concept of biological organization. *Proc Natl. Acad Sci USA* **86**:8837–8841.

Kolasa J, Drake JA, Huxel GR, Hewitt CL. 1996. Hierarchy underlies patterns of variability in species inhabiting natural microcosms. *Oikos* **77**:259–266.

Kossinets G, Watts DJ. 2006. Empirical analysis of an evolving social network. *Science* **311**:88–90.

Lavallee D, Lovejoy S, Schertzer D, Ladoy P. 1993. Nonlinear variability of landscape topography: Multifractal analysis and simulation. pp 158–192. In Lam NS, Cola LD (Eds.), *Fractals in Geography*. Prentice Hall, Englewood Cliffs, NJ.

Levin SA. 1998. Ecosystems and the biosphere as complex adaptive systems. *Ecosystems* **1**:431–436.

Li H, Wu J. 2006. Uncertainty analysis in ecological studies: An overview. In Wu J, Jones KB, Li H, Loucks OL (Eds.), *Scaling and Uncertainty Analysis in Ecology: Methods and Applications*. Springer, Dordrecht, pp. 45–66.

Li H, Wu J. 2007. Landscape pattern analysis: Key issues and challenges. In Wu J, Hobbs R (Eds.), *Key Topics in Landscape Ecology*. Cambridge University Press, Cambridge, UK, pp. 39–61.

Li HB, Reynolds JF. 1994. A simulation experiment to quantify spatial heterogeneity in categorical maps. *Ecology* **75**:2446–2455.

Lobo A, Moloney K, Chic O, Chiariello N. 1998. Analysis of fine-scale spatial pattern of a grassland from remotely-sensed imagery and field collected data. *Landscape Ecol* **13**:111–131.

Ludwig JA, Wiens JA, Tongway DJ. 2000. A scaling rule for landscape patches and how it applies to conserving soil resources in savannas. *Ecosystems* **3**:84–97.

Lunetta RS, Knight JF, Ediriwickrema J, Lyon JG, Worthy LD. 2006. Land-cover change detection using multi-temporal MODIS NDVI data. *Remote Sensing Environ* **105**:142–154.

MacArthur R, Wilson E. 1967. *The Theory of Island Biogeography*. Princeton University Press, Princeton, NJ.

Malamud BD, Morein G, Turcotte DL. 1998. Forest fires: An example of self-organized critical behavior. *Science* **281**:1840–1842.

Mandelbrot BB. 1983. *The Fractal Geometry of Nature*. W.H. Freeman and Co., San Francisco, CA.

Mandelbrot BB. 1998. Is nature fractal? *Science* **279**:783–784.

Mandelbrot BB. 1999. *Multifractals and 1/f Noise: Wild Self-Affinity in Physics*. Springer-Verlag, New York.

McGarigal K, Marks BJ. 1995. *FRAGSTATS: Spatial Pattern Analysis Program for Quantifying Landscape Structure*. USDA Forest Service General Technical Report PNW-GTR-351.

McPherson JM, Jetz W, Rogers DJ. 2006. Using coarse-grained occurrence data to predict species distributions at finer spatial resolutions-possibilities and limitations. *Ecol Modelling* **192**:499–522.

Miller JR, Turner MG, Smithwick EAH, Dent CL, Stanley EH. 2004. Spatial extrapolation: The science of predicting ecological patterns and processes. *Bioscience* **54**:310–320.

Montoya JM, Pimm SL, Sole RV. 2006. Ecological networks and their fragility. *Nature* **442**:259–264.

Moody A, Gopal S, Strahler AH. 1996. Artificial neural network response to mixed pixels in coarse-resolution satellite data. *Remote Sensing Environ* **58**:329–343.

Naiman RJ, Pinay G, Johnston CA, Pastor J. 1994. Beaver influences on the long-term biogeochemical characteristics of boreal forest drainage networks. *Ecology* **75**:905–921.

Newman MEJ, Watts DJ. 1999. Renormalization group analysis of the small-world network model. *Physics Lett A* **263**:341–346.

Nikora VI. 1994. On self-similarity and self-affinity of drainage basins. *Water Resources Res* **30**:133–137.

Nikora VI, Pearson CP, Shankar U. 1999. Scaling properties in landscape patterns: New Zealand experience. *Landscape Ecol* **14**:17–33.

O'Neill RV, DeAngelis DL, Waide JB, Allen TFH. 1986. *A Hierarchical Concept of Ecosystems*. Princeton University Press, Princeton, NJ.

Parker VT, Pickett STA. 1998. Historical contingencies and multiple scales of dynamics within plant communities. In Peterson DL, Parker VT (Eds.), *Ecological Scale: Theory and Applications*. Columbia University Press, New York, pp. 171–192.

Phillips JD. 1993. Interpreting the fractal dimension of river networks. In Lam NS, Cola LD (Eds.), *Fractals in Geography*. Prentice Hall, Englewood Bluffs, pp. 142–157.

Pickett STA, Cadenasso ML. 1995. Landscape ecology: Spatial heterogeneity in ecological-systems. *Science* **269**:331–334.

Pickett STA, Collins SL, Armesto JJ. 1987. A hierarchical consideration of causes and mechanisms of succession. *Vegetation* **69**:109–114.

Plotnick RE, Prestegaard K. 1993. Fractal analysis of geologic time series. In Lam NS, Cola LD (Eds.), *Fractals in Geography*. Prentice Hall, Englewood Cliffs, NJ, pp. 193–210.

Prigogine I, Stengers I. 1984. *Order Out of Chaos*. Bantam Books, New York.

Ricotta C, Avena G, Marchetti M. 1999. The flaming sandpile: Self-organized criticality and wildfires. *Ecol Modelling* **119**:73–77.

Rigon R, Rinaldo A, Rodrigueziturbe I. 1994. On landscape self-organization. *J Geophys Res Solid Earth* **99**:11971–11993.

Riitters KH, Oneill RV, Hunsaker CT, Wickham JD, D. Yankee H, Timmins SP, Jones KB, Jackson BL. 1995. A factor-analysis of landscape pattern and structure metrics. *Landscape Ecol* **10**:23–39.

Rinaldo A, Rodriguez-Iturbe I, Rigon R, Ijjaszvasquez E, an Bras RL. 1993. Self-organized fractal river networks. *Phys Rev Lett* **70**:822–825.

Rodriguez-Iturbe I, Rinaldo A, Rigon R, Bras RL, Ijjaszvasquez E, Marani A. 1992. Fractal structures as least energy patterns: The case of river networks. *Geophys Res Lett* **19**:889–892.

Rossi RE, Mulla DJ, Journel AG, Franz EH. 1992. Geostatistical tools for modeling and interpreting ecological spatial dependence. *Ecol Monogr* **62**:277–314.

Sabol DE, Adams JB, Smith MO. 1992. Quantitative subpixel spectral detection of targets in multispectral images. *J Geophys Res Planets* **97**:2659–2672.

Sarkar S, Kafatos M. 2004. Interannual variability of vegetation over the Indian sub-continent and its relation to the different meteorological parameters. *Remote Sensing Environ* **90**:268–280.

Scheffer M, Carpenter SR. 2003. Catastrophic regime shifts in ecosystems: Linking theory to observation. *Trends Ecol Evol* **18**:648–656.

Scheffer M, Carpenter S, Foley JA, Folke C, Walker B. 2001. Catastrophic shifts in ecosystems. *Nature* **413**:591–596.

Schmidli J, Goodess CM, Frei C, Haylock MR, Hundecha Y, Ribalaygua J, Schmith T. 2007. Statistical and dynamical downscaling of precipitation: An evaluation and comparison of scenarios for the European Alps. *J Geophys Res Atmos* doi:10.1029/2005JD007026.

Schneider DC. 2001. Spatial allometry: Theory and application to experimental and natural aquatic ecosystems. In Gardner RH, Kemp WM, Kennedy VS, Peterson JE (Eds.), *Scaling Relations in Experimental Ecology*. Columbia University Press, New York, pp. 113–153.

Shen W, Jenerette GD, Wu J, Gardner RH. 2004. Scaling properties of simulated landscapes. *Ecography* **27**:459–469.

Simon HA. 1996. *The Sciences of the Artificial*, third edition. MIT Press, Cambridge, MA.

Sole RV, Bascompte J, Manrubia SC. 1996. Extinction: Bad genes or weak chaos? *Proc R Soc London Ser B Biol Sci*. **263**:1407–1413.

Sole RV, Manrubia SC, Benton M, Kauffman S, Bak P. 1999. Criticality and scaling in evolutionary ecology. *Trends Ecol Evol* **14**:156–160.

Sponseller RA, Fisher SG. 2006. Drainage size, stream intermittency, and ecosystem function in a Sonoran Desert landscape. *Ecosystems* **9**:344–356.

Steele BM, Reddy SK, Nemani RR. 2005. A regression strategy for analyzing environmental data generated by spatio-temporal processes. *Ecol Modelling* **181**:93–108.

Stolum HH. 1996. River meandering as a self-organization process. *Science* **271**:1710–1713.

Turcotte DL, Malamud BD, Guzzetti F, Reichenbach P. 2002. Self-organization, the cascade model, and natural hazards. *Proc Natl Acad Sci USA* **99**:2530–2537.

Turcotte DL, Rundle JB. 2002. Self-organized complexity in the physical, biological, and social sciences. *Proc Natl Acad Sci USA* **99**:2463–2465.

Turner BL, R. Kasperson E, Matson PA, McCarthy JJ, Corell RW, Christensen L, Eckley N, Kasperson JX, Luers A, Martello ML, Polsky C, Pulsipher A, Schiller A. 2003. A framework for vulnerability analysis in sustainability science. *Proc Natl Acad Sci USA* **100**:8074–8079.

Turner MG, O'Neill RV, Gardner RH, Milne BT. 1989. Effects of changing spatial scale on the analysis of landscape pattern. *Landscape Ecol* **3**:153–162.

Urban DL, O'Neill RV, Shugart HH. 1987. Landscape ecology. *Bioscience* **37**:119–127.

Urban D, Keitt T. 2001. Landscape connectivity: A graph-theoretic perspective. *Ecology* **82**:1205–1218.

Urban DL. 2005. Modeling ecological processes across scales. *Ecology* **86**:1996–2006.

Vitousek PM. 1994. Beyond global warming: Ecology and global change. *Ecology* **75**:1861–1876.

Walker BH. 1994. Landscape to regional-scale responses of terrestrial ecosystems to global change. *Ambio* **23**:67–73.

Waltho N, Kolasa J. 1994. Organization of instabilities in multispecies systems, a test of hierarchy theory. *Proc Natl Acad Sci USA* **91**:1682–1685.

Werner BT. 1999. Complexity in natural landform patterns. *Science* **284**:102–104.

West GB, Brown JH, Enquist BJ. 1997. A general model for the origin of allometric scaling laws in biology. *Science* **276**:122–126.

West GB, Savage VM, Gillooly J, Enquist BJ, Woodruff WH, Brown JH. 2003. Why does metabolic rate scale with body size? *Nature* **421**:713–713.

Wickham JD, Riitters KH. 1995. Sensitivity of landscape metrics to pixel size. *Int J Remote Sensing* **16**:3585–3594.

Wilby RL, Wigley TML. 1997. Downscaling general circulation model output: A review of methods and limitations. *Prog Phys Geogr* **21**:530–548.

Wu JG. 1999. Hierarchy and scaling: Extrapolating information along a scaling ladder. *Can J Remote Sensing* **25**:367–380.

Wu J. 2004. Effects of changing scale on landscape pattern analysis: Scaling relations. *Landscape Ecol* **19**:125–138.

Wu J, Li H. 2006a. Concepts of scale and scaling. In Wu J, Jones KB, Li H, Loucks OL (Eds.), *Scaling and Uncertainty Analysis in Ecology: Methods and Applications*. Springer, Dordrecht, pp. 3–15.

Wu J, Li H. 2006b. Perspectives and methods of scaling. In Wu J, Jones KB, Li H, Loucks OL (Eds.), *Scaling and Uncertainty Analysis in Ecology: Methods and Applications*. Springer, Dordrecht, pp. 17–44.

Wu JG, David JL. 2002. A spatially explicit hierarchical approach to modeling complex ecological systems: Theory and applications. *Ecol Modelling* **153**:7–26.

Wu J, Hobbs R. 2002. Key issues and research priorities in landscape ecology: An idiosyncratic synthesis. *Landscape Ecology* **17**:355–365.

Wu JG, Loucks OL. 1995. From balance of nature to hierarchical patch dynamics: A paradigm shift in ecology. *Q Rev Biol* **70**:439–466.

Wu J, Shen W, Sun W, Tueller PT. 2002. Pattern and scale: Effects of changing grain size and extent on landscape metrics. *Landscape Ecol* **17**:761–782.

Wu J, Li H, Jones KB, Loucks OL. 2006. Scaling with known uncertainty: A synthesis. In Wu J, Jones KB, Li H, Loucks OL (Eds.), *Scaling and Uncertainty Analysis in Ecology: Methods and Applications*. Springer, Dordrecht, pp. 329–346.

6

BAYESIAN MODELS IN ASSESSMENT AND MANAGEMENT

S. Jannicke Moe

Bayesian statistical methods are becoming increasingly popular in environmental risk assessment (Fox 2006), but risk assessors and managers could still benefit from wider use of this methodology. The purpose of this chapter is to give a brief presentation of Bayesian modeling and statistical methods and to show how these methods can be useful for environmental risk assessment and management from a landscape perspective. Bayesian methods are not only a set of methods, but a whole framework for modeling and statistical inference, and it offers an alternative to the classical or frequentist statistical framework. Although Bayesian theory has been known for 250 years, the application of Bayesian statistics is still seen as controversial in parts of the scientific community. This has resulted in strong disagreements between "Bayesians" and "frequentists" (Dennis 1996, Clark 2005).

Many people find it hard to correctly interpret the main concepts of classical statistics, such as p-value and confidence intervals. A good reason for this confusion is that classical statistics is constructed in a nonintuitive way. For example, the p-value of classical statistics represents the probability of observing the data (although they are actually observed), given that the null hypothesis is true (although you try to reject it). Bayesian statistics do the opposite: It calculates the probability of the hypothesis, given the data or other evidence. The outcome of this calculation is a probability distribution for the hypothesis or variable of interest, which is easier to interpret and more directly useful in a risk assessment than the output of classical statistics

(Ellison 1996, Johnson 1999). In the Bayesian framework, everything can be treated as probability distributions—both the hypotheses (including the model parameters) and the evidence (data or other types of information). This means that uncertainty can be accounted for and integrated from all steps of the model. This property should make the Bayesian approach especially valuable for environmental risk assessment and decision-making, where uncertainty from various sources is an important issue.

The use of probabilistic methods in risk assessment is already well established. For an overview of the of use of probabilistic methods in geo-referenced ecological risk assessment, see Verdonck (2003). A key property of Bayesian modeling, however, is to use new evidence to update the probability of a hypothesis. The prior probability distribution of the hypothesis can be combined with the probability distribution of the evidence (also called the likelihood function), and the product is the updated or posterior probability distribution (conditional on the evidence). The name "Bayesian" comes from the use of the Reverend Thomas Bayes' theorem for calculating posterior probabilities, which can be stated in a simplified version as

$$\text{Pr}_{\text{(hypothesis given new evidence)}} = \text{Pr}_{\text{(hypothesis)}} \times \text{Pr}_{\text{(new evidence)}}$$

[posterior probability]   [prior probability]   [likelihood function]

An important strength of Bayesian methodology is that different types of information can be combined, including data, model predictions, and expert judgment. This makes the approach very flexible, and particularly useful in situations with scarce data or high uncertainty. On the other hand, the fact that subjective opinion can be incorporated in a model (as prior probability distributions) has been a main reason for the controversy around Bayesian modeling (Dennis 1996). However, subjective opinion will often influence a model or assessment in various ways also in non-Bayesian approaches. Besides, with Bayesian methods the subjective influence can be more transparent and traceable than with classical methods (Clark 2005). Therefore, Bayesian methods are now considered to provide legitimate ways of incorporating subjective belief or expert opinion in the form of prior probability distributions (Fox 2006). Another reason why Bayesian methods have not been more in use historically is that the parameter estimation requires much computational power. However, this problem is now largely solved by modern powerful computers and freely available software, which I describe at the end of this chapter.

A literature search indicates that different branches of natural science have used Bayesian methods to very different extent. The following examples of applications of Bayesian methods have relevance for environmental risk assessment and management, particularly in a landscape context:

1. *Fisheries Management and Whaling.* This science has been forced to deal with scarce data, high uncertainty, multiple pressures (including harvest) and many stakeholders with diverging interest. Bayesian methods have made it possible to deal with many of these challenges (Punt and Kennedy 1997, Varis and Kuikka 1999, Maunder 2004, Peterman 2004). However, spatial aspects have not usually played a major role in this field.

2. *Water Management.* Management of freshwater resources also often involves many stakeholders, multiple pressures, and complex management problems. Bayesian network methodology has proved useful in this field, not least as a tool for communication between researchers and stakeholders (Varis and Kuikka 1999, Tattari et al. 2003, Borsuk et al. 2004, Henriksen et al. 2004, 2007; Malve 2007). A recent European Union research project (http://www. merit-eu.net/) has resulted in a special issue on Bayesian networks in water resource modeling and management (Henriksen et al. 2007, Pollino et al. 2007, Ticehurst et al. 2007, Castelletti and Soncini-Sessa 2007a, 2007b, 2007c; Martín de Santa Olalla et al. 2007). For coastal and estuarine waters, however, Bayesian methods as well as ERA in general appear to have been less applied (Newman et al. 2007).

3. *Forestry.* Forestry management, like water management, involves socially and ecologically complex problems. Outcome of Bayesian methods has proved more suitable for communication with stakeholders than has outcome of classical statistical methods (Ghazoul and McAllister 2003).

4. *Conservation Biology and Wildlife Management.* Prediction of probabilities is central for conservation biology (Wade 2000). Moreover, this field often deals with rare or threatened species, where data can be limited or data collection can be restricted. The possibility of combining different sources of information and judgment in a Bayesian model can therefore be particularly important in conservation biology. The importance of landscape properties is particularly evident for metapopulations; for examples of Bayesian applications see McCarthy et al. (2001), O'Hara et al. (2002), ter Braak and Etienne (2003), and Ovaskainen and Hanski (2004). Other examples of Bayesian approaches to population assessments that involve spatial aspects include Johnston and Forman (1989), Kangas and Kurki (2000), Rowland et al. (2003), Ree and McCarthy (2005), Sargeant et al. (2006), Steventon et al. (2006), and Gibbs (2007).

5. *Population Genetics.* Landscape features can constrain the movements of individuals within a population and thereby create subpopulations with reduced genetic diversity and increased risk of local extinction. Analysis of population genetics can therefore contribute to landscape-level risk assessment. So-called assignment tests use the observed allele frequencies to calculate the likelihood of each genotype in each of the populations. Bayesian assignment methods have been applied to assess the gene flow of wolverine populations (Cegelski et al. 2003) and ant populations (Mäki-Petäys et al. 2005). In both cases, it was found that increased habitat fragmentation had reduced the gene flow and might threaten the long-term persistence of these populations.

6. *Epidemiology.* The combination of geostatistical and Bayesian methods appear to be particularly advanced within spatial epidemiology and disease mapping (see reviews by Best et al. (2005) and Brooker (2007).)

7. *Natural Catastrophes.* Models for extreme events such as earthquakes and floods must obviously deal with high uncertainty. Natvig and Tvete (2007) discuss the use of Bayesian hierarchical space–time models for predicting areas of high earthquake risk, as an alternative to more standard extreme value models. Coles and Tawn (2005) show how modern statistical modeling techniques in combination with extreme value theory can improve flood risk assessment. They argue that Bayesian methodology is preferable to more conventional analyses, because it enables a risk-based interpretation of the results.

In the following, I will focus on certain methods that may be particularly relevant for environmental risk assessment and management in a landscape perspective: (1) Bayesian networks, (2) hierarchical Bayesian models, and (3) Bayesian geostatistical models. The principles and usefulness of these methods will be illustrated by examples mainly from water management and from epidemiology.

BAYESIAN NETWORKS

In this section, I introduce the concepts and properties of Bayesian Networks (BN). This is followed by an example of development and application of BNs for ecosystem modeling and for decision analysis. I conclude with a summary of the strengths and weaknesses in the use of BNs for ecological risk assessment.

Properties of BNs

Bayesian network models are gaining increasing popularity in environmental modeling, especially within fields such as natural resource management (Varis and Kuikka 1999, Tattari et al. 2003, Borsuk et al. 2004, Henriksen et al. 2004, 2007; Malve 2007) and conservation biology (Wade 2000, Marcot et al. 2001). A BN model is composed of probability distributions, but it can easily be represented as a graphical model (Fig. 6.1). Thus, a BN can function both as a conceptual model and as a mathematical model. The graphical version facilitates development of and communication about a complex environmental model, and it makes BN models suitable for involving stakeholders in the process of analyzing and improving the model.

The nodes of a BN model represent the random variables, which are formulated as probability distributions (or, equally, frequency distributions). The nodes are usually defined by discrete intervals, even if they represent a continuous variable. The arrows represent the causal links, which are formulated as conditional probability tables (CPTs). A "child node" (output node) descends from a "parent node" (input node) by an arrow directed toward the child node. The probability distribution of the children nodes are calculated as the combined probabilities of all combinations of all of their parent nodes, while the probability distribution of the parent node must be specified as a "marginal probability distribution" (Fig. 6.1). The marginal probability distributions of parent nodes and the CPTs can be based on different types of information, such as data, expert opinions, or estimates or predictions from other models

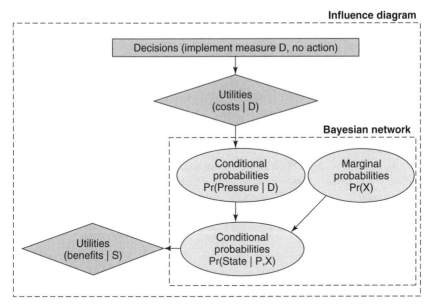

Figure 6.1. Principles of BN and influence diagrams, applied to an environmental manage-ment situation. The BN itself consists of random variables only (oval nodes). The state (*S*) of the environment is modeled by a probability distribution, Pr(State|*P*, *X*), which is conditional on the pressure (*P*) on the environment as well as on some other environmental variable (*X*). The variable *X* must be specified as a marginal (unconditional) probability distribution, which can be based on different sources. This BN can thus be used to explore the effects of varying pressure levels (and other environmental variables) on the state of the environment. The probability distribution of the pressure variable, Pr(Pressure|*D*), is in turn conditional on a decision (*D*) on whether to implement an abatement measure. The decision alternatives (here: yes/no) have different costs, which are modelled by a utility node (cost|*D*). The different levels of the state of the environment are also evaluated by a utility node, (benefits|*D*). The utility nodes can be used to compare the cost–efficiency or cost–benefit ratio of different management decisions.

(Borsuk et al. 2004, Castelletti and Soncini-Sessa 2007a, 2007b). This ability to inte-grate different types of information and models with associated uncertainties can make BNs useful for ecological risk assessment, which must often be based on information of various types and with high uncertainty. A BN model does not necessarily give the most realistic representation of the ecosystem, but rather a representation of our knowledge of the system (Borsuk et al. 2004).

Bayesian networks can also include special nodes for decisions and for valuation. Such a network is called an influence diagram (Fig. 6.1), and it can be a useful tool for cost–benefit analysis and for decision support.

An interesting property of BN models is structure learning: The network structure (causal links) need not be specified in advanced, but can be estimated (or learned) from the data. However, automatic structure learning can often result in illogical relationships, not least for environmental data, which are often noisy and correlated. For the purpose of environmental risk assessment, it is probably better to develop the network structure step by step in collaboration with experts and stakeholders.

Use of BNs in Ecosystem Modeling and Ecological Risk Assessment

The construction and use of a BN model will be illustrated with an example: a network that was developed for assessing the risk of eutrophication and algal blooms in Lake Vansjø in Norway (Fig. 6.2, Fig. 6.3). Of particular interest is the amount of cyanobacteria, because these can produce toxins that are harmful to humans and various other organisms. The model was developed and analyzed by researchers with different expertise (modeling, limnology, ecology, hydrology, and economics) in collaboration (Barton et al. 2008a, 2008b). The aim of the model was to answer questions such as:

- Which variable is the better predictor of the risk of cyanobacterial blooms; phosphorus concentration or algal biomass?
- How much reduction of phosphorus loading is required to keep the risk of cyanobacterial blooms at an acceptable level?
- Among the alternative abatement measures, which is the most cost-efficient in reaching the management target of good ecological status?

A submodel representing the ecological processes (Fig. 6.3) was used for testing several variants of the network, including other environmental and biological nodes, before this relatively simple version was agreed. The data used for parameterizing the conditional probability tables (CPTs), which provide the links between the nodes, consist of 1326 observations on total phosphorus, chlorophyll *a*, and percentage cyanobacteria (proportion out of total algal biomass) from a national lake eutrophication survey in Norway in 1988 (Oredalen and Faafeng 2002). The chlorophyll *a* concentration is modeled as conditional on the total phosphorus load and on the fraction of biologically available phosphorus (RBio_load). The percentage cyanobacteria is modeled as conditional on the chlorophyll *a* concentration only. The lake status (good or not) is set to be dependent on the percentage cyanobacteria. A separate node for the limit value of percentage cyanobacteria makes it possible to vary this criterion for good status.

A common way to use a BN is to fix the value of one of the nodes (set 100% probability for the selected interval) and to study the impacts on other nodes. For example, in order to assess the impact of various total phosphorus load levels, we can fix the total phosphorus node at a selected concentration interval and explore how this affects the probability distributions for algal abundances and for cyanobacterial blooms (see Fig. 6.3). A particularly interesting attribute of BNs is that not only input nodes, but also output nodes, can be fixed. This means that the network can be analyzed in any direction. We can use the network as a "normal" simulation model, to predict the probability distribution of output nodes based on given values of input nodes. But we

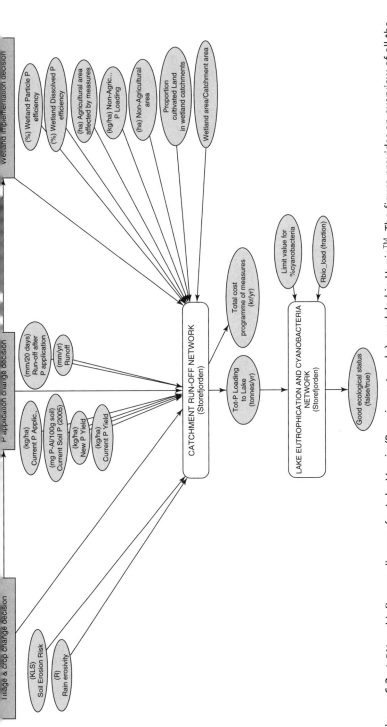

Figure 6.2. A BN and influence diagram for Lake Vansjø (Bayesian networks), modeled in Hugin™. The figure provides an overview of all the unconditional nodes that may be catchment-specific, the underlying sub-networks, and the conditional nodes showing policy-relevant results such as whether "good ecological status" is obtained and the "total cost of the program of measures." The model contains decision nodes for three different abatement measures: tillage and crop change; P (phosphorus) application change, and wetland implementation. The model also contains two sub-networks (shown as rounded rectangles), which are modeled as separate modules. The effects of these three decision and the associated costs are integrated in the first sub-network, "Catchment run-off network" (details not shown here). The resulting output variable "total phosphorus loading to lake" is the input variable for the next sub-network.

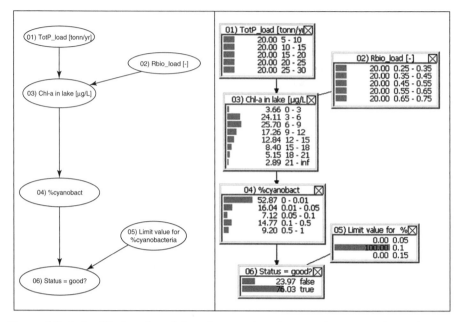

Figure 6.3. A simplified version of the sub-network "Lake eutrophication and cyanobacteria network." The left-hand side shows the ordinary display version, while the right-hand side shows the probability distribution for each node (probability for each possible state of the node).

can also use the model to project scenarios "backwards": to predict the probability distribution of input nodes based on given values of output nodes. For example, if we select a maximum allowed level for the algal abundance (output node), then what is the maximum phosphorus load (input node) that can be received by the lake? The answer will be given as a probability distribution, which also represent the uncertainty associated with this assessment.

Uncertainties in the ecological model can be represented both in the variables (nodes) and in the causal links (conditional probability tables) of a BN: Higher uncertainty is represented by a more even distribution in nodes or in CPTs. Note that the "uncertainty" represented by these probability distributions can represent natural, irreducible variability as well as conceptual uncertainty due to limited knowledge (which is potentially reducible). Therefore it is not necessarily an aim to minimize the uncertainty in such a model, but to explicitly describe the uncertainty as far as possible in all steps of the model construction and analysis.

A natural consequence of incorporating high uncertainties in a BN model (broad probability distributions) is that the model becomes less responsive to manipulation of the nodes. This is also what we experienced during our work with the BN for Lake Vansjø: Reducing the input phosphorus node to a minimum did not result in a correspondingly strong reduction of the algal abundance or the risk of toxic algal bloom. It is possible that more information about the ecological system can reduce some of this uncertainty (give narrower probability distributions) and thus amplify the response of the model. However, the "weak" relationships in the CPTs can also appropriately represent natural variability in the cause-effect relationship. We used two alternative versions of the CPT that link the cyanobacteria node to the total phosphorus concentrations node. In version 1 ("empirical relationship"), the values of the CPT was calculated directly as the joint frequency distributions of total phosphorus concentrations and cyanobacterial abundance in the dataset. In version 2 ("statistical relationship"), the CPT values were derived from a statistical regression model of this relationship, with associated standard error (more details in Barton et al. 2008a). The CPT based on the statistical relationship had lower variability (narrower distribution), which gave a stronger response in the network. But the empirical version of the CPT may nevertheless have given a more correct picture of the true variability in the system. Thus, a BN where natural variability is properly accounted for may give more correct predictions than a BN where the variability is underestimated (or a deterministic model where the variability is ignored).

Bayesian network models have been used in several other cases of water management, for modeling of nutrient load pressures as well as for other management issues. For example, a BN was used to integrate different models, representing the various processes involved in eutrophication in the Neuse River, USA (Borsuk et al. 2004). The sub-networks were based on different sources such as expert opinion, process-based models, multivariate regression models, and data from the literature. Their network model demonstrated that ecological improvement is likely to result from nitrogen reductions. However, like Barton et al. (2008b), they also experienced high predictive uncertainty, arising both from natural variation and from lack of knowledge.

Landscape properties can be incorporated into Bayesian networks in different ways. The BN models for Lake Vansjø and Neuse River were not spatially explicit, but landscape properties were included as by nodes that represented for example, habitat characteristics. A BN can also be constructed in a hierarchical way. A network can thus consist of several sub-networks (e.g., representing local patches or waterbodies) that are linked in a meta-network (e.g., representing a larger region or watershed). A hierarchical, spatially explicit BN was used to evaluate potential effects of federal land management alternatives on trends of salmonids and their habitats in the Columbia River (Rieman et al. 2001). This model consisted of sub-networks representing more than 6000 sub-watersheds. Like the BN for the Neuse River (Borsuk et al. 2004), the BN for the Columbia River was composed of several types of information: expert opinion, empirical observations, statistical models, and process-based models. Landscape properties were represented by nodes both on a regional scale (e.g., road density, slope steepness, grazing) and on more local scale (connectivity, habitat capacity, corridor conditions). In general, the aquatic habitat network reflected a collective belief that conditions of riparian corridors and their management would strongly influence the condition of habitats. The importance of landscape became obvious in their results: the management scenarios could play out differently, depending on where in the landscape the effects were considered. In this example, the formalization of a huge and complex ecological system allowed the representation of key relationships implemented with spatial detail, without the full complexity of a process-based model. However, such a meta-network can easily get complicated and requires substantial amounts of information for parameterization. In contrast, a Bayesian statistical model can more easily be constructed in a hierarchical manner (discussed later in this chapter).

When data are lacking completely, it is still possible to develop a BN entirely based on expert opinion. A BN was developed for game management of wildlife populations, built upon judgments of eight experts on wildlife ecology (Pellikka et al. 2005). The model was used to analyze uncertainties and inconsistencies among the judgments. Another BN based on expert judgment only was developed to assess the impact of buffer zones on water protection and biodiversity (Tattari et al. 2003). The model included factors on different spatial scales (site properties, buffer zone properties, field properties) as well as management measures and concern endpoints (including landscape effects). The uncertainties of the water experts' models were clearly higher at fine spatial resolution (site properties) than at coarser spatial scale (field properties). In risk assessment situations where no data are available, this approach can still give some information—for example, on which components are considered to be the most important by experts, and about which factors there is most uncertainty or disagreement.

In the examples given so far, the BNs have been used to model an ecological system. Furthermore, a BN can also be used to model other components of the risk assessment or the decision-making procedure. A BN was developed as a decision-analysis tool for evaluating a conservation plan for hundreds of species in northwestern United States (Marcot et al. 2001). The viability of these species was potentially at risk from multiple stressors, particularly disruption of their habitat and reduction and isolation of their population caused by various land management activities. The landscape and habitat properties were incorporated by various nodes such as distribution within habitat and geographic range. The BN models helped to detect logical flaws, bias, and vagueness in the guidelines and provided a way of representing inconsistent guidelines in a consistent structure.

Bayesian Influence Diagrams

A BN can be used more actively as a decision-analysis tool under uncertainty when it is developed into an influence diagram (Varis and Kuikka 1999). An influence diagram contains decision nodes and utility nodes (or valuation nodes) in addition to the ordinary probability nodes (see Fig. 6.1). Such a network can be used to rank management options according to a cost-efficiency or benefit–cost analysis. A decision node contains alternative decision, such as "action" or "no action." Utility nodes can represent both costs and benefits of the chosen strategy: the cost of the decisions and the valuation that the decision-maker places on the resulting environmental consequences. Nodes that are included in a decision-analysis model should typically be (1) potentially manageable, (2) predictable from available data or expert opinion, or (3) observable at the scale of interest for the management problem or risk assessment. In addition, nodes can be included if they are (4) of interest to the decision-makers or stake-holder or (5) helpful for assessing probability distributions for other variables that are of interest (Labiosa et al. 2005).

The influence diagram for Lake Vansjø contains decision nodes for three alternative abatement measures for reducing phosphorus loading: changes in fertilizer application, tillage practice, and wetland use (represented by rectangular nodes in Fig. 6.3). One purpose of the influence diagram was to rank the cost-effectiveness of the alternative abatement measures, according to the impact of the measures on the risk of cyanobacterial blooms. The different measures had rather diffuse effects in the model, when all uncertainties were accounted for (Barton et al. 2008b). Another version of the influence diagram for Lake Vansjø included benefit–cost analysis for a program of measures (Barton et al. 2005). This analysis concluded that the predicted benefits of the actions were not sufficient to justify the cost: For most abatement measures, the expected net value of "no action" exceeded the expected value net of implementation. This lack of responsiveness may indeed be an appropriate representation of the situation in this lake: Various abatement measures have successfully reduced the amount of phosphorus load over the last years, but the algal abundances and cyanobacterial blooms have nevertheless remained high (Bjørndalen et al. 2006). However, the lack of response in the models may also be due to nonoptimal model design, and the work on testing and improving the application of BNs for Lake Vansjø will continue in collaboration with lake managers.

Another example of an influence diagram (Labiosa et al. 2005) describes mercury loading from mines and uptake in fish tissue in the Sulphur Creek watershed. More influence diagrams for fisheries and water management are described by Varis and Kuikka (1999).

Strengths and Weaknesses of BNs

Certain properties of BN models can be particularly useful for EcoRA in a landscape setting:

- The modular structure of the network means that it can easily be updated with new submodels and new information.
- Spatial and landscape components can be included as separate nodes in the model.
- A BN can be used as a tool for communicating complex environmental problems among scientific experts, managers, and stakeholders.
- A BN can be used to integrate model of different types (Castelletti and Soncini-Sessa 2007b).
- Influence diagrams can be used for economic analysis and as a decision-analysis tool.

Nevertheless, BNs are a relatively new methodology and there are still many limitations considering use of BNs for ecological modeling and ecological risk assessment:

- BNs cannot be used as dynamic models (simulation over time steps). A possible solution is to link several replicated BNs by time steps, but this is not an efficient way for model simulation.

- BNs are by definition directed acyclic graphs, which means that feedback loops cannot be included. In cases where there are feedbacks among variables, the arrows must represent a net effect of the processes. Alternatively, feedback can be implemented in replicated BNs that are linked by time steps.
- The variables must usually be discretized (from continuous to discrete distributions); the choice of discretization can influence the outcome of the analysis. If a node has too few intervals, then patterns and relationships can be blurred. On the other hand, nodes with many intervals will require more information for specification of the causal links (parameterization of the CPTs). A CART analysis (Classification and Regression Tree, Breiman et al. 1984) can give a useful indication of how variables should be discretized in order to maximize the responsiveness of the network (Malve 2007, Barton et al. 2008a).
- BNs are not optimal for statistical inference such as parameter estimation, although it is possible.

The most important challenges for using BNs are not the technical issues, but to (a) properly comprehend the methodology as well as the environmental problem to be solved and (b) communicate this with all people involved in the process (Henriksen et al. 2007). Much has been learned about communication of Bayesian modeling results from applying Bayesian decision analysis to environmental and natural resource management (Varis and Kuikka 1999). Among their most important lessons are that working with an unconventional approach requires time, imagination, and ability to see analogies. Moreover, it is even more demanding to get other people to understand and accept such a new approach.

HIERARCHICAL BAYESIAN MODELING

In this section, I introduce Bayesian statistical inference as the foundation for hierarchical Bayesian modeling (HBM) in risk assessment. I close this section with a summary of the strengths and weaknesses of the approach.

Bayesian Statistical Inference

The statistical models that are typically used for hypothesis testing and parameter estimation in classical statistics, such as linear regression or analysis of variance (ANOVA), can also be constructed and analyzed in a Bayesian framework (Gelman et al. 1995, Clark and Lavine 2001). However, the procedure for parameter estimation differs fundamentally for the two approaches. Bayesian inference is based on summary statistics for a large number of simulations from a specified probability distribution; usually Monte-Carlo Markov chain (MCMC) simulations. In simple cases, such as linear regression, classical and Bayesian parameter estimations yield the same results. Environmental problems, however, are typically complex, especially when multiple stressors and spatial scales are considered. Therefore, when hypothesis testing or parameter estimation is needed for environmental risk assessment, there are advantages of Bayesian statistical models over that of classical statistics:

- The result of a Bayesian analysis can be more relevant and easier to interpret for a risk manager than the result of a classical analysis: the probability of a hypothesis being true, instead of the probability of the evidence being true.
- Bayesian statistics can more easily include more than two alternative hypotheses in a model, and it can rank the hypotheses according to the estimated probabilities of each hypothesis.
- Data or knowledge from different sources can more easily be combined, and the model can be updated as new information is obtained.
- Uncertainty can be modeled explicitly, for both evidence and hypotheses.
- It is possible to analyze models with complex structure, such as hierarchical models, which would get very complicated with classical statistics (Clark 2005).

An example of Bayesian inference in risk assessment is reported by Williams et al. (2005, 2006). They used a Bayesian logistic regression model to analyze the influence of landscape attributes on the probability of extirpation of localized plant populations in Australia. They found that attributes of the surrounding landscape (such as road density) and habitat quality were more influential than local-scale patch attributes (such as patch area and isolation) in their study. Other examples of Bayesian statistical inference that combine environmental risk and spatial aspects can be found in conservation biology (Tufto et al. 2000, Ree and McCarthy 2005) and wildlife management (Etterson and Bennett 2006).

HBM in EcoRA

Hierarchical modeling is one the most important advances of Bayesian statistics (Gelman et al. 1995). The fact that Bayesian statistical models can easily be constructed with a hierarchical structure can also make them very suitable for EcoRA in a landscape perspective. A thorough description of HBM and applications in ecological studies is given by Richardson and Best (2003). Here, the hierarchical approach is illustrated (Fig. 6.4) by an example from Finnish lake management (Malve and Qian 2006).

The management problem is similar to the Lake Vansjø case (introduced under the section "Bayesian Networks"): How much must phosphorus loadings be reduced in order to obtain a good ecological status of the lake and minimize the risk of algal blooms? The management target is to ensure that the algal concentration does not exceed a given limit. The relationship between phosphorus and algal abundance can be affected by various environmental factors, and a lake manager will therefore need a reliable estimation of the phosphorus–algae relationship for this particular lake. However, the amount of data on nutrients and algae for a specific lake is often limited, therefore using lake-specific data only will result in estimates with low precision (high uncertainty; Fig. 6.5, left plot). To reduce the uncertainty, data from other lakes

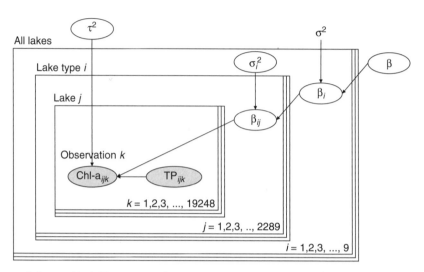

Figure 6.4. Graphical illustration of a HBM for estimating the relationship between total phosphorus (TP) and chlorophyll a (Chl-a) in Finnish lakes. The model has three levels: individual lakes, lake type (group of similar lakes), and all lakes. The observations of Chl-a concentrations (Chl-a$_{ijk}$) are modeled conditional on lake-specific parameter values (β_{ij}), with model error variance (τ^2). The lake-specific parameter values (β_{ij}) are in turn modeled conditional on lake-type-specific parameters (β_i), with between-lake variance within lake type (σ_i^2). Finally, the lake-type-specific parameters are modeled conditional on a parameter distribution of all lakes in Finland (β), with between-lake-type variance (σ^2).

can be included in the analysis. However, because environmental factors may vary across spatial scales, the inclusion of such data may introduce bias in the parameter estimates and make them less applicable for the target lake (Malve 2007). The dilemma is thus whether to use only lake-specific data and be "roughly right" (low precision, high accuracy) or to use all available data and be "precisely wrong" (high precision, low accuracy). A possible solution is to use a hierarchical model, where one can assume that all lakes in the dataset have some properties in common, while lakes that belong to the same region (or some other kind of grouping) have more in common than lakes from different regions. One can in this way "borrow information" about the phosphorus–algae relationship from other lakes in the dataset. A three-level Bayesian regression model was used for analysis of algal abundance in the Finnish lakes (Malve and Qian 2006): target lake, lake type (group of lakes with similar geological properties), and all lakes (Fig. 6.4). In cases with few data for the target lake, the precision improved considerably when data for all lakes were included, but the estimates were now biased ("precisely wrong") (Fig. 6.5, middle plot). Furthermore, the bias was considerably reduced when the data were used in a hierarchical manner (Fig. 6.5, right plot).

As described for BNs, an HBM can be run both "forwards and backwards": It can be used to predict output variables as well as input variables. For example, in the Finnish lakes case (Malve and Qian 2006), one can predict the risk of algal blooms

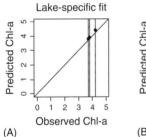
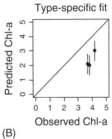
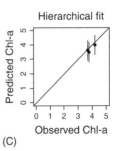

Figure 6.5. Comparison of predicted and observed values of Chl-a (μg/L, log-transformed) from three variants of a Bayesian regression for relationship between total phosphorus and Chl-a in Finnish lakes, illustrating the effect of hierarchical modeling on bias and uncertainty. **(A)** Lake-specific fit: Only data from the target lake is used ($N = 3$). The fit has low bias, but uncertainty is very high. **(B)** Type-specific fit. Data from all lakes are used ($N = 19,248$), and lake type is used as a covariate (but the identity of the target lake is not specified in the model). Uncertainty is reduced, but the fit is biased. **(C)** Hierarchical fit. Data from all lakes are used, and both lake type and target lake identity is specified. This variant has both low bias and low uncertainty. Modified from Malve and Qian (2006).

that must be expected for a given phosphorus load, or, vice versa one can predict the phosphorus load that must not be exceeded for a given risk of algal blooms. This "reversal" is possible in a Bayesian model, because any of the variables can be either specified (parameterized by a probability distribution) or unspecified (and thus be estimated by the data). Such a "reversal" would probably not be possible with a hierarchical model in a classical framework.

Hierarchical Bayesian models have been used in several other cases of water management. For example, Borsuk et al. (2001) used a hierarchical model to estimate the relationship between algal production and oxygen demand of river sediments in the Neuse River. With the hierarchical approach, they could account for the partial commonality in parameter values across different estuarine systems. Hierarchical models were also applied to analyze multi-stock data for pink salmon (Peterman 2004, Su et al. 2004). The hierarchical approach improved the estimates of probability distributions for model parameters compared with those derived through single-stock analyses.

A hierarchical Bayesian Poisson regression model was used to analyze the effects of patch attributes (area and isolation), as well as species–environment relations, on the species richness of amphibian assemblages in Australia (Parris 2006). A hierarchical model was chosen to account for the spatial structure in the data: clusters of ponds with varying size and degree of isolation. Both patch attributes and habitat quality were found to influence the species richness.

Many examples of risk assessment with HBM are found in the sciences of epidemiology and veterinary medicine. The importance of the spatial aspect in risk models is perhaps most obvious for infectious diseases such as tuberculosis (Souza et al. 2007) or the "mad-cow disease" (Abrial et al. 2005, Allepuz et al. 2007). However, geographical patterns are important also in predicting risks for noninfectious diseases such as cancer (Xia et al. 1997, Greco et al. 2005, Wakefield 2007) and diabetes (Ranta and Penttinen 2000). One benefit of the hierarchical approach is that one can obtain reliable risk estimates for small areas while using information from a larger geographic scale (Xia et al. 1997). Moreover, health data are often reported at an aggregate level, whereas risk factors are measured with higher resolution. This creates a spatial misalignment for exposure and effect data, and hierarchical modeling is suggested as a natural approach to this problem (Greco et al. 2005).

Hierarchical risk models that are developed for epidemiology have many components that can be relevant for landscape-based EcoRA. For example, HBMs can be used when: the exposure is heterogeneous and depends on geographical patterns; receptors exhibit spatial aggregation and movements; the spatial processes for exposure and for receptor have different scales; or information on these processes have different spatial resolution.

Strengths and Weaknesses of HBM

There are many benefits reported from the use of HBMs on complex problems involving health and environment effects (Richardson and Best 2003). These benefits should apply equally to other applications in EcoRAs:

- Modular model elaboration; if new information becomes available, this can be modeled in a separate module, which can subsequently be linked to the existing model.
- Integration of different sources of information; the modular properties of a HBM enable the simultaneous integration of different sources of information.
- Coherent propagation of uncertainty; the joint hierarchical model leads automatically to a correct propagation of all sources of uncertainty that have been quantified in each module onto the estimation of the parameters of interest.
- "Borrowing of strength"; the hierarchical structure allows for borrowing of strength between different datasets, thus leading to improved and more stable estimates of the parameters of interest. (This is an active area of research at the moment, with natural extensions to space–time models.)
- Integrated treatment of information at different levels; an HBM can in principle accommodate data that are observed at different spatial scales.

Application of HBMs obviously has limitations as well. Some limitations on use of this method for natural resource management are identified by Peterman (2004):

- A hierarchical model will not be beneficial for estimating parameters unless they are at least somewhat similar across datasets.
- When there are only a few datasets on different populations or locations, the approach will not be particularly advantageous.
- Nonlinear relationships, which are common in ecological systems, may cause problems for estimation with MCMC methods (convergence problems).

- It is often difficult to specify noninformative priors, and care should be taken to choose appropriate scale.

BAYESIAN GEOSTATISTICAL MODELS

The HBM presented previously can account for spatial heterogeneity mostly in a qualitative way—that is, by assigning units to groups with different landscape/habitat properties. However, this approach does not incorporate spatial gradients in a quantitative way. Spatial information can be incorporated more directly by coupling Bayesian methods with geographical information system (GIS) platforms and with remote sensing of environmental features. This combination has only been developed quite recently, but has proved very fruitful in risk analysis, particularly for disease mapping (analysis and prediction of spatial distributions of diseases). The combined Bayesian and GIS approach is used for various of types of human diseases [see, for example, Ranta and Penttinen (2000) and Best et al. (2005); see, for example, Louie and Kolaczyk (2006) and Brooker (2007)] as well as animal diseases (Thompson et al. 2005). A reason why this approach has been used so widely within epidemiology is that data on geographical distribution of diseases are often sparse, especially in developing countries, while there is a strong need for rational decision-making to help maximize cost-effective resource allocation. Analysis and management of uncertainty is therefore of vital importance in this field.

A benefit of using a Bayesian platform for geostatistical analyses is that the model can take into account the spatial variability in the environmental and exposure data by specifying probability distributions for the spatial processes and their covariation (Brooker 2007). The geostatistical analysis can help determine whether patterns are due to random processes or caused by specific variables such as environmental heterogeneity. A useful product of Bayesian geostatistical modeling is a risk map, which shows the different levels of risk as geographical patterns. In addition, maps of prediction error can highlight which areas need further investigations, in order to make the risk maps uniformly reliable (Raso et al. 2006).

A GIS-based, watershed-level assessment with Bayesian weights-of-evidence (WOE) and weighted logistic regression (WLR) was developed by Kapo and Burton (2006). They show how this technique provides a method to determine and compare potential environmental stressors in river ecosystems and to create predictive models of general or species-specific biological impairment across numerous spatial scales based on limited existing sample data. The WOE/WLR technique used in their study is a data-driven, probabilistic approach that was conceptualized in epidemiological research. Extrapolation of this method to a case-study watershed assessment (Ohio, USA) produced a quantitative determination of physical and chemical watershed stressor associations with biological impairment, along with a predicted comparative probability of biological impairment at a spatial resolution of 0.5 km^2 over the watershed study region. Habitat stressors showed the greatest spatial association with biological impairment in low-order streams, whereas water chemistry, particularly that of wastewater effluent, was associated most strongly with biological impairment in high-order reaches. Significant potential stressors varied by land-use and stream order as well as by species. This WOE/WLR method provides a highly useful "tier 1" watershed risk assessment product through the integration of various existing data sources, and it produces a clear visual communication of areas favorable for biological impairment and a quantitative ranking of candidate stressors and associated uncertainty (Kapo and Burton 2006).

Bayesian risk mapping has been particularly useful for risk-mapping of parasite-borne diseases such as schistosomiasis (Clements et al. 2006, Raso et al. 2006, Beck-Worner et al. 2007, Brooker 2007). The parasite is a trematode (fluke), which uses a freshwater snail as an intermediate host. Beck-Worner et al. (2007) have used one of the latest remote sensing products, a high-resolution (90 m) digital elevation model (DEM), to produce detailed topographic maps of watersheds in Côte d'Ivoire. The launch of the Shuttle Radar Topography Mission 2 in 2000 has facilitated the production of this model, and digital topographic data of high quality are now freely available. Spatial correlation of infections was modeled both as stationary (dependent on distance only) and as nonstationary (dependent on both distance and location). Finally, Bayesian kriging (interpolation between points of observation) was used to generate smooth risk maps for infection prevalence with covariates from both multi-variate stationary and nonstationary models. Landscape features such as stream order and flow velocity turned out to be important explanatory variables for the risk of schistomiasis infection. The explanation is that streams with high velocity are less suitable as habitat for the host snail on which the parasite depends.

The approach used by Beck-Worner et al. (2007) to map the risk of parasite trans-mission and infection might be applicable also for risks of other biological pressures, such as invasive species. In a similar way, knowledge of the invasive species' habi-tat requirements (food, territory, migration corridors, competing species, etc.) could be coupled with digital maps that provide relevant information about the habitat. The authors also encourage other researchers to use this "unprecedented near-global high-resolution elevation dataset" and to adopt and further develop their spatial risk-profiling approach.

Brooker (2007) suggests three additional features of Bayesian risk mapping of that deserves further scientific study, based on his review of spatial risk models for schistosomiasis:

- Studies investigating the importance of different risk factors at varying spatial scales, and their relative importance, are necessary.
- The extent of the spatial correlation in data will be influenced by local character-istics and can therefore be expected to differ in different parts of a geographical region. Modeling of such nonstationary spatial processes has received little attention, but it might be possible to use a method similar to ARIMA (auto-regressive integrated moving average) models used in time-series analysis.
- Risk models have considered the epidemiology of single parasite species in isolation, whereas human infections typically involve multiple species (Raso et al. 2006).

Although these suggestions are based on risk mapping for epidemiology, they may also apply to risk mapping in environmental risk assessment more generally.

SUMMARY AND RECOMMENDATIONS

This chapter has given examples of application Bayesian methods in environmental risk assessment and management as well as in other related fields of environmental science. Although Bayesian methods are already being used increasingly in EcoRA (Fox 2006), there are considerably more examples of advanced Bayesian methodology in other fields, in particular for methodology that includes spatial aspects. Therefore, landscape-based EcoRA can benefit greatly from adopting Bayesian techniques that are developed within fields such as fisheries, water management, and epidemiology.

Strengths and weaknesses of the different Bayesian methods presented here have already been pointed out for each method. Certain potential benefits to EcoRA that these methods have in common can be highlighted:

- These methodologies are under rapid development in several branches of the scientific community.
- The graphical representation of the method (particularly of Bayesian networks) facilitates communication and model development in collaboration with other parties involved in the EcoRA.
- The models can be used for different purposes: statistical inference (parameter estimation), model prediction (simulation), and exploration of scenarios.
- The models can be updated continuously as new information becomes available.
- The model output is always probability distributions, which can be directly useful for EcoRA.
- Landscape features can be incorporated in all types of models; most explicitly in the geostatistical Bayesian models.

The use of Bayesian methods also poses many challenges:

- Because subjective opinion can be incorporated in the model, special caution must be used in the formulation of priors.
- It can be difficult for non-scientists involved in the ERA to fully understand the concepts of probability distributions and conditional probabilities.
- Although the Bayesian approach can be seen as more intuitive, it requires a new way of reasoning and model formulation compared to the classical approach. Since Bayesian statistics is usually not taught in introductory statistics courses, researchers are often not familiar with this way of reasoning.
- Performing a Bayesian analysis requires not only a good understanding of the Bayesian theory and the environmental problem, but also a good understanding of the statistical software. Indeed, the manual for the Bayesian software BUGS (Bayesian inference Using Gibbs Sampling) states in red letters: "Beware: MCMC sampling can be dangerous!"

The most commonly used software for Bayesian inference are the freely available WinBUGS (interactive Windows version of BUGS) and the open-source version OpenBUGS (both available from http://www.mrc-bsu.cam.ac.uk/bugs/) is the most commonly used software for Bayesian inference. The BUGS project also has a web resource site (http://www.mrc-bsu.cam.ac.uk/bugs/weblinks/webresource.shtml), which contains links to web pages describing ongoing research within a wide range of fields: social science, actuarial science, cost-effectiveness analysis, pharmacokinetics, analysis of imperfect diagnostic tests, complex epidemiological population genetics, and archaeology. New users are recommended to attend a course in Bayesian statistics before using the software. The web resource site also contains links to other resources: tutorial material, usergroups, and books; sites on particular application areas; and software to run in conjunction with BUGS.

A version of BUGS for geostatistics, GeoBUGS, has been developed by a team at the Department of Epidemiology and Public Health of Imperial College at St Mary's Hospital London. It is an add-on to WinBUGS that fits spatial models and produces a range of maps as output.

Bayesian networks cannot be run in WinBUGS, but different programs are available—for example, Hugin™, Analytica™, and Netica™. Courses in use of Hugin for Bayesian Networks are arranged frequently (http://www.hugin.com).

More information of Bayesian networks and other Bayesian methods can be found in Wikipedia, for example:

- http://en.wikipedia.org/wiki/Bayesian_network
- http://en.wikipedia.org/wiki/Influence_diagrams
- http://en.wikipedia.org/wiki/Bayesian_inference
- http://en.wikipedia.org/wiki/Hierarchical_Bayes_model

ACKNOWLEDGMENTS

The author was supported by the Research Council of Norway through a grant to the NIVA project Model-SIP (contract no. 172 708/S30).

REFERENCES

Abrial D, Calavas D, Jarrige N, Ducrot C. 2005. Poultry, pig and the risk of BSE following the feed ban in France—A spatial analysis. *Vet Res* **36**:615–628.

Allepuz A, Lopez-Quilez A, Forte A, Fernandez G, Casal J. 2007. Spatial analysis of bovine spongiform encephalopathy in Galicia, Spain (2000–2005). *Preventive Vet Med* **79**:174–185.

Barton DN, Saloranta T, Bakken TH, Solheim AL, Moe J, Selvik JR, Vagstad N. 2005. Using Bayesian network models to incorporate uncertainty in the economic analysis of pollution abatement measures under the Water Framework Directive. *Water Sci Technol: Water Supply* **5**:95–104.

Barton D, Bechmann M, Eggestad HO, Moe J, Saloranta T, Kuikka S, Haygarth P. 2008a. EUTROBAYES—Integration of nutrient loading and lake eutrophication models in cost-effectiveness analysis of abatement measures. OR-5555, NIVA.

Barton DN, Saloranta T, Moe SJ, Eggestad HO, Kuikka S. 2008b. Bayesian belief networks as a meta-modelling tool in integrated river basin management—Pros and cons in evaluating nutrient abatement decisions under uncertainty in a Norwegian river basin. *Ecol Econ* **66**:91–104.

Beck-Worner C, Raso G, Vounatsou P, N'Goran EK, Rigo G, Parlow E, Utzinger J. 2007. Bayesian spatial risk prediction of *Schistosoma mansoni* infection in western Cote d'Ivoire using a remotely-sensed digital elevation model. *Am J Trop Med Hygiene* **76**:956–963.

Best N, Richardson S, Thomson A. 2005. A comparison of Bayesian spatial models for disease mapping. *Stat Methods Med Res* **14**:35–59.

Biggeri A, Dreassi E, Catelan D, Rinaldi L, Lagazio C, Cringoli G. 2006. Disease mapping in veterinary epidemiology: a Bayesian geostatistical approach. *Stat Meth Med Res* **15**:337–352.

Bjørndalen K, Andersen T, Bechmann M, Borgvang S, Brabrand AA, Delstra J, Gunnarsdottir H, Hobæk A, Saloranta T, Skarbøvik E, Solheim AL. 2006. Utredninger Vansjø 2005—Sammendrag og anbefalinger (in Norwegian). 5146, NIVA.

Borsuk ME, Higdon D, Stowa CA, Reckhow KH. 2001. A Bayesian hierarchical model to predict benthic oxygen demand from organic matter loading in estuaries and coastal zones. *Ecol Model* **143**:165–181.

Borsuk ME, Stow CA, Reckhow KH. 2004. A Bayesian network of eutrophication models for synthesis, prediction, and uncertainty analysis. *Ecol Model* **173**:219–239.

Breiman L, Friedman JH, Olshen RA, Stone CJ. 1984. *Classification and Regression Trees*. Wadsworth International Group, Belmont, CA.

Brooker S. 2007. Spatial epidemiology of human schistosomiasis in Africa: Risk models, transmission dynamics and control. *Trans R Soc Trop Med Hygiene* **101**:1–8.

Castelletti A, Soncini-Sessa R. 2007a. Bayesian networks in water resource modelling and management. *Environ Model Software* **22**:1073–1074.

Castelletti A, Soncini-Sessa R. 2007b. Bayesian networks and participatory modelling in water resource management. *Environ Model Software* **22**:1075–1088.

Castelletti A, Soncini-Sessa R. 2007c. Coupling real-time control and socio-economic issues in participatory river basin planning. *Environ Model Software* **22**:1114–1128.

Cegelski CC, Waits LP, Anderson NJ. 2003. Assessing population structure and gene flow in Montana wolverines (*Gulo gulo*) using assignment-based approaches. *Mole Ecol* **12**:2907–2918.

Clark JS. 2005. Why environmental scientists are becoming Bayesians. *Ecol Lett* **8**:2–14.

Clark JS, Lavine M. 2001. Bayesian statistics. In Scheiner SM, Gurevitch J (Eds.), *Design and Analysis of Ecological Experiments, second edition*. Oxford University Press, Oxford, UK, pp. 327–346.

Clements ACA, Moyeed R, Brooker S. 2006. Bayesian geostatistical prediction of the intensity of infection with *Schistosoma mansoni* in East Africa. *Parsitology* **133**:711–719.

Coles S, Tawn J. 2005. Bayesian modelling of extreme surges on the UK east coast. *Philos Trans Rl Soc A Math Phys Eng Sci* **363**:1387–1406.

Dennis B. 1996. Discussion: Should ecologists become bayesians? *Ecol Appl* **6**:1095–1103.

Ellison AM. 1996. An introduction to Bayesian inference for ecological research and environmental decision-making. *Ecol Appl* **6**:1036–1046.

Etterson MA, Bennett RS. 2006. On the use of published demographic data for population-level risk assessment in birds. *Hum Ecol Risk Assess* **12**:1074–1093.

Fox DR. 2006. Statistical issues in ecological risk assessment. *Hum Ecol Risk Assess* **12**:120–129.

Gelman A, Carlin JB, Stern HS, Rubin DB. 1995. *Bayesian Data Analysis*. Chapman & Hall, London.

Ghazoul J, McAllister M. 2003. Communicating complexity and uncertainty in decision making contexts: Bayesian approaches to forest research. *Int Forest Rev* **5**:9–19.

Gibbs MT. 2007. Assessing the risk of an aquaculture development on shorebirds using a Bayesian belief model. *Hum Ecol Risk Assess* **13**:156–179.

Greco FP, Lawson AB, Cocchi D, Temples T. 2005. Some interpolation estimators in environmental risk assessment for spatially misaligned health data. *Environ Ecol Stat* **12**:379–395.

Henriksen HJ, Rasmussen P, Brandt G, Bülow Dv, Jensen FV. 2004. Engaging stakeholders in construction and validation of Bayesian belief networks for groundwater protection. IFAC Workshop on Modelling and Control for Participatory Planning and Managing Water Systems, Venezia, Italy, 29.09.2004–01.10.2004.

Henriksen HJ, Rasmussen P, Brandt G, von Bulow D, Jensen FV. 2007. Public participation modelling using Bayesian networks in management of groundwater contamination. *Environ Model Software* **22**:1101–1113.

Johnson DH. 1999. The insignificance of statistical significance testing. *J Wildlife Manage* **63**:763–772.

Johnston KM, Forman RTT. 1989. Scale-dependent proximity of wildlife habitat in a spatially-neutral Bayesian model. *Landscape Ecol* **2**:101–110.

Kangas A, Kurki S. 2000. Predicting the future of capercaillie (*Tetrao urogallus*) in Finland. *Ecol Model* **134**:73–87.

Kapo KE, Burton GA. 2006. A geographic information systems-based, weights-of-evidence approach for diagnosing aquatic ecosystem impairment. *Environ Toxicol Chem* **25**:2237–2249.

Labiosa W, Leckie J, Shachter R, Freyberg D, Rytuba J. 2005. Incorporating uncertainty in watershed management decision-making: A mercury TMDL case study. ASCE Watershed Management Conference, Managing Watersheds for Human and Natural Impacts: Engineering, Ecological, and Economic Challenges, Williamsburg, VA, USA, 19–22 July 2005.

Louie MM, Kolaczyk ED. 2006. A multiscale method for disease mapping in spatial epidemiology. *Stat Med* **25**:1287–1306.

Mäki-Petäys H, Zakharov A, Viljakainen L, Corander J, Pamilo P. 2005. Genetic changes associated to declining populations of *Formica* ants in fragmented forest landscape. *Mol Ecol* **14**:733–742.

Malve O. 2007. *Water Quality Prediction for River Basin Management*. PhD thesis. Helsinki University of Technology, Helsinki, Finland.

Malve O, Qian SS. 2006. Estimating nutrients and chlorophyll *a* relationships in Finnish lakes. *Environ Sci Technol* **40**:7848–7853.

Marcot BG, Holthausen RS, Raphael MG, Rowland MM, Wisdom MJ. 2001. Using Bayesian belief networks to evaluate fish and wildlife population viability under land management alternatives from an environmental impact statement. *Forest Ecol Manage* **153**:29–42.

Martín de Santa Olalla F, Dominguez A, Ortega F, Artigao A, Fabeiro C. 2007. Bayesian networks in planning a large aquifer in Eastern Mancha, Spain. *Environ Model Software* **22**:1089–1100.

Maunder MN. 2004. Population viability analysis based on combining Bayesian, integrated, and hierarchical analyses. *Acta Oecol* **26**:85–94.

McCarthy MA, Lindenmayer DB, Possingham HP. 2001. Assessing spatial PVA models of arboreal marsupials using significance tests and Bayesian statistics. *Biol Conserv* **98**:191–200.

Natvig B, Tvete IF. 2007. Bayesian hierarchical space-time modeling of earthquake data. *Method Comput Appl Prob* **9**:89–114.

Newman M, Zhao Y, Carriger J. 2007. Coastal and estuarine ecological risk assessment: The need for a more formal approach to stressor identification. *Hydrobiologia* **577**:31–40.

O'Hara RB, Arjas E, Toivonen H, Hanski I. 2002. Bayesian analysis of metapopulation data. *Ecology* **83**:2408–2415.

Oredalen TJ, Faafeng B. 2002. Landsomfattende undersøkelse av trofitilstanden i norske innsjøer. Datarapport 2001. (National survey of eutrofication status in Norwegian lakes. Data report 2001). OR-4570, NIVA.

Ovaskainen O, Hanski I. 2004. From individual behavior to metapopulation dynamics: Unifying the patchy population and classic metapopulation models. *Am Nat* **164**:364–377.

Parris KM. 2006. Urban amphibian assemblages as metacommunities. *J Animal Ecol* **75**:757–764.

Pellikka J, Kuikka S, Lindén H, Varis O. 2005. The role of game management in wildlife populations: uncertainty analysis of expert knowledge. *Eur J Wildlife Res* **51**:48–59.

Peterman RM. 2004. Possible solutions to some challenges facing fisheries scientists and managers. *Ices J Marine Sci* **61**:1331–1343.

Pollino CA, Woodberry O, Nicholson A, Korb K, Hart BT. 2007. Parameterisation and evaluation of a Bayesian network for use in an ecological risk assessment. *Environ Model Software* **22**:1140–1152.

Punt AE, Kennedy RB. 1997. Population modelling of Tasmanian rock lobster, *Jasus edwardsii*, resources. *Marine Freshwater Res* **48**:967–980.

Ranta J, Penttinen A. 2000. Probabilistic small area risk assessment using GIS-based data: A case study on Finnish childhood diabetes. *Stat Med* **19**:2345–2359.

Raso G, Vounatsou P, Singer BH, N'Goran EK, Tanner M, Utzinger J. 2006. An integrated approach for risk profiling and spatial prediction of *Schistosoma mansoni* –hookworm coinfection. *Proc Natl Acad Sci USA* **103**:6934–6939.

Ree R, McCarthy MA. 2005. Inferring persistence of indigenous mammals in response to urbanisation. *Animal Conserv* **8**:309–319.

Richardson S, Best N. 2003. Bayesian hierarchical models in ecological studies of health-environment effects. *Environmetrics* **14**:129–147.

Rieman B, Peterson JT, Clayton J, Howell P, Thurow R, Thompson W, Lee D. 2001. Evaluation of potential effects of federal land management alternatives on trends of salmonids and their habitats in the interior Columbia River basin. *Forest Ecol Manage* **153**:43–62.

Rowland MM, Wisdom MJ, Johnson DH, Wales BC, Copeland JP, Edelmann FB. 2003. Evaluation of landscape models for wolverines in the interior Northwest, United States of America. *J Mammal* **84**:92–105.

Sargeant GA, Sovada MA, Slivinski CC, Johnson DH. 2006. Markov Chain Monte Carlo estimation of species distributions: A case study of the swift fox in western Kansas. *J Wildl Manage* **69**:483–497.

Souza WV, Carvalho MS, Albuquerque M, Barcellos CC, Ximenes RAA. 2007. Tuberculosis in intra-urban settings: A Bayesian approach. *Trop Med Intern Health* **12**:323–330.

Steventon JD, Sutherland GD, Arcese P. 2006. A population-viability-based risk assessment of Marbled Murrelet nesting habitat policy in British Columbia. *Can J For Res-Rev Can Rech For* **36**:3075–3086.

Su Z, Peterman RM, Haesker SL. 2004. Spatial hierarchical Bayesian models for stock-recruitment analysis of pink salmon (*Oncorhynchus gorbuscha*). *Can J Fish Aquat Sci* **61**:2471–2486.

Tattari S, Schultz T, Kuussaari M. 2003. Use of belief network modelling to assess the impact of buffer zones on water protection and biodiversity. *Agric Ecosyst Environ* **96**:119–132.

ter Braak CJF, Etienne RS. 2003. Improved Bayesian analysis of metapopulation data with an application to a tree frog metapopulation. *Ecology* **84**:231–241.

Thompson JA, Brown SE, Riddle WT Seahorn JC, Cohen ND. 2005. Use of a Bayesian risk-mapping technique to estimate spatial risks for mare reproductive loss syndrome in Kentucky. *Amer J Veterinary Res* **66**:17–20.

Ticehurst JL, Newham LTH, Rissik D, Letcher RA, Jakeman AJ. 2007. A Bayesian network approach for assessing the sustainability of coastal lakes in New South Wales, Australia. *Environ Model Software* **22**:1129–1139.

Tufto J, Sæther B-E, Engen S, Arcese P, Jerstad K, Rostad OW, Smith JNM. 2000. Bayesian meta-analysis of demographic parameters in three small, temperate passerines. *Oikos* **88**:273–281.

Varis O, Kuikka S. 1999. Learning Bayesian decision analysis by doing: Lessons from environmental and natural resources management. *Ecol Model* **119**:177–195.

Verdonck F. 2003. *Geo-referenced Probabilistic Ecological Risk Assessment*. Ghent University, Belgium.

Wade PR. 2000. Bayesian methods in conservation biology. *Conserv Biol* **14**:1308–1316.

Wakefield J. 2007. Disease mapping and spatial regression with count data. *Biostatistics* **8**:158–183.

Williams NSG, Morgan JW, McDonnell MJ, McCarthy MA. 2005. Plant traits and local extinctions in natural grasslands along an urban–rural gradient. *J Ecol* **93**:1203–1213.

Williams NSG, Morgan JW, McCarthy MA, McDonnell MJ. 2006. Local extinction of grassland plants: The landscape matrix is more important than patch attributes. *Ecology* **87**:3000–3006.

Xia H, Carlin BP, Waller LA. 1997. Hierarchical models for mapping Ohio lung cancer rates. *Environmetrics* **8**:107–120.

7

LINKING REGIONAL AND LOCAL RISK ASSESSMENT

Rosana Moraes and Sverker Molander

Physical or biological processes that affect the fluxes of organisms or stressors can functionally define regions, such as wide-scale watersheds or physiographic provinces (Hunsaker et al. 1990). Stressors, sources, and their associated risks exist along a continuum of spatial scales, and it is impossible to draw an objective line between what is a regional and what is a local environmental situation. For instance, regional change can be caused either by a local phenomenon (stressor source or primary effect) that has a regional consequence or by multiple local sources that, when combined, create changes that qualify as regional or by diffuse sources—very many, very small. Despite the lack of a clear demarcation between local and regional, we distinguish two categories (i.e., regional and local) of ecological risk assessment (EcoRA), acknowledging that this is an arbitrary categorization that provides a practical way to distinguish local and regional risk assessment methods.

Local risk assessments cover, for example, evaluation of changes in the water quality of a reach of a particular river related to discharges from one or many specific industries. The scale in this case is small and the organisms are exposed in the area. In general, risk assessments at a small scale can comprise more refined analyses than assessments made at a regional scale. On the other end of the scale, a regional risk assessment is more general and coarse-grained. Details are left out in order to aggregate information, sometimes in order to prioritize where more detailed assessment, on a local scale, should be performed.

Environmental Risk and Management from a Landscape Perspective, edited by Kapustka and Landis
Copyright © 2010 John Wiley & Sons, Inc.

AIMS AND CHALLENGES OF REGIONAL RISK ASSESSMENTS

This discussion of regional risk assessment begins with a view of scale. Two additional subsections describe the consideration of multiple stressors and the challenges of dealing with voids in regional-specific spatial data.

Scale

One of the purposes of EcoRA is to inform environmental management. Managers must in many cases make decisions concerning environmental issues at a regional scale (e.g., threats to natural reserves or large watersheds) rather than in single sites or small geographic areas. Regional issues are not only larger in scale than local ones, but are also larger in scope, since they may involve multiple stressors, multiple receptors, and spatially heterogeneous natural systems. The evaluation of ecological risks posed by environmental problems over large geographical areas requires regional risk assessments. In these circumstances the applicability of traditional site-specific EcoRA procedures, such as the ones designed for assessing contaminated sites published by agencies in different jurisdictions such as Canada (CCME 1996, EC 1994), the United States (US EPA 1998), Australia (A NEPC, 1999a, b), the Netherlands (Rutgers et al. 2001), the United Kingdom (UK EA 2008), and Basque Country (IHOBE 2003) may have limited applicability.

Participants of the "Workshop on Characterizing Ecological Risk at the Watershed Scale" (US EPA, 2000) indeed recognized the problem of dealing with large geographical areas. Since the event of that workshop, several papers have been published on watershed risk assessments methods and case studies (e.g., US EPA 2002a, 2000b; Wickham and Wade 2002, Serveiss 2002, Serveiss et al. 2004). In Europe, recent changes in environmental directives and strategies (e.g., Water Framework Directive 2000/60/EC, CEC 2000) will require that currently used approaches of risk assessment shift from small scale to the watershed level and will need a better understanding of different media and habitats across many spatial and temporal scales (Apitz et al. 2006). Regarding prospective investigations the European legislation requires studies of Strategic Environmental Assessments on larger scales according to the directive 2001/42/EC on the assessment of the effects of certain plans and programs on the environment (CEC 2001).

Even studies concerning effects on small geographic areas should consider effects linked to broader scales than the exposed area because of fluxes of organisms or stressors from the exposed area to the surrounding landscape and vice versa (Gardner 1998, McLaughlin and Landis 2000). For instance, impacted sites whose communities have been extinguished by acute exposures may be recolonized by immigrants from noncontaminated areas (Johnson 2002). On the other hand, migratory species that have traveled from exposed sites may exhibit effects when they live in a nonexposed area. None of the EcoRA procedures listed above (US EPA, CCME, UK EPA, Australian NEPC, Basque Country IHOBE, etc.) explicitly incorporates this issue. Moreover, air and water pollutants can travel long distances, and the understanding of those pollutants' sources is relevant, even if they are outside of the boundary of the studied site (Efroymson et al. 2005). A further aspect that has been put forward by Colnar and Landis (2007) is effects caused by invasive species where the sources of these species are outside the studied area. It is clear that in order to cover significant influences on the assessment endpoint, also the connection to sources of various stressors occurring outside the study area need to be included. The point is to keep flexibility of what to include so that the relevant linkages from stressor sources to endpoints are captured.

Multiple Kinds of Stressors

Another case in which traditional EcoRA procedures have limited application is associated with the assessment of multiple kinds of stressors. The importance of identifying and evaluating the adverse effects of biological, physical, and chemical stressors in ecosystems and their components is explicitly mentioned in the US EPA *Guidelines for Ecological Risk Assessment*, but that procedure does not offer specific guidance on the quantitative assessment of multiple kinds of stressors in the context of an EcoRA. Even though a deterioration of natural ecosystems is typically the result of cumulative impacts of chemical, physical, and biological origin, most published EcoRA studies analyze the effects of only one kind of stressor. This is clearly unsatisfactory but understandable due to practical constraints.

In most cases the stressors examined are exclusively chemicals, such as PCBs, metals, PAHs, or pesticides. EcoRA studies less frequently concern physical stressors—that is, stressors that lead to reduction in availability and accessibility of receptor-specific, structurally defined habitat (Hope 2005). Examples of physical stressors include habitat fragmentation, sound pollution, and visual pollution. Even rarer are EcoRA studies concerning exclusively biological stressors, such as presence of pathogens, or the removal of resources needed, such as dead wood, which is a prerequisite for woodpeckers serving both as substrate for food resources and nesting (Butler et al. 2004).

All three types of stressors may interact in such ways that modify a chemical stressor's contribution to the overall risk to a given receptor (Hope 2005). Furthermore, the interaction of different types of stressors are unlikely to be additive. The development of quantitative and qualitative assessments of multiple kinds of stressors and their incorporation in EcoRA procedures still remain a difficult challenge.

Lack of Region Specific Spatial Data

Few models have been suggested for the purpose of dealing with risks at regional scale; and most of these are constrained in terms of data availability, because the cost of data collection at large scales is often high. For instance, Graham et al. (1991) published a prototype assessment to evaluate risks associated with elevated ozone in a forested region in the United States. The regional risk assessment used a stochastic spatial simulation model of land-cover change linked with a water quality regression model. Positive aspects of the model include consideration of spatial heterogeneity and the linkages between terrestrial and aquatic systems. Limitations comprise the need of spatially explicitly data such as water quality data for a larger region.

Another example is the multiple regression model relating anthropogenic activities in watersheds in Ohio, USA, and several stream conditions variables related to biological quality proposed by Gordon and Majumder (2000). Useful accomplishments of the model include a broad view of the degree and significance of the factors impacting ecological resources on a regional scale. Once again, limitations are related to needs of large input dataset. Just as an illustration, the authors ran the model using data from 276 locations across 25 watersheds in Ohio.

In contrast, the procedure proposed for risk assessments by the Commission of European Communities (CEC 1996) does include evaluations of risk at both the regional and local scale, but is designed for generic environments, not actual sites, and is restricted to assessments of chemical stressors, one chemical at a time. The model used can be adapted to represent aspects of a local environment, but in that case, locally gathered data will be needed.

THE RELATIVE RISK MODEL (RRM) FOR REGIONAL RISK ASSESSMENTS

An alternative approach used to overcome some of the limitations described in the paragraphs above is the Relative Risk Model (RRM) developed by Landis and Weigers (1997). The merits of the RRM are not only related to low demand for input data, and consequently low cost, but also to the transparency of its assessment models, plus the fact that its outputs are easy to use in risk communication and for creating priorities for management actions or for the identification of locations for local risk assessments.

While traditional EcoRAs estimate the level of exposure and effects to calculate risk to a specific receptor at a local site (see Chapter 2), the assessment at regional level, deals with a level of complexity that is not the same. At a regional level, numerous existing sources in a certain region (e.g., industries, farms) can release different types of stressors (e.g., chemicals, invasive species, soil erosion); multiple receptors (e.g., wildlife, fish) present in the various habitats (e.g., forest, wetland, rivers) in the region can be affected by those stressors. The evaluation of exposure in this case will involve different types of measurements [concentrations of chemicals in water (mg/L), number of individuals of invasive species per m^2 of wetland, surface of eroded soil in m^2]. Although the three stressors can lead to the same effects (e.g., decline of wetland indigenous species), the exposure levels are not easily comparable. In order to address this issue, Landis and Weigers (1997) suggested the establishment of ranks and weighting factors to evaluate the risk. The model is based on the community conditioning hypothesis (Matthews et al. 1996), which eliminates dependency of a reference site and therefore opens up for relative risk assessment.

The basic steps of the RRM are:

- To identify management goals based on stakeholders' values and regulatory needs and associate those goals into a spatial context

- To locate sources of stressors and habitats relevant to management goals on a map
- To delineate sub-areas based on combinations of identified management goals and the location of stressors and habitats
- To build a conceptual model linking the multiple sources, stressors, habitats, and endpoints
- To assign a ranking scheme for each source, stressor, and habitat to allow the calculation of relative risk to the assessment endpoint (details in the next pages)
- To calculate the relative risk based on the relationship between sources, habitat, and impact to assessment endpoints (details in the next pages)
- To evaluate the uncertainty and sensitivity analysis of the relative rankings
- To generate testable hypotheses for future investigations to reduce uncertainties and to confirm the risk rankings
- To test those hypotheses
- To communicate the risks.

There have been some recent applications of the RRM model demonstrating its adaptability to a variety of scales and ecosystems (Landis and Wiegers 1997, Wiegers et al. 1998, Landis et al. 2000, Walker et al. 2001, Moraes et al. 2002, Obery and Landis 2002, Hamamé 2002, Hart Hayes and Landis 2004, Landis et al. 2004, Colnar and Landis 2007). Most of those case studies were compiled in the book *Regional Scale Ecological Risk Assessment Using the Relative Risk Model* (Landis 2005). A reflection over the 10 years (1997–2007) of the Relative Risk Model and Regional Scale Ecological Risk Assessment was recently published by Landis and Wiegers (2007).

Here we describe a procedure that links regional and local risk assessments. The linkage is soft in that the regional assessment is used for identifying stressors, sources, and habitats with potentially affected endpoints in certain areas. This is apart from the conceptually different approach of Colnar and Landis (2007), who used hierarchical modeling (Wu and David 2002, Allen and Starr 1982) where local scale is linked to regional by an aggregative approach.

PETAR: A PROCEDURE TO INTEGRATE REGIONAL AND LOCAL RISK ASSESSMENTS

In 2004, we proposed a procedure for EcoRA that incorporates qualitative and quantitative methods for assessing multiple kinds of stressors related to environmental problems across a range of scales (Moraes and Molander 2004). It also includes Landis and Weigers' RRM and takes into account restrictions on data availability and open up for smaller-scale studies within the framework. The proposed procedure was given the name "PETAR" as an acronym for Procedure for Ecological Tiered Assessment of Risks. The acronym also pays homage to the place in which the preliminary ideas for the procedure were first conceived and put into practice, the Parque Estadual Turistico do Alto Ribeira (PETAR), a Brazilian rain forest reserve. Results of several studies leading up to the proposed procedure are also found in Moraes (2002), Moraes et al. (2003), and Gerhard et al. (2004).

The PETAR procedure consists of a three-tiered evaluation of risks (Fig. 7.1), with successive tiers requiring more detailed investigations. The procedure, which starts with large and ends with small geographical areas, recognizes restrictions in data availability and acquisition. It is based on a number of complementary methods selected from the scientific literature regarding assessments of multiple kinds of stressors and for establishing causality between exposure and effects. The three main phases of the assessment (preliminary, regional, and local) will be detailed in the next sections.

Preliminary Assessment

A preliminary assessment aims at a qualitative evaluation of the stressors sources, ecosystems potentially at risk, characteristics of the stressors, and expected ecological effects (Fig. 7.2). Based on the agreed goals and scope of the planning process, the assessor should gather information on all existing stressor sources in the study area and ecosystems potentially at risk. The study area should include a broader spatial scale than the stressor sources of concern (e.g., a municipal district) or the specific exposed area of interest (e.g., a bay, a natural reserve, a fiord, a river). The boundaries of the study area may preferentially be of physical (e.g., airsheds or watersheds) or biological nature (e.g., physiographic provinces) because these features are known to influence the flux of stressors and organisms from a source area to an exposed area and to its neighborhood, and vice versa (Hunsaker et al. 1990, Efroymson et al. 2005).

Potential stressors should be listed on the basis of existing sources in the region. Physical, chemical, or biological characteristics of the stressors and ecosystem that may influence their fate and distribution in the environment as well as their potential hazard should be compiled from the scientific literature (e.g., existing toxicological databases and studies done at other sites). These data are important for evaluation of expected ecological effects.

The outcome of this preliminary assessment regarding the initial survey of stressors sources, ecosystems potentially at risk, characteristics of stressors, and expected ecological effects will guide the problem formulation of the regional risk assessment tier.

Regional Risk Assessment

The aim is to identify stressors having the greatest potential for ecological impact, the habitats most at risk, and the sub-areas inside the study area most likely to be affected by impacts. A semiquantitative evaluation of ecological risks over large geographical areas is the result of the regional risk assessment.

In the PETAR approach, the RRM of Landis and Weigers is the suggested tool to evaluate risks at the regional level. The next paragraphs provide a general understanding of the RRM and its application within the PETAR procedure. The reader looking

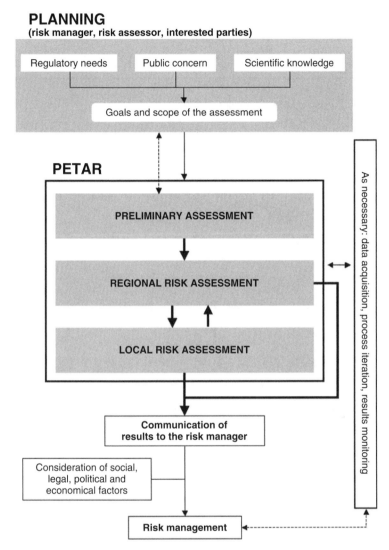

Figure 7.1. Overall process of the Procedure for Ecological Tiered Assessment of Risks (PETAR) and related activities (planning process, risk communication, and risk management).

for a more detailed description of the RRM method and examples of its application should refer to Landis (2005).

Like traditional local risk assessments, the overall process of the regional risk assessment includes three main steps: problem formulation, analysis phase, and risk characterization (Figure 7.3). During the problem formulation stage, assessment endpoints are selected, conceptual models are created, and an analysis plan for the evaluation of exposure and effects at regional scale is designed.

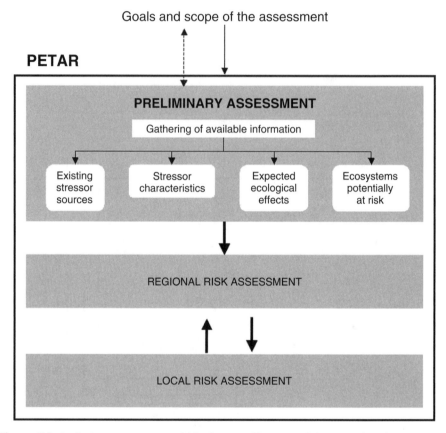

Figure 7.2. Preliminary assessment within the overall process of the Procedure for Ecological Tiered Assessment of Risks (PETAR). White rectangles represent steps in the process and rounded rectangles represent products of the preliminary assessment.

The major criteria to be used when assessment endpoints are selected are their ecological relevance, their susceptibility to the known or potential stressors, and their representation of management goals (US EPA 1992, Barton and Sergeant 1998). According to Suter (1990), assessment endpoints for regional risk assessments may differ from those traditionally selected for local ecological risk assessments. In regional risk assessments, environmental characteristics valued by the public and by decision-makers should be given great weight in the choice of the assessment endpoint. The societal values are usually linked to the region's ability to provide food, clean water and air, aesthetic experiences, recreation, and so on. It is also possible to relate the choice of the assessment endpoint to the ecosystem services following the terminology of the Millennium Ecosystem Assessment (2005).

Suitable types of endpoints can then be connected to valued populations such as endangered, commercial, or recreational species in the region, provided that the status

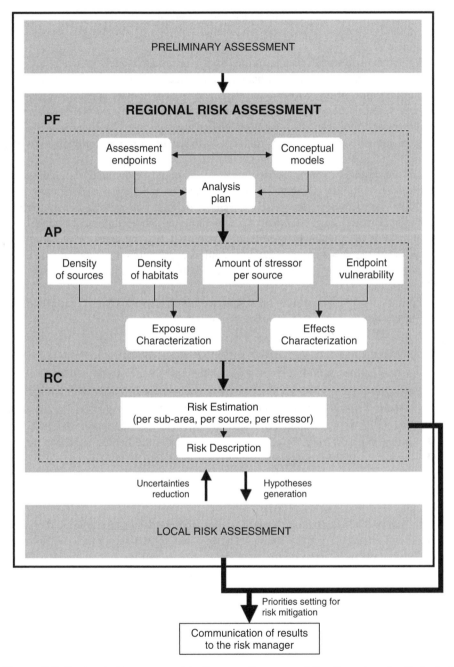

Figure 7.3. Regional risk assessment in the overall process of the Procedure for Ecological Tiered Assessment of Risks (PETAR): PF, problem formulation; AP, analysis phase; and RC, risk characterization. White rectangles represent steps in the process, and rounded rectangles represent products of the assessment.

of the selected populations integrates physical, chemical, and biological disturbances. At the community level, changes in community type can be suitable for assessments at a regional scale.

The problem formulation also produces a conceptual model, which is the link between stressor sources, stressors, habitats, and effects linkages (e.g., pool of industries, air emissions, primary forest, and decline of sensitive plant species). A strategy for creating clear and complete conceptual models for complex EcoRAs with multiple stressors and multiple assessment endpoints was presented by Suter (1999). The conceptual model can also be constructed as a hierarchical model following Wu and David (2002) and Colnar and Landis (2007).

The final product of the problem formulation should be the analysis plan, which includes criteria for the delimitation of sub-areas of the region under study. The criteria may be analytical factors (e.g., watershed limits) as well as human-derived organizational factors (e.g., political, administrative or socioeconomical boundaries). With the use of Geographical Information Systems (GIS), the assessor should create maps showing the sub-areas and, within them, the distribution of sources, habitats, and assessment endpoints. Afterwards, the density of sources and habitats per sub-area can be estimated.

The analysis phase consists of a semiquantitative evaluation of exposure and effects at the regional scale. It includes the design of:

1. A ranking based on the density of sources per subarea (rank sources, RS)
2. A ranking based on the density of habitats per subarea (rank habitats RH)
3. Weighting factors for stressors (WS) to account for differences in amounts of a stressor that can be released from diverse sources
4. Weighting factors for effects (WE) to evaluate the likelihood and extent to which exposure to different stressors may harm the habitat as a function of the system's sensitivity and its ability to adapt to new conditions

For an illustration of how designing of ranking systems and weighting factors can be conducted, see Moraes et al. (2002).

The use of RRM may require simplifying assumptions because empirical data on exposure and effects are usually limited. The designing of weighting factors also relies to a great degree on inferred preferences and data availability. To evaluate the likelihood and extent of harm to the habitat, the assessor may use, for instance, information on the number of potential stressors of each type (e.g., number of organic compounds and their toxicity relative to the number of metals and their toxicity included in each "type" of stressor) and the temporal and spatial variability of the stressor. Relevant information for design of effect weighting factors includes the possibility of previous exposure by natural sources in the region (e.g., metal containing ores), the duration of the exposure (chronic or acute), and consequent opportunity for organisms to develop tolerance and adaptation mechanisms.

Risk characterization is based on the three following assumptions of the RRM: First, the greater the density of sources in a sub-area, the amount of stressor released by the source, and the density of habitats where the endpoint resides in a sub-area, the greater the potential for exposure; second, the greater the vulnerability of an endpoint to a stressor, the greater the potential for effects; and third, the greater the potential combination of exposure and effects, the greater the risk.

The risk of adverse effects in sub-area i for the habitat h is determined by

$$\text{Risk\_Sub-area}_{ih} = \sum \text{Risk\_Sub-area}_{ijh} \qquad (1)$$

$$\text{Risk\_Sub-area}_{ijh} = \text{Exposure}_{ijh} \times \text{Effect}_{jh} \qquad (2)$$

where i represents the sub-area, j represents the stressor, and h represents the habitat.

The likelihood of the endpoint to be exposed to a stressor depends on the abundance of sources (RS) that release such a stressor, the amount of stressors released in the environment by each source (WS), and the abundance of habitats where the endpoint can be found in a certain area (RH). The exposure factor in area i to stressor j for the habitat h was determined by

$$\text{Exposure}_{ijh} = \sum_{k=\text{source}_{1...n}} (\text{RS}_{ki} \times \text{WS}_{jk}) \times \text{RH}_{hi} \qquad (3)$$

where k represents the sources of a particular type (e.g., industries or landfills); RS_{ki} is the rank chosen for the sub-area i according to the density of the source k (e.g., ranks determined by the number of industries in the sub-area); WS_{jk} is the weighting factor for the stressor j according to the amounts produced by source k; and RH_{hi} is the rank chosen for the area i according to the density of the habitat h.

The likelihood of undesired effects will depend on the vulnerability of the endpoint to the stressor (WE). The effect factor to stressor j for the habitat h was determined by

$$\text{Effect}_{jh} = \text{WE}_{jh} \qquad (4)$$

where WE_{jh} is the effect-weighting factor based on the vulnerability of endpoints in habitat h to stressor j.

The risk of adverse effects due to the source k for the habitat h is determined by adding the risks of adverse effects of each stressor released by the source k:

$$\text{Risk\_Source}_{kh} = \sum_{j=\text{stressor}_{1...n}} \text{Risk\_Stressor}_{jh} \qquad (5)$$

The risk of adverse effects due to the stressor j in the habitat h was determined by

$$\text{Risk\_Stressor}_{jh} = \text{Exposure}_{jh} \times \text{Effect}_{jh} \qquad (6)$$

The probability of exposure to stressor j to the endpoints living in the habitat h is a product of the abundance of the different sources (RS) that release that stressor, the amount of that stressor released by each source (WS), and the abundance of the habitat h (RH) where that endpoint lives:

$$\text{Exposure\_Stressor}_{jh} = \sum_{i=\text{sub-area}_{1...n}} \left[\sum_{k=\text{source}_{1...n}} (\text{RS}_{ki} \times \text{WS}_{jk}) \times \text{RH}_{hi} \right] \quad (7)$$

The likelihood of undesired effects will depend on the vulnerability of the endpoint to the stressor (WE). The effect factor to stressor j for the habitat h was determined by Eq. (4).

The application of the RMM in retrospective studies results in (a) a ranking of stressors having the greatest potential for ecological impact, (b) habitats most at risk, and (c) sub-areas inside the study area that are more likely to be impacted. The results can be used for risk mitigation in specific sub-areas and, furthermore, to generate testable hypotheses and direct investigations for the local risk assessment.

Many uncertainties are involved in the estimations of effects, which typically occur in any EcoRA. The RRM is no exception. The uncertainties, overestimating or underestimating the risk, can be divided into two groups. In the first group are uncertainties related to the model structure, which are the result of a possible disregard of key elements in the conceptual models due to oversimplifications or lack of knowledge. The second group consists of uncertainties related to assigning the input values to the model. This type of uncertainty is a consequence of extrapolations, assumptions, and lack of knowledge. Follow-up studies can add knowledge and, to a certain degree, reduce uncertainties by filling data gaps with site-specific data. Methods for the collection, analysis, and interpretation of site-specific data are presented in the next section.

Local Risk Assessment

The final tier of the PETAR procedure is a more detailed quantitative and site-specific risk assessment, designed for relatively small geographical areas. The starting point of the local risk assessment is the hypotheses generated based on the regional risk assessment regarding stressors with greatest potential for impact, habitats most at risk, and sub-areas more likely to be impacted. One of the aims of the site-specific investigations is to verify the occurrence of the ecological effects in specific habitats and in specific sub-areas and to relate observed effects to specific stressors.

Probably the most difficult task of retrospective risk assessments at the local scale is to provide data sufficient to conclude that a particular human-induced stressor causes an observed ecological effect in natural systems. Without knowing the cause, decision-makers are not able to properly tackle an environmental problem. The approach suggested here involves the construction of several lines of evidence (comparisons between measured concentrations and toxicity endpoints from toxicity tests, biomarkers, biological surveys, etc.). The different lines of evidence should then be evaluated in a formal and systematic way with a weight-of-evidence approach (Suter and Barnthouse 1993, Menzie et al. 1996, Hull and Swanson 2006) that uses a collection of epidemiological criteria to evaluate causality (Hill 1965, Fox 1991, Suter et al. 2002).

The problem formulation of the local risk assessment has three main outputs: assessment endpoints, a conceptual model, and an analysis plan (Fig. 7.4). The main difference between the regional scale and the local scale assessments is the level of detail in which the site-specific data are incorporated.

In the case of the local risk assessment, it is particularly desirable for the assessment endpoint to have a biological significance. In particular, that an effect on the level of biological organization to which the assessment endpoint belongs has implications for the next higher level of biological organization (Suter 1990). For the EcoRA procedure proposed here, the assessment endpoint for the local risk assessment may (but not necessarily) be the same as the one selected for the regional risk assessment. Generally speaking, population-level assessment endpoints may be the most useful in local risk assessments because (a) populations of many species have economical, recreational, aesthetic, and biological significance that is easily appreciated by the public and is likely to influence management decisions and (b) population responses are well-defined and easier to predict with available data and methods (see, for instance, Munkittrick et al. 2000) in comparison to community and ecosystem responses (Suter 1990). Regarding the choice of ecological entity, individual or sub-individual levels of biological organization often have low societal value and are of controversial biological significance. However, observations on these levels of organization can be used as evidence underpinning particular cause–effect relationships (Moraes et al. 2003).

The problem formulation also results in a detailed conceptual model constructed on the basis of hypothesized cause–effect chains and experience from the regional risk assessment. The model should describe (1) the stressors pathways from sources to receptors, considered to have the greatest potential for ecological impact, and (2) expected effects on the assessment endpoint. It should include all known stressors existent in the study area that can cause similar effects, including stressors of both anthropogenic and natural origin. Included potential effects is suggested to cover effects at the same and lower levels of biological organization to which the assessment endpoint belongs, since lower-level effects can be the controller of, or the constraint on, the effects on the assessment endpoint. In addition, the conceptual model should also consider potential implications for the next higher level of biological organization—for example, from the species level to the community level, since ecological and societal relevance may be perceived as larger on higher levels of organization.

The process of evaluation of the hypothesized cause–effect chains to be made during the analysis phase characterizes receptor, ecosystem, exposure, and effects. The characterization of the receptor organism, population, or community is the evaluation of the life history characteristics (e.g., habitat use, behavior, trophic relations) of the endpoint that can influence the exposure pathways and the response. The characterization of receptors at several locations is also vital to distinguish between patterns of disturbed and undisturbed receptors in order to recognize differences between observed variations caused by natural variability (e.g., differences in habitat complexity between sampling sites) and those caused by anthropogenic stressors.

Ecosystem characterization consists of the qualitative or quantitative evaluation of ecosystem features, related to both site-specific and larger-scale phenomena, that may influence stressors' fate and distribution or the receptor sensitivity to the stressors

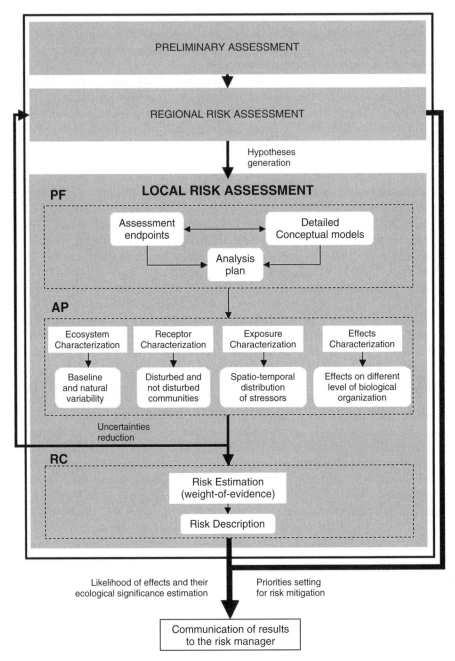

Figure 7.4. Local risk assessment in the overall process of the Procedure for Ecological Tiered Assessment of Risks (PETAR): PF, problem formulation; AP, analysis phase; and RC, risk characterization. White rectangles represent steps in the process, and rounded rectangles represent products of the assessment.

(e.g., water chemical and physical parameters). Some features can even be used as indicators of responses to human-induced alterations (e.g., pH, nutrient concentration). The ecosystem characterization may be based on available maps and reports together with site-specific observations in field studies. Detailed site-specific investigations for analysis of exposure and effects should only start after a careful selection of reference sites, which is based on evaluation of baseline conditions of ecosystems and a preliminary distinction between disturbed and undisturbed receptors.

Exposure characterization may include evaluation of number, size, and distribution of natural and anthropogenic sources by digital processing of maps on land use, aerial photographs, or satellite images. In addition, exposure characterization should include assessment of spatial and temporal distribution of stressors by an evaluation of literature data and databases or (preferentially) by field measurements on different media. Evaluation of pathways from the source to the receptor and quantification of stressors' concentration and distribution can also, for chemical stressors, be based on transport and fate models, or on other assessment tools for biological and physical stressors.

Effects characterization comprises complementary methodologies to be used for inference or measures of biological effects in natural systems. The selection of methods to use, and the levels of organization to be investigated, will depend on resource availability. However, it is important to note that adding different methodologies and studies on effects at different levels of biological organization will greatly improve the quality of the assessment. A single community sample (e.g., fish) can be used for several studies—for instance, body burden, enzyme activity, condition factor, reproductive stage, diversity, and so on [see, e.g., Moraes et al. (2003)].

The analysis phase of a local risk assessment is based on an assessment of the baseline condition of the ecosystem and its natural range of variation that precedes any detailed site-specific study for establishing cause-effect relations. Considerable variability in the distribution and intensity of factors such as river flow, temperature, and nutrient concentration occurs naturally. Human interventions in their patterns of variability may cause important ecological effects. As environmental factors already show a range of natural variability, it is difficult (but critical) to predict when human-induced changes result in significant alterations outside the boundaries of the baseline range for a particular ecosystem (Dorward-King et al. 2001). The evaluation of the baseline conditions of ecosystems, together with the distinction between disturbed and undisturbed communities, is an important element in a careful selection of sampling sites. A failure to have baseline data on the natural variability of ecosystems may thus cast doubt on the causal links between stressor, exposure, and effects, and the probability to establish a cause–effect relation will be higher at sites where ecosystem baseline conditions are comparable to both previously and currently unexposed similar sites.

Detection of effects on different levels of biological organization can be achieved by using complementary methods. The greater the amount of evidence indicating the occurrence (or not) of biological effects at different levels of high specificity and levels of high ecological relevance, the better the quality of the effect assessment.

The final stage of the local risk assessment is the risk characterization during which evidence that a particular human-induced stressor causes observed biological effects at specific sites is analyzed. Evaluation of causal relationships between exposure and effects suggested by the PETAR procedure consists of two steps: elimination of potential causes (Suter et al. 2002) followed by a risk estimation step using a weight-of-evidence approach for the remaining causes.

The weight-of-evidence approach suggested by the PETAR procedure is largely similar to the approach suggested by Suter et al. (2000). Lines of evidence can be results of laboratory studies, manipulations of field situations, and results of biological surveys. The main difference between the PETAR procedure and other published EcoRA approaches is that measures of effects at different levels of biological organization are divided into individual lines of evidence. So when analyzed, for example, biochemical, physiological, organism, reproductive biomarkers, population, or community characteristics are evaluated separately. Epidemiological criteria for an evaluation of each line of evidence may include: strength of the association, consistency and specificity of the association; biological gradient and plausibility; and temporality (Adams 2003). For an illustration of how to establish causality between effects and stressors using the suggested weight-of-evidence, [see e.g., Moraes et al. (2003)].

CONCLUSIONS

The EcoRA procedure presented here integrates methods for assessing multiple types of stressors related to environmental problems at both a regional and local scale. It also includes methods for establishing causal relationships between exposures and effects. Furthermore, three additional characteristics of the PETAR procedure can be considered to be improvements over existing EcoRA procedures: the way cause–effect chains are incorporated in the detailed conceptual model, the integration of a method for the selection of sites for exposure and effect analyses, and an adaptation of the weight-of-evidence approach for establishing causality.

While getting as close as possible to a corroboration, significant simplifications are performed since interaction between entities in ecosystems is very complex. Simplifications imply uncertainties in assessments and leave room for much criticism (Power and Adams 1997, Holdway 1997). The extent to which extrapolations, assumptions, and lack of knowledge can influence the scientific validity of the assessment will vary from case to case, and in particular this is evident for the regional risk assessment in PETAR. Therefore it is essential to include uncertainty and sensitivity analyses in the regional risk assessment (Landis and Weigers 1997) and, when needed, iterate the procedure and invest in further data gathering or model refinements. Despite possible criticism regarding simplifications that the PETAR and other EcoRA procedures may encounter, decisions regarding environmental problems will continue to be made regardless of our current (and endlessly) incomplete scientific knowledge of ecosystems. Nevertheless, it is also important to recognize that the majority of decision-makers do not require the degree of certainty that scientists seek to achieve to make decisions (US EPA 2000).

The further development of the method will probably be related to the further development of remote sensing techniques. For instance, the characterizations of land cover (Ju et al. 2005) may, in conjunction with methods for identification of human activities from satellite images and the establishment of links between these and environmental effects, make it possible to perform regional risk assessments that cover wide areas at low costs. Such methods may also, assisted by various models, contribute to prospective regional risk assessments that can serve as input to strategic environmental assessments. However, what might be even more important is the development of methods to estimate the uncertainty, which in these studies is more related to model than to measurements.

REFERENCES

Adams SM. 2003. Establishing causality between environmental stressors and effects on aquatic ecosystems. *Hum Ecol Risk Assess* **9**:17–35.

Allen TFH, Starr TB. 1982. *Hierarchy—Perspectives for Ecological Complexity*. University of Chicago Press, Chicago, IL.

A NEPC (Australian National Environment Protection Council). 1999a. *Guideline on Ecological Risk Assessment. Schedule B5*. National Environment Protection (Assessment of Site Contamination) Measure. Canberra, Australia, 50 pp.

A NEPC (Australian National Environment Protection Council). 1999b. *National Environmental Protection Measure for Assessment of Site Contamination, Draft Guideline 5: Ecological Risk Assessment*. National Environment Protection Council, Canberra, Australia, 68 pp.

Apitz SE, Elliott M, Fountain, M, Galloway TS. 2006. European environmental management: Moving to an ecosystem approach. *Integr Environ Assess Manage* **2**:80–85.

Barton A, Sergeant A. 1998. Policy before the ecological risk assessment: What are we trying to protect? *Hum Ecol Risk Assess* **4**:787–795.

Butler R, Angelstam P, Ekelund P, Schlaeffer R. 2004. Dead wood threshold values for the three-toed woodpecker presence in boreal and sub-Alpine forest. *Biol Conserv* **119**:305–318.

CCME (Canadian Council of Ministers of the Environment). 1996. *A Framework for Ecological Risk Assessment at Contaminated Sites: General Guidance*. CCME Subcommittee on Environmental Quality Criteria for Contaminated Sites, March 1996. Available c/o Manitoba Statutory Publications, 200 Vaughn St., Winnipeg, Manitoba, R3C-1T5.

CEC (Commission of the European Communities). 1996. Technical guidance documents in support of the Commission Directive 93/67/EEC on risk assessment for new substances and the Commission regulation (EC). Commission of the European Communities, Brussels, Belgium. 1488/94.

CEC (Commission of the European Communities). 2000. Directive 2000/60/EC of the European Parliament and of the Council of 23 October 2000 establishing a framework for Community action in the field of water policy (available at http://ec.europa.eu/environment/water/water-framework/index_en.html as accessed on 27 June 2009).

CEC (Commission of the European Communities). 2001. Directive 2001/42/EC of the European Parliament and of the Council of 27 June 2001 on the assessment of the effects of certain plans and programmes on the environment (available at http://www.europa.eu.int/eur-lex/pri/en/oj/dat/2001/l_197/l_19720010721en00300037.pdf as accessed on 27 June 2009).

Colnar AM, Landis WG. 2007. Conceptual model development for invasive species and a regional risk assessment case study: The European Green Crab, *Carcinus maenas*, at Cherry Point, Washington, USA. *Hum Ecol Risk Assess* **10**:299–325.

Dorward-King EJ, Suter GW, Kapustka LA, Mount DR, Reed-Judkins DK, Cormier SM, Dwyer SD, Luxon MG, Parrish R, Burton A. 2001. Distinguishing among factors that influence ecosystems. In Baird DJ, Burton GA (Eds). *Ecological Variability: Separating Natural from Anthropogenic Causes of Ecosystem Impairment*. SETAC Press, Pensacola, FL, Chapter 1, pp. 1—26.

Efroymson RA, Dale VH, Baskaran LM, Chang M, Aldridge M, Berry MW. 2005. Planning transboundary ecological risk assessments at military installations. *Hum Ecol Risk Assess* **11**:1193–1215.

EC (Environment Canada). 1994. A framework for ecological risk assessment at contaminated sites in Canada: Review and recommendations. Ecosystem Conservation Directorate, National Contaminated Sites Remediation Program, Guidance Document, Ottawa, Ontario.

Fox GA. 1991. Practical causal inference for ecoepidemiologists. *J Toxicol Environ Health* **33**:359–373.

Gardner RH. 1998. Patter, process, and the analysis of spatial scales. In Peterson DL, Parker VT (Eds.), *Ecological Scale: Theory and Applications*. Columbia University Press, New York, pp. 18–34.

Gerhard P, Moraes R, Molander S. 2004. Stream fish communities and their associations to habitat variables in a rain forest reserve in southeastern Brazil. *Environ Bio Fishes* **71**: 321–340.

Gordon S, Majumder S. 2000. Empirical stressor-response relationships for prospective risk analysis. *Environ Toxicol Chem* **19**:1106–1112.

Graham R, Hunsaker C, Oneill R, Jackson B. 1991. Ecological risk assessment at the regional scale. *Ecol Appl* **1**:196–206.

Hamamé M. 2002. Regional risk assessment in northern Chile. Environmental Systems Analysis Report 2002:1, Chalmers University of Technology, Gothenburg, Sweden, p. 52.

Hart Hayes E, Landis WG. 2004. Regional ecological risk assessment of a near shore marine environment: Cherry Point, WA. *Hum Ecol Risk Assess* **10**:299–325.

Hart Hayes E, Landis WG. 2005. Ecological risk assessment using the relative risk model and incorporating a Monte Carlo uncertainty analysis. In Landis WG (Ed.), *Regional Scale Ecological Risk Assessment Using the Relative Risk Model*. CRC Press, Boca Raton, FL, pp. 257–290.

Hill AB. 1965. The environment and disease: Association or causation. *Proc R Soc Med* **58**:295–300.

Holdway D. 1997. Truth and validation in ecological risk assessment. *Environ Manage* **27**:816–819.

Hope BK. 2005. Performing spatially and temporally explicit ecological exposure assessment involving multiple stressors. *Hum Ecol Risk Assess* **11**:539–565.

Hull RN, Swanson S. 2006. Sequential analysis of lines of evidence—An advanced weight-of-evidence approach for ecological risk assessment. *Integr Environ Assess Manage* **2**:302–311.

Hunsaker CT, Graham R L, Suter GW, Oneill RV, Barnthouse LW, Gardner RH. 1990. Assessing ecological risk on a regional scale. *Environ Manage* **14**:325–332.

IHOBE (Sociedad Pública de Gestión Ambiental del Gobierno Vasco). 2003. *Investigation de la Contaminacion de la contamination del suelo. Guia Metodologico: Analisis de riesgos para la salud humana y los ecosistemas*. Departamento de Ordenación del Territorio Vivienda y Medio Ambiente, Gobierno Vasco, Victoria-Gasteiz. 119 pp.

Johnson AR. 2002. Landscape ecotoxicology and assessment of risk at multiple scales. *Hum Ecol Risk Assess* **8**:127–146.

Ju J, Gopal S, Kolaczyk ED. 2005. On the choice of spatial and categorical scale in remote sensing land cover characterization. *Remote Sensing Environ* **96**:62–77.

Landis WG. (Ed.) 2005. *Regional Scale Ecological Risk Assessment Using the Relative Risk Model*. CRC Press, Boca Raton, FL, 286 pp.

Landis WG, Weigers JA. 1997. Design considerations and a suggested approach for regional and comparative ecological risk assessment. *Hum Ecol Risk Assess* **3**:287–297.

Landis WG, Weigers JK. 2007. Ten years of the relative risk model and regional scale ecological risk assessment. *Hum Ecol Risk Assess* **13**:1–14.

Landis WG, Luxon M, Bodensteiner LR. 2000. Design of a relative rank method regional-scale risk assessment with confirmational sampling for the Willamette and McKenzie rivers, Oregon. In Price T, Brix KV, Lane NK (Eds.), *Ninth Symposium on Environmental Toxicology and Risk Assessment: Recent Achievements in Environmental Fate and Transport*. ASTM STP1381 F, American Society for Testing and Materials, West Conshohocken, PA, pp. 67–88.

Landis WG, Duncan PB, Hart-Hayes E, Markiewicz AJ, Thomas JF. 2004. A regional assessment of the potential stressors causing the decline of the Cherry Point Pacific herring run and alternative management endpoints for the Cherry Point Reserve (Washington). *Hum Ecol Risk Assess* **10**:271–297.

Matthews RA, Landis WG, Matthews GB. 1996. The community conditioning hypothesis and its application to environmental toxicology. *Environ Toxicol Chem* **15**:597–603.

McLaughlin JF. Landis WG. 2000. Effects of environmental contaminants in spatially structured environments. In Albers P, Heinz G, Ohlendorf H. (Eds.), *Environmental Contaminants and Terrestrial Vertebrates: Effects on Populations, Communities, and Ecosystems*. SETAC Press, Pensacola, FL, pp. 245–276.

Menzie C, Henning MH, Cura J, Finkelstein K, Gentile JH, Maughan J, Mitchell D, Petron S, Potocki B, Svirsky S, Tyler P. 1996. A weight-of-evidence approach for evaluating ecological risks: Report of the Massachusetts Weight-of-Evidence Work Group. *Hum Ecol Risk Assess* **2**:277–304.

Millennium Ecosystem Assessment. 2005. *Ecosystems and Human Well-Being; Synthesis*. Island Press, Washington, DC.

Moraes R. 2002. *A Procedure for Ecological Tiered Assessment of Risks (PETAR)*. Doctoral Thesis. Environmental Systems Analysis, Chalmers University of Technology, Gothenburg, Sweden.

Moraes R, Molander S. 2004. A procedure for ecological tiered assessment of risks (PETAR). *Hum Ecol Risk Assess* **10**:49–371.

Moraes R, Landis WG, Molander S. 2002. Regional risk assessment of a Brazilian rain forest reserve. *Hum Ecol Risk Assess* **8**:1779–1803.

Moraes R, Gerhard P, Andersson L, Sturve J, Rauch S, Molander S. 2003. Establishing causality between exposure to metals and effects on fish. *Hum Ecol Risk Assess* **9**:149–169.

Munkittrick K R, McMaster M E, Van Der Kraak G, Portt C, Gibbons W N, Farwell A, Gray M. 2000. *Development of Methods for Effects-Driven Cumulative Effects Assessment Using Fish Populations: Moose River Project (1991–1999). Final Report*. Society of Environmental Toxicology and Chemistry, Pensacola, FL, 236 pp.

Obery A, Landis WG. 2002. Application of the relative risk model for Codorus Creek watershed relative ecological risk assessment: An approach for multiple stressors. *Hum Ecol Risk Assess* **8**:405–428.

Power M, Adams S. 1997. Charlatan or sage? A dichotomy of views on ecological risk assessment. *Environ Manage* **21**:803–830.

Rutgers M, Faber J, Postma J, Eijsackers H. 2001. *Site-Specific Ecological Risk: A Basic Approach to Function-Specific Assessment of Soil Pollution.* Netherlands Integrated Soil Research Program, Wageningen, The Netherlands.

Serveiss VB. 2002. Applying ecological risk principles to watershed assessment and management. *Environ Manage* **29**:2145–2154.

Serveiss VB, Bowen JL, Dow B, Valiela I. 2004. Using ecological risk assessment to identify the major anthropogenic stressor in the Waquoit Bay Watershed, Cape Cod, Massachusetts. *Environ Manage* **33**:730–740.

Suter GW. 1990. Endpoints for regional ecological risk assessments. *Environ Manage* **14**:9–23.

Suter GW. 1999. Developing conceptual models for complex ecological risk assessments. *Hum Ecol Risk Assess* **5**:397–413.

Suter GW, Barnthouse LW. 1993. *Ecological Risk Assessment.* Lewis Publishers, Boca Raton, FL.

Suter GW, Efroymson RA, Sample BE, Jones DS. 2000. *Ecological Risk Assessment of Contaminated Sites.* CRC Press LLC, Boca Raton, FL.

Suter GW, Norton SB, Cormier SM. 2002. A methodology for inferring the causes of observed impairments in aquatic ecosystems. *Environ Toxicol Chem* **21**:1101–1111.

US EPA (United States Environmental Protection Agency). 1992. Framework for ecological risk assessment. Risk Assessment Forum, U.S. Environmental Protection Agency, Washington, DC. EPA/630/R-92/001.

US EPA (United States Environmental Protection Agency). 1998. Guidelines for ecological risk assessment. Risk Assessment Forum U.S. Environmental Protection Agency, Washington, DC. EPA/630/R-95/002F.

US EPA (United States Environmental Protection Agency). 2000. *Workshop Report on Characterizing Ecological Risk at the Watershed Scale.* National Center for Environmental Assessment, Office of Research and Development, Washington, DC. EPA/600/R-99/111.

US EPA (United States Environmental Protection Agency). 2002a. *Clinch and Powell Valley Watershed Ecological Risk Assessment.* National Center for Environmental Assessment. Office of Research and Development. U.S. Environmental Protection Agency. Washington, DC. EPA/600/R-01/050.

US EPA (United States Environmental Protection Agency). 2002b. *Waquoit Bay Watershed Ecological Risk Assessment: The effect of land-derived nitrogen loads on estuarine eutrophication.* National Center for Environmental Assessment. Office of Research and Development. U.S. Environmental Protection Agency. Washington, DC. EPA/600/R-02/079.

UK EA (United Kingdom Environment Agency). 2008. *An Ecological Risk Assessment Framework for Contaminants in Soil.* Science report SC070009 UK Environment Agency, Bristol, UK.

Walker R, Landis W, Brown P. 2001. Developing a regional ecological risk assessment: A case study of a Tasmanian agricultural catchment. *Hum Ecol Risk Assess* **7**:417–439.

Wickham JD, Wade TG. 2002. Watershed level risk assessment of nitrogen and phosphorus export. *Comput Electron Agric* **37**:15–24.

Wiegers JK, Feder HM, Mortensen LS, Shaw DG, Wilson VJ, Landis WG. 1998. A regional multiple-stressor rank-based ecological risk assessment for the fjord of Port Valdez, Alaska. *Hum Ecol Risk Assess* **4**:1125–1173.

Wu J, David JL. 2002. A spatially explicit hierarchical approach to modeling complex ecological systems: Theory and applications. *Ecol Model* **153**:7–26.

8

INTEGRATING HEALTH IN ENVIRONMENTAL RISK ASSESSMENTS

Kenneth L. Froese and Marla Orenstein

> Health is a state of complete physical, mental,
> and social well-being, and not merely the
> absence of disease or infirmity.
>
> World Health Organization

Can we integrate human health into landscape-level environmental risk assessment? This is an important and timely question for several reasons. Health is a high-priority concept for a broad range of audiences, such as governments, regulatory agencies, NGOs, communities, and individuals. However, the approach to health that has been developed for use in Environmental Impact Assessments and human health risk assessments has become extremely narrow and disconnected from the more holistic concept of health generally envisioned by affected populations. A landscape perspective coupled with systems thinking (addressed in Chapters 16, 17, and 18) may provide a way for risk assessment practitioners to incorporate health interests in a more meaningful way into assessment modalities. In this chapter, we review what is meant by "health" in current assessment paradigms; discuss ways of bringing a landscape perspective into the assessment of health; and present two case studies that have attempted to integrate to health in a meaningful way into the environmental risk assessment process.

Environmental Risk and Management from a Landscape Perspective, edited by Kapustka and Landis
Copyright © 2010 John Wiley & Sons, Inc.

WHAT IS HEALTH?

In contemporary risk assessment practice, health is defined narrowly to reflect physiological conditions of individual organisms relative to chemical exposures or physical insults. Accordingly, health (impact) assessments, whether of humans or wildlife, are required components of Environmental Impact Assessments (EIAs) in Canada and the United States. What this commonly translates to in practice is a chemical-based human health risk assessment (HHRA), in which the practitioner estimates the risks of possible adverse health effects related to specified chemical (e.g., NO_x, SO_x, metals) or physical (e.g., dust, fine particulate matter) emissions and the estimated subsequent exposures. EIAs may also focus on a few other areas considered to be important to health, including noise and public safety (e.g., traffic, emergency planning).

In this paradigm, health is seen as a cause-and-effect, exposure-and-outcome process. Each risk factor and associated disease can be identified individually and categorized, and the overall risk to the health of an individual is arrived at by a process of aggregating particular disease-specific processes or mechanisms.

For many years, this definition of health has sufficed. Although it is limited, it is workable and thus remains appealing for risk practitioners to use. However, there is a problem with approaching the assessment of health in this way. There is good evidence that this definition of health is not broad enough to satisfy many stakeholder groups, an increasing number of regulatory bodies, and the health practitioner community. There is increasing pressure to evaluate the full spectrum of health—not only the limited number of outcomes such as cancer and respiratory disease commonly examined in HHRAs, but also health outcomes such as alcoholism, injury, and diet. This pressure is coming primarily from Aboriginal groups and governments in the northern regions of Canada and the United States (Noble and Bronson 2006, Wernham 2007), certain corporations that are conscious of and concerned about the broader impact of their projects (Birley 2005), and European governments and the World Bank (Mercier 2003). Interestingly, this push has not been coming from the primary provincial and federal government agencies that originally passed legislation to require consideration of health within the established EIA process.

In 1946, the World Health Organization (WHO) proposed a definition of health that encompasses not only biophysical health and the absence of disease, but also physical, mental, and social well-being (WHO 1985). This definition has since gained currency with a wide range of organizations, and has become the accepted standard worldwide. By focusing on a broad range of illnesses, well-being, and health states, the definition is an improvement on the current paradigm. This definition, however, can be difficult to put into practice in terms of impact assessment, as methods for probing mental, emotional, and social well-being are not as well developed as those that assess biomedical disease outcomes. Additionally, the multifactorial nature of many health outcomes (e.g., cardiovascular disease or overall mortality) means that attributing cause can be difficult.

While the exposure-and-outcome model still has great traction in the research community due primarily to its simplicity for epidemiologic modeling, the health world has moved toward a framework that more broadly reflects the WHO model of

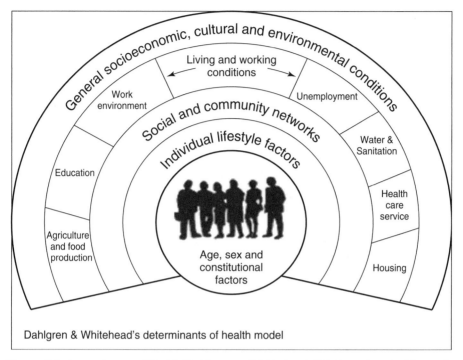

Figure 8.1. Determinants of Health [Dahlgren and Whitehead (1991), reprinted with permission.]

health. Instead of focusing on the biological risk factors for individual diseases, many health researchers have now chosen to focus on the "determinants of health"—social, economic, physical, and environmental factors that are correlated with individual and population health outcomes. Determinants of health include factors such as education, housing, income, employment, social class, environmental contamination, and public infrastructure, as well as many others. A focus on the determinants of health shifts assessment and analysis away from individual-level disease and risk factors to the root causes of health status at a population level (Fig. 8.1). It is a systems-based approach that is predicated on a complex network of causation that cannot be disaggregated very easily.

A number of models of the determinants of health have been developed. Two of the most popular are the Health Canada (Public Health Agency of Canada 2001) model that lists 12 categories of health determinants (income and social status; employment; education; social environments; physical environments; healthy child development; personal health practices and coping skills; health services; social support networks; biology and genetic endowment; gender; culture) and the Dahlgren and Whitehead model (Dahlgren and Whitehead 1991). It is likely that new models will continue to be generated, because there is a very large number of health determinants, and the selection of which to focus on and how to group them differs according to purpose.

To help translate this concept for the ecological discipline, there are a number of parallels in the determinants of health approach with the habitat suitability approaches that have been developed for ecosystems (Kapustka 2005). The fundamental parallel between these concepts is that by including in an assessment the interrelated building blocks of a nurturing landscape, we are more able to reflect a holistic realism in the assessment, which in turn can lead to more informed decision-making in a risk management process. As with the determinants of health for a community, habitat determinants for a particular receptor community in an ecosystem are affected to varying degrees by the direct actions and indirect influences of a particular project, including but not limited to its chemical releases.

By using a determinants of health framework, practitioners are moving to a higher systems level to enable meaningful health assessment and health risk management. As described in Chapters 16, 17, and 18, "Underlying systems thinking is the premise that systems behave as a whole and that such behavior cannot be explained solely in terms that simply aggregate the individual elements." The determinants of health approach enables a systems level view that defines health as exactly this kind of complex system that cannot be extracted from its context nor broken down into separable pieces.

Toward a Systems Approach for Integrating Health in EnRA

There appears to be a disconnect between the way in which affected people conceive of health and the way in which health is defined for the purposes of health risk analysis. The problem can be considered as a difference in scale, or system hierarchy. The health concerns identified by individuals and groups of people affected by contaminated sites or large developments encompass not only specific diseases such as cancer, but also overall wellness and community well-being including the condition of ecological resources—a landscape perspective that links humans to their surroundings. However, scientists and regulators have consistently responded to stakeholders' human landscape perspective concerns with site- or molecular-spatial scale technical and mechanistic arguments. That is, the spatial scale at which questions are being raised is entirely different than that at which responses are given, effectively stifling communication and dialogue.

The molecular-scale approach represents a "bottom-up" philosophy, much like that described by Suter (2004). It is founded on a science paradigm that employs a reductionist perspective in which the belief is that problems will be solved by delving deeper and deeper into mechanistic detail. That is, the more we understand at the molecular scale about fate, transport, absorption, eco-toxicity mechanisms, and so on, the more capable we will be to address human health effects caused by chemical pollutants. Such a perspective is demonstrated in this WHO/IPCS statement (Vermeire et al. 2007):

> By weighing the evidence from conventional mammalian toxicology, ecotoxicology, human epidemiology, and eco-epidemiology, risk assessors could better characterize mechanisms of action and the forms of the relationships of exposures to responses.

Such an approach does have merit. It also has limitations. Valuable progress has been made in elucidating mechanisms of ecotoxicology using this approach, but substantial uncertainty remains in many aspects of toxicology and exposure science as they relate to human health (e.g., cross-species comparisons; whether measured biochemical responses truly lead to health effects, for example, CYP450 induction or T4 hormone induction; and the fundamental assumptions behind assessing well-characterized compounds). We fail to recognize in these instances that human health and the variables that affect it are many, complex, and inextricably interrelated.

Recognition of this problem at the regulator level is surfacing, along with recognition that the whole approach to environmental assessment and risk assessment must be shifted from solely a bottom-up approach to one that includes a top-down approach, where the top can be defined as the landscape perspective questions that require attention. A quote from the Terms of Reference for the Gahcho Kué diamond mine project in Canada's Northwest Territories demonstrates the direction in which responsible regulators are moving on the approach and performance of impact assessments (MVEIRB 2007):

> Previous environmental impact assessments on diamond mines in the NWT generally analyzed and predicted impacts on a subject-specific basis. [Affected] communities have expressed that their primary concerns often are broad and holistic, dealing with interconnecting systems of the land and the people who depend on it, instead of the more narrow subjects often studied by conventional scientific specialists. The Panel made efforts to structure this document [terms of reference] to meet the needs of both types of reviewers.

The reference to two types of reviewers appears to reflect the distinctions between top-down and bottom-up perspectives. Ultimately, the developer was required to frame descriptions of potential effects to key receptors and resources in comprehensive integrated analyses, not merely a series of linear treatments common to most conventional environmental impact assessments. The problem of scale was recognized by Suter et al. (2003) in their discussion of an integrated framework for health and ecological risk assessment. Their discussion emphasized integration of the chemical and toxicological aspects (molecular scale) of traditional risk assessment; however, they acknowledge the importance of nature and the "services of nature" on human well-being (human community scale), including mental health. Suter et al. also recognize, as with MVEIRB on the Gahcho Kué project, that conventional chemical-based risk assessments cannot answer stakeholders' holistic questions of health and welfare.

Contemporary risk assessment practices tend to push the "boundaries" of health sciences and ecological sciences in opposite directions. By boundaries, we contend that risk assessments often are tasked with providing answers that are at the edge or beyond the capabilities of contemporary science. In systems hierarchy terms, assessments that address population-based public health "macro" wellness drivers and place-based, rather than chemical or emissions-based assessments push the science toward a higher system hierarchy. Conversely, when assessments are approached via detailed mechanistic studies of chemical mixtures and biomolecular and toxicological relationships (molecular scale) between human and ecological systems [e.g., the direction furthered by Vermiere et al. (2007)], the scientific boundaries are pushed downward in system hierarchy.

A clear division in the current primary literature appears when we examine this issue from a systems-based approach: population health-level health impact assessment at one extreme (e.g., Bhatia and Wernham 2008, Cole et al. 2005, Wernham 2007), and toxico-mechanistic based health and ecological risk at the other (e.g., Callahan and Sexton 2007, Cormier and Suter 2008, Gohlke and Portier 2007). This division effectively blocks integration of human health and wellness risk assessment with ecological risk assessment, because communication does not occur between the different systems levels of risk inquiry. Beginning at the problem formulation stage of an assessment, the key questions that best capture the risk issue to be addressed must be asked at the correct systems level, and subsequent assessment must proceed along a path that allows those questions to be addressed. A macro-level question cannot be addressed satisfactorily with a micro-level investigation without first translating the macro-level question to a form that is consistent with the lower system scale.

Health Impact Assessment and Integrated Assessment

There are two methods that have been developed relatively recently to bring a more holistic approach to the analysis of human health effects. These are *Health Impact Assessment* (HIA) and *Integrated Assessment* (IA, or sometimes IIA—*Integrated Impact Assessment*).

Health Impact Assessment examines the potential health consequences of a project or policy on nearby populations. It is often performed as a complement to EIA and Socioeconomic Impact Assessment. An HIA follows the same sequence of screening, scoping, assessment of impacts, risk characterization, and formulation of mitigation and monitoring strategies—an evaluation familiar to ecological risk assessors. An HIA is intended to be a participatory process that actively engages local communities in most stages of the assessment. It uses a broader approach to health than that taken in EIA or HHRA, and it assesses the impact of the project or policy on a very broad range of physical and mental health outcomes, and also on relevant determinants of health. The HIA has become mandatory in some regions (the United Kingdom, Australia, the Netherlands, Québec) and is a required element for projects funded by the World Bank under the equator principles (International Finance Corporation 2006).

The move toward Integrated Assessment is an attempt to improve the quality of assessments and subsequent decision-making by simultaneously addressing biophysical, ecological, social, economic, cultural, and health impacts of a project or policy (Bhatia and Wernham 2008, Gohlke and Portier 2007, Kwiatkowski and Ooi 2003). An IA also links to other knowledge streams (such as traditional knowledge studies and stakeholder engagement) and integrates with the decision-making process and timeline. It is the antithesis of the concept that integration can be achieved at the end of the EA process, as an afterthought. At this point, there has been more published on the theory of integrated assessment than examples of IAs themselves; nonetheless, IA holds promise as a way of joining assessment tools with a landscape perspective of humans and their environment.

IMPLEMENTATION BARRIERS

Numerous authors have identified and discussed the barriers to effective implementation of HIA within the EIA context in Canada and the United States (Bhatia and Wernham 2008, McCaig 2005, Noble 2006, Noble and Bronson 2005, Petticrew et al. 2007, WHO 2001). These include limits in the health requirements in EIA legislation; differences in interpretation of the scope of health and expectations of EIAs to address health; and lack of communication or outright dissent among government institutions, health practitioners, risk assessors, environmental assessment specialists, and project proponents. There are positive indications that some of the barriers are being reduced (Bhatia and Wernham 2008, Dannenberg et al. 2008, Wernham 2007).

Those who are reluctant or hesitant to embrace a holistic integration of health in environmental assessment and environmental risk assessment often point to difficulties in evaluating nonquantitative data, such as the social, cultural, and psychological determinants of wellness, in conjunction with their "quantitative" evaluation of human activities on the environment. This is a real concern, particularly because much of the social, cultural, and psychological determinants and what specifically affects them cannot be quantified in the same manner. However, there is sufficient public health evidence to allow practitioners to evaluate planned events, anticipate health determinant outcomes, and proactively implement risk management initiatives. For example, the development of a new mine is often associated with a population of external workers moving into a worker camp for the construction phase. Risk factors include age and gender of workers (often males in their early twenties), disposable income (high), lack of broader social support (remote location), and access to alcohol and drugs (high). This situation commonly results in a range of health problems in both local communities and the in-migrant worker population, including increases in sexually transmitted infections, alcohol and substance abuse, violence, mental health problems, and traffic collisions. A top systems level (qualitative) evaluation of this scenario can result in generalized and specific risk management activities being implemented at the planning stages of the project, without the requirement for detailed quantitative evaluation of the hazards, effects, and magnitudes of risk.

We all recognize that EnRA is an inherently multidisciplinary activity, whether one is doing ecological, human health, integrated HHRA-EcoRA risk assessment, or developing a holistic risk strategy. The range and complexity of the multidisciplinary effort are different for each of these, and should not be underestimated but often is. One of the primary challenges of multidisciplinary work is finding the common language and concepts to undertake the work. Furthermore, system levels at which we develop the key questions should be considered and defined within the context of the different disciplines.

In preparing this chapter, we observed a critical lack of communication among the leaders championing integrated, holistic risk assessment. For example, let us presume that G. Suter II, K. Sexton, W. Munns, and T. Vermiere are key figures in theory and practice from the ecological side of integrated risk assessment, including support of the US EPA's 2003 Guidance on Cumulative Risk Assessment (US EPA 2003), and that A. Dannenberg, R. Bhatia, J. Bronson, B. Cole, K. McCaig, B. Noble, J. Kemm, and A. Wernham are key figures in the theory and practice of the public health side of health impact assessment and its integration with environmental assessments. Using the Web of Science tools, we entered the analysis starting with two recent articles from each of the members of the disciplinary groups (Bhatia and Wernham 2008, Callahan and Sexton 2007, Cormier and Suter 2008, Dannenberg et al. 2006, McCaig 2005, Wernham 2007). None of the key figures from either group were cited directly by the other group in these articles. The Web of Science also enabled us to look up related references (articles that have one or more cited references in common with the selected paper). With the Dannenberg (Dannenberg et al. 2006), Wernham (Wernham 2007) and McCaig (McCaig 2005) papers at the time of writing this chapter (spring 2009), 1924 articles were identified as being related to these three papers. Of those 1924 articles, only one related article was co-authored by one of the key figures in the ecological risk assessment group (Murdock et al. 2005). Similarly, of the 3911 related articles found for the Callahan and Sexton (Callahan and Sexton 2007) and Cormier and Suter (Cormier and Suter 2008) papers, only one related article was authored by one of the key people in the HIA group (Corburn and Bhatia 2007). In total, 38 common related records were found between the two groups from a combined 5835 related records. No common guidance documents were cited between the two groups.

This lack of interaction at the primary literature level is, in our opinion, the single greatest barrier to moving forward effectively in truly addressing holistic questions in environmental health and ecology. This demonstrates the lack of dialogue and professional exchange of knowledge and ideas between influential human health and ecological risk leaders. If there is effectively no acknowledgment of alternative views in the primary literature, where can practitioners, regulatory bodies, and environmental and public health organizations find the basis for undertaking the necessary multidisciplinary effort? How can practitioners begin to build the required support at the fundamental organizational level when attempting to do such work?

EXAMPLES OF HEALTH BEING INCORPORATED INTO HIA AND IA

In this section, we review two examples of integrated assessment and discuss their successes and failures in terms of (a) addressing health holistically and (b) functionally integrating with environmental and social impact assessment disciplines at compatible systems levels.

BHP Diamonds, Northwest Territories, Canada

In 1994, BHP Diamonds Inc. (now BHP Billiton) proposed the development of Canada's first diamond mine, the Ekati mine, located approximately 300 km northeast of Yellowknife in the Northwest Territories (NWT). The Ekati mine was the first of several diamond mines that came on line in the NWT over the following decade. It was sited in an area with a predominantly Aboriginal population and little previous industrial development. Two publications (Bronson and Noble 2006, Kwiatkowski and Ooi 2003) amply review (a) the process that was used to arrive at the terms by which the EIA was conducted and reviewed and (b) the consultation that was undertaken with various stakeholder groups. Those details are not repeated here.

There are several aspects of the health risk assessment that make the Ekati project notable. The first is the inclusion of health in the EIA in a way that is not only broad and holistic, drawing on both environmental and socioeconomic determinants of health, but also in a manner relevant to a northern, subsistence-based culture.

The communities in the area near the Ekati mine are predominantly Aboriginal (Dene); and a large part of the population still retains many aspects of a traditional lifestyle, including a traditional diet (food from the land with caribou extremely important), hunting and trapping as a significant economic activity, and many traditional cultural features. Consultation with local groups during the scoping process identified a number of concerns that are consistent with those raised by Aboriginal groups within Canada and other northern areas regarding development projects. Specific health concerns included disruption of traditional ways of life, including hunting and fishing; problems associated with the introduction of a money-based economy; social, economic, and cultural stress; racism; social diseases (e.g., alcoholism and sexually transmitted diseases); personal development; self-esteem and confidence; mental health; and families left with one or no parents as a result of employment opportunities outside the community (Bronson and Noble 2006, Kwiatkowski and Ooi 2003). Effects on wildlife, water, plants, and other environmental receptors were also a primary concern. The EIA as mandated addressed the potential effects of the project on this broad array of health-related topics, and mitigation measures were developed in a partnership between industry and community.

Another significant aspect of the Ekati project identified by Bronson and Noble (2006) is the development and implementation of a socioeconomic effects monitoring partnership between BHPB Ekati and the Government of the Northwest Territories. This monitoring partnership has also been established with the proponents of subsequent mines in the area. As part of the agreement, the government of the Northwest Territories publishes "Communities and Diamonds," an annual analysis of socioeconomic indicators in communities affected by diamond mining (GNWT 2009). The report examines several indicators specific to health (injuries, suicides, communicable diseases, potential years of life lost) as well as a number of social and economic determinants of health, including family and community well-being, crime, housing, cultural well-being, traditional economy, nontraditional economy and employment, and education. The analysis compares the effects originally predicted by the diamond mining companies with the ongoing effects observed in territorial data banks. The latest analysis (2008) indicates a trend of improvement in housing and crowding, domestic violence, income, disparity, educational attainment, injuries, and in some communities potential years of life lost. At the same time, the report describes an increase in sexually transmitted infections (a common consequence of resource development).

The socioeconomic monitoring efforts fall short of an ideal health evaluation for a number of reasons. For one, the outcomes reflect many factors and not just the effect of the diamond mines. For another, the choice of indicators is not ideal for examining health; a comprehensive health monitoring program would also need to include health outcome measures such as mortality, reproductive outcomes, diet-related disease, and respiratory problems such as asthma and acute respiratory infections.

Nonetheless, the significance of this monitoring approach is that the use of broad parameters of health reflects the community's original concerns related to health. It also allows for the demonstration of positive health outcomes—something that is not usually possible when considering only those health risks that are linked with environmental contaminants or other ecosystem disturbance.

Oil and Gas Development in the National Petroleum Reserve, Alaska

An HIA was commissioned by the North Slope Borough (the regional government on Alaska's North Slope) through a cooperating agency relationship with the US Bureau of Land Management, and was led by Dr. Aaron Wernham (Wernham 2007). It represents a significant development in project-based impact assessments in North America. Wernham was asked by the Alaska Inter-Tribal Council to look at the health effects of proposed oil and gas development within the North Slope Borough in Alaska. The HIA was requested in response to the perception that EIS statements were being produced without sufficient health input. This assessment is significant in that it represents the first formal effort to undertake an HIA within the legal framework of the National Environmental Policy Act (NEPA).

> The inclusion of a broad, systematic analysis of health within a US EIS is unprecedented...and has implications for the implementation of NEPA elsewhere in the US [Wernham 2007].

The study area for the Northeast NPR-A project contains eight remote Inupiat villages ranging in population from 250 to over 4000 people. There is a strong subsistence culture, and the local population relies heavily on hunting, fishing, and whaling for food, livelihoods, and culture. Residents have expressed concern about potential health effects associated with oil development in the area. Stated health concerns include possible effects of contaminants, increased social dysfunction (alcoholism, suicide, domestic violence), and general fears over loss of culture. In other words, their health concerns are *holistic* in nature and are often expressed in terms of ecosystem-based impacts generalized over the entire area.

The assessment method did not rely on a quantitative modeling approach; rather, Wernham employed a logic framework model that identified and validated relationships between oil and gas activities, health determinants, and health outcomes. The assessment supported the development of mitigation measures recommended for incorporation in future oil and gas development.

This HIA was successful in several respects. First, the approach successfully represented both the specific health concerns and the overall view of health determinants supported by the local community. Intensive, high-level participation by the Inupiat communities was key both to supporting the analysis of potential impacts and also in convincing regulatory agencies of the relevance of the work. The work resulted in successful partnership with the regulatory agency to embed results/recommendations in future development.

Since this HIA/EIS was completed, there is evidence that the approach is being adopted by other federal agencies with increasing frequency. The Alaska Native Tribal Health Consortium (a state-wide tribal public health agency) has initiated a state-wide HIA program led by Wernham. This group has partnered with the North Slope Borough on an HIA/EIS for offshore oil and gas leasing and has worked with the US EPA and a regional tribal health agency (Maniilaq Association) to complete an HIA integrated into an EIS for a large zinc mine, and the US EPA has now offered to subcontract work on another HIA for a large coal mine in Alaska. Outside Alaska, the US EPA (responsible for reviewing other agencies' EIS practice) has now called for HIAs for a variety of regulatory, EIS-based federal actions.

There are also a few shortcomings of the assessment. Most notably, it was completed without the presence or participation of oil and gas development proponents. This means that the interesting and intensive discussion fomented during this process/approach may not inform EIS of specific projects and may or may not be acknowledged by project developers, which can limit the ability to develop project-specific creative solutions in cooperation with local community. However, this HIA was part of an EIS that focused on leasing, and project-specific assessments are expected for future developments.

CONCLUSIONS

The question we posed at the outset of this chapter is, Can we integrate human health into landscape-level environmental risk assessment? The answer we have arrived at is, yes, we can.

It will be critical to ensure that we are starting from a common understanding of what health is and what factors in the biophysical and social environment can affect it. Assessments will need to identify the appropriate spatial and temporal scales for asking key questions, and it will need to tailor the assessment approach for those scales in terms of grain and extent (see Chapter 4).

The approach we offer is to instill a systems-thinking paradigm and to begin planning holistic assessments from a high organizational level—for example, the essence of the "top-down" approach as put forward by Suter (2004)—and ask specific questions to focus targeted "bottom-up" inquiries within the overall framework. Practitioners from diverse disciplines, including public health, epidemiology, social sciences, economics, natural sciences, and especially engineering, must view more broadly the complex relationships between humans and their environmental, social, cultural, and economic landscapes and the fundamental determinants of health that connect different aspects of those landscape views.

We must evaluate effects on wellness with a clear goal of developing management strategies that target an achievable range of balance among the key indicators of health/wellbeing. Management strategies that, in some way or another, are "public actions that simultaneously strengthen people's power, expand people's choices, institute norms of widespread accountability, and ultimately transform adverse living conditions along with the health indices that they engender" (Milstein 2008).

The fundamental message that comes from this chapter is that risk assessment needs to become more integrated in both research and discussions; that is, cross-disciplinary dialogue between risk practitioners. It needs to develop a concise vision of holistic impact assessment that builds on the existing legislated requirements for HIA and EIA. A holistic, systems-based risk strategy will enable health to be included in environmental risk assessments in ways that are more in keeping with emerging scientific and stakeholder views.

REFERENCES

Bhatia R, Wernham A. 2008. Integrating human health into environmental impact assessment: An unrealized opportunity for environmental health and justice. *Environ Health Perspect* **116**:991–1000.

Birley M. 2005. Health impact assessment in multinationals: A case study of the Royal Dutch/Shell group. *Environ Impact Assess Rev* **25**:702–713.

Bronson J, Noble BF. 2006. Health determinants in Canadian northern environmental impact assessment. *Polar Record* **42**:315–324.

Callahan MA, Sexton K. 2007. If cumulative risk assessment is the answer, what is the question? *Environ Health Perspect* **115**:799–806.

Cole BL, Shimkhada R, Fielding JE, Kominski G, Morgenstern H. 2005. Methodologies for realizing the potential of health impact assessment. *Am J Prevent Med* **28**:382–389.

Corburn J, Bhatia R. 2007. Health impact assessment in San Francisco: Incorporating the social determinants of health into environmental planning. *J Environ Plan Manage* **50**:323–341.

Cormier SM, Suter GW. 2008. A framework for fully integrating environmental assessment. *Environ Manage* **42**:543–556.

Dahlgren G, Whitehead M. 1991. *Policies and Strategies to Promote Social Equality in Health*. Institute of Future Studies, Stockholm, Sweden.

Dannenberg AL, Bhatia R, Cole BL, Dora C, Fielding JE, Kraft K, McClymont-Peace D, Mindell J, Onyekere C, Roberts JA, Ross CL, Rutt CD, Scott-Samuel A, Tilson HH. 2006. Growing the field of health impact assessment in the United States: An agenda for research and practice. *Am J Public Health* **96**:262–270.

Dannenberg AL, Bhatia R, Cole BL, Heaton SK, Feldman JD, Rutt CD. 2008. Use of health impact assessment in the US—27 case studies, 1999–2007. *Am J Prevent Med* **34**:241–256.

GNWT. 2009. Communities and Diamonds: Socio-economic impacts in the communities of Behchoko, Gemeti, Whati, Wekweeti, Dettah, Ndilo, Lutselk'e, and Yellowknife. 2008 Annual Report of the Government of the Northwest Territories Under the BHP Billiton, Diavik and De Beers Socio-economic Agreements. August 2007

Gohlke JM, Portier CJ. 2007. The forest for the trees: A systems approach to human health research. *Environ Health Perspect* **115**:1261–1263.

International Finance Corporation. 2006. *Policy and Performance Standards on Social & Environmental Sustainability*. World Bank Group, Washington, DC.

Kapustka LA. 2005. Assessing ecological risks at the landscape scale: Opportunities and technical limitations. *Ecol Soc* **10**:11. [Online] URL: http://www.ecologyandsociety.org/vol10/iss2/art11/

Kwiatkowski RE, Ooi M. 2003. Integrated environmental impact assessment: A Canadian example. *Bull World Health Org* **81**:434–438.

McCaig K. 2005. Canadian insights: The challenges of an integrated environmental assessment framework. *Environ Impact Assess Rev* **25**:737–746.

Mercier J. 2003. Health impact assessment in international development assistance: The World Bank experience. *Bull World Health Org* **81**:2.

Milstein B. 2008. *Hygeia's Constellation—Navigating Health Futures in a Dynamic and Democratic World*. The Centers for Disease Control and Prevention, Syndemics Prevention Network, Atlanta, GA.

Murdock BS, Wiessner C, Sexton K. 2005. Stakeholder participation in voluntary environmental agreements: Analysis of 10 project xl case studies. *Sci Technol Human Values* **30**:223–250.

MVEIRB (MacKenzie Valley Environmental Impact Review Board). 2007. Terms of reference for the Gahcho Kue environmental impact statement. Yellowknife, NWT.

Noble B. 2006. Human health in environmental impact assessment. *Arctic* **59**:234–235.

Noble B, Bronson J. 2006. Practitioner survey of the state of health integration in environmental assessment: The case of northern Canada. *Environ Impact Assess Rev* **26**:410–424.

Noble BF, Bronson JE. 2005. Integrating human health into environmental impact assessment: Case studies of Canada's northern mining resource sector. *Arctic* **58**:395–405.

Petticrew M, Cummins S, Sparks L, Findlay A. 2007. Validating health impact assessment: Prediction is difficult (especially about the future). *Environ Impact Assess Rev* **27**:101–107.

Public Health Agency of Canada. 2001. *What Determines Health?* Health Canada, Ottawa, ON.

Suter GW. 2004. Bottom-up and top-down integration of human and ecological risk assessment. *J Toxicol Environ Health Part a—Current Issues* **67**:779–790.

Suter GW, Vermeire T, Munns WR, Sekizawa J. 2003. Framework for the integration of health and ecological risk assessment. *Hum Ecol Risk Assess* **9**:281–301.

US EPA (United States Environmental Protection Agency). 2003. *Framework for Cumulative Risk Assessment*. Office of Research and Development NCfEA, Washington Office, U.S. Environmental Protection Agency, Washington, DC.

Vermeire T, Munns WR, Sekizawa J, Suter G, Van der Kraak G. 2007. An assessment of integrated risk assessment. *Hum Ecol Risk Assess* **13**:339–354.

Wernham A. 2007. Inupiat health and proposed Alaskan oil development: Results of the first integrated health impact assessment/environmental impact statement for proposed oil development on Alaska's north slope. *Ecohealth* **4**:500–513.

WHO (World Health Organization). 1985. *Basic Documents*, 35th edition. World Health Organization, Geneva, Switzerland.

WHO (World Health Organization). 2001. *Report on Integrated Risk Assessment*. Safety IPoC, World Health Organization, Geneva, Switzerland.

ADDITIONAL SOURCES

Health Impact Assessment and Integrated Impact Assessment

Overview Publications

Baines JT, Morgan B, Researcher D. 2006. Getting on with integrated impact assessment: One set of guiding principles—many methods.In *Annual Conference of the Environment Institute of Australia and New Zealand*, Adelaide, September 18, 2006, Vol. 18.

Birley M. 2003. Health impact assessment, integration and critical appraisal, *Impact Assess Project Appraisal* **21**:313–321.

Krieger N. 2003. Assessing health impact assessment: multidisciplinary and international perspectives. *J Epidemiol Community Health* **57**:659–662.

Lee N. 2006. Bridging the gap between theory and practice in integrated assessment. *Environ Impact Assess Rev* **26**:57–78.

Guidelines

Minister of Health. 2004. *Canadian Handbook on Health Impact Assessment* (4 volumes). Canadian Ministry of Health. Ottawa, Ontario.

Krieger G. 2009. *Introduction to Health Impact Assessment*. International Finance Corporation, World Bank Group, Washington, DC.

Scott-Samuel A, Birley M, Ardern K. 1998/2001. *The Merseyside Guidelines for Health Impact Assessment, second edition*. Liverpool: Merseyside Health Impact Assessment Steering Group. Reissued by International Health Impact Assessment Consortium.

Quigley R, den Broeder L, Furu P, Bond A, Cave B, Bos R. 2006. *Health Impact Assessment International Best Practice Principles. Special Publ Series No. 5*. International Association for Impact Assessment, Fargo, ND.

Repositories of Completed HIAs, Tools and Evidence Sources

HIA Gateway: http://www.apho.org.uk/default.aspx?QN=P_HIA

HIA Connect: http://www.hiaconnect.edu.au/

World Health Organization: http://www.who.int/hia/

Netherlands RIVM HIA Database: https://webcollect.rivm.nl/hiadatabase/

HIA Community Wiki: http://www.seedwiki.com/wiki/health_impact_assessment_hia_community_wiki

9

VALUING WILDLANDS

Rebecca A. Efroymson, Henriette I. Jager,
and William W. Hargrove

One of the central problems of land and water management is "the way in which scarce resources are allocated among alternative uses and users. The question is, of course, fundamental to economic thinking . . . and it is for this reason that we have seen the introduction of essentially economic models and modes of thought in ecology" (Rapport and Turner 1977). Many questions that are at the heart of environmental management may be answered not only through the use and advancement of landscape ecology and EcoRAs (the primary topics of this book), but also through resource valuation. The value of wildlands is derived from human use of resources, as well as ecological functions such as provision of habitat, that support nonuse or existence values of organisms, populations, communities, and ecosystems. Ecological valuation entails both the description of valued attributes of the environment, as well as quantitative methods for comparing these attributes and alternative scenarios. The valuation of wildlands can support several types of decisions, such as which lands to conserve, which lands to develop, which waters to impound, how much flow to leave in rivers, which lands or waters to remediate, and which lands or waters to set aside for research. Moreover, various US federal agencies are increasingly required to evaluate benefits of conservation and environmental research programs, both of which rely on valuation methods. For example, the US Department of Agriculture evaluates benefits of its Conservation Reserve Program (USDA 2004), and the US Department of Defense is increasingly interested in valuing its lands that are exclusion zones or buffer areas for military training or testing (R. Pinkham, Booz Allen Hamilton, personal communication, September 2006).

Wildlands

The use of the term "wildland" implies that value is somehow derived from wildness. Wildlands are lands and waters where natural processes dominate and human impact is minimized. The term "wilderness" can be a synonym for wildlands, but is more narrowly defined by law, though the US Wilderness Act of 1964 took the rather broad definition "area where the earth and its community of life are untrammeled by man, where man himself is a visitor who does not remain" (Public Law 88–577). For the purpose of this chapter, we assume a gradient of "wildness" or lack of human impact, and only completely exclude from discussion areas of extensive urbanization, industrial development, intense resource extraction (e.g., oil and gas development, agriculture, timber extraction), and stream impoundment. Thus, most forests, grasslands, rangelands, streams, and natural lakes would fit our definition of wildlands, as would small natural areas such as riparian zones that are surrounded by urban, suburban, or industrial development. Although some readers would dispute that powerline rights-of-way are wildlands, for example, those that are managed for dense scrub vegetation provide substantial pollination services (Russell et al. 2005). Similarly, many military installations have large wildland communities that serve as reservoirs for protected species, despite the proximate disturbances from training (Tazik and Martin 2002).

For the purposes of this chapter, we include aquatic ecosystems within the definition of wildlands. In the United States, some rivers are designated *Wild and Scenic Rivers*: "certain selected rivers of the Nation which, with their immediate environments, possess outstandingly remarkable scenic, recreational, geologic, fish and wildlife, historic, cultural or other similar values, shall be preserved in free-flowing condition, and ... they and their immediate environments shall be protected for the benefit and enjoyment of present and future generations" (16 U.S.C. §§ 1271–1287). "Free-flowing" is defined as "existing or flowing in a natural condition without impoundment, diversion, straightening, rip-rapping, or other modification of the waterway." Dams upstream of Wild and Scenic portions of rivers are typically required to maintain natural flow regimes (Jager and Bevelhimer, 2007). The "wild and scenic rivers" designation recognizes the public's interest in maintaining a subset of rivers in a relatively pristine state.

Similarly, lakes without shoreline development have enhanced value as wildlands. In 1965, the US Congress established the Land and Water Conservation Fund (16 U.S.C. §§ 4601–4 to 4601–11) to purchase and protect undeveloped shoreline along critical lakes and streams. These lands are often placed in the custody of the USDA Forest Service.

Types of Value

The value of wildlands is not derived primarily from human extractive use, even where hunting, fishing, and timber harvesting are common. Although game fish and wildlife are sometimes classified as market entities (e.g., US EPA 2006), most people who engage in these activities are not recouping their travel or other costs from sales. These activities are valued because of the cultural experience and environment as well as the resource product. Human use values of wildlands include recreational and aesthetic value. They also include other ecological service values, many of which are not well-quantified or well-monetized: supportive functions such as nutrient cycling and pollination, regulating services such as climate modulation and soil retention, provisioning services such as water supply, and cultural services such as historical or spiritual symbolism (see the *Millennium Ecosystem Assessment*, WRI 2005, for more detail).

Nonuse values are existence values or bequest values that are unrelated to use of or visits to wildlands. For example, we value rare species just because they exist. Likewise, we value the fact that we could visit the African Plains even if we never travel there. Option value is an additional type of value related to preserving the opportunity of possible future use of the resource (e.g., for genes or medicines), but it may also be viewed as belonging to the nonuse category of values. This taxonomy of ecological valuation is described in more detail at http://www.ecosystemvaluation.org/ (viewed January 2010).

Preservation value (a combination of option value for recreational use, existence value, and bequest value) contributes most of the value of wildlands, but willingness to pay for preservation declines as the number of protected resources becomes large. For example, in a study of the protection of rivers in the US Rocky Mountains in Colorado. Sanders et al. (1990) found that preservation value was higher than recreational use value, but declined as the number of protected rivers increased. Consequently, total value reached a peak at an intermediate number of protected rivers (Fig. 9.1).

In its *Ecological Benefits Assessment Strategic Plan*, the US EPA defines "indirect-use" values as those that indirectly benefit society though the "support [of] offsite ecological resources or [maintenance of] the biological ... or biochemical processes

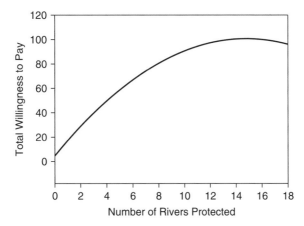

Figure 9.1. Total benefits of river preservation, including preservation values for protection of wild and scenic rivers, Colorado, 1983. [Redrawn from Sanders et al. (1990).]

required for life support." In this definition, EPA includes many "benefits" that are relevant to wildlands, including maintenance of biodiversity, protection of habitat, pollination, seed dispersal, flood protection, water supply (quantity), water purification, pest and pathogen control, and energy and nutrient flow (US EPA 2006). Many of these benefits are not well-quantified.

Many environmentalists are reluctant to value natural environments from an economic or even an ecological service perspective. For example, McCauley (2006) argues that conservation must be a moral or ethical enterprise and that "Nature has an intrinsic value that makes it priceless." While this cultural belief is valid, it does not help environmental managers choose which lands to conserve or which to restore first.

ECOLOGICAL VALUATION

We believe that the future of wildland valuation will be driven by the increased incorporation of ecological relationships. Ecologists can quantify many economic concepts that are at the heart of valuation, such as rarity, complementarity (i.e., value in context), and substitutability. Values of wildlands depend on spatial relationships, temporal systems dynamics, and thresholds. Ecological models can be used to transfer estimated value from one previously valued (e.g., by surveys) entity to a related, unvalued one (such as a predator, forage, habitat, etc.). It is unlikely that new economic methods of valuation of direct-use benefits, such as recreation, will advance the science of valuation as much as ecology. Therefore, we believe that a discussion of the future of wildland valuation is a discussion of the future of *ecological* valuation, involving valuation of populations and their habitats, communities, and ecosystem function. (See also Chapters 16, 17, and 18, all of which address economic ecology.)

The science of ecological valuation is moving in two directions at once—toward increased simplification and toward increased complexity. Simple approaches include several nonmonetary valuation methods: (2006) semiquantitative lists of valued attributes, such as aspects of habitat value; (2005) environmental benefit indicators; (2005) environmental benefit indices (aggregations of indicators); and (1999) areal equivalencies for ecological services. Simple approaches are often chosen when funding is not available to monetize, direct measurement of value is important, all relevant ecological benefits cannot be monetized, monetization is not in the interest of the land owner or manager (e.g., if a high value might prompt a sale of land that is not desired by all stakeholders), or valuation is being used primarily as a communication tool (e.g., if users want maps of value). More complex approaches use dynamic models that include feedbacks between ecology and economics. These are typically used when adequate funding is available to support a large valuation effort, value can be monetized, and mechanistic relationships are understood.

EXAMPLE APPROACHES TO ECOLOGICAL VALUATION

We now review some of the methods that are available for quantitative and semiquantitative valuation of wildlands. These include simple models of value (e.g., habitat valuation metrics, and indicators/indices of environmental benefits) and more complex models of value (integrated models, mechanistic models of ecology). We also discuss the use of ecological values that are derived using these methods in optimizations to address objectives that combine ecological and nonecological values.

Simple Models of Value

Ecological value can be decomposed into measurable characteristics. One of the important questions is, What makes wildlands wild? Remoteness is a characteristic of wildlands that is valued by many hunters, fishers, hikers, and wide-ranging vertebrate species. Remoteness is often correlated with valued ecological services and attributes of habitat. For example, bird densities are reduced near automobile traffic (Reijnen et al. 1995). One could represent remoteness by using a simple measure such as average road density in the area (the value would be sensitive to the area chosen) or distance to closest road. By the latter measure, R. T. Forman asserts that the most remote location in the eastern United States would be somewhere in the Florida Everglades (Cromie 2001), coinciding with prime Florida panther (*Puma concolor coryi*) and American alligator (*Alligator mississippiensis*) habitat. However, this quality of remoteness raises the dilemma presented in Banzhaf and Boyd (2005): If an ecosystem benefit is enjoyed by many, rather than a few, is a higher level of ecological service being provided? Is ecological value higher?

Wildness also implies a lack of disturbance from other stressors, not just roads and their vehicles. Therefore, measures of extent or intensity of disturbance might be viewed as other broad indicators of wildness, or more precisely, a lack of wildness. However, the term *disturbance* has many meanings, sometimes representing exposure to physical (e.g., noise, erosion) and biological (e.g., invasive plant species) agents and sometimes biological effect. Disturbance is not easily measured as a broad value metric, but descriptions of specific disturbances have been used in valuation studies. For example, in a habitat valuation study, Efroymson et al. (2008a, 2008b) included examples of disturbances or management practices as part of the site descriptions that were used in the analysis of habitat complexity, land cover, and ecological corridors: presence of invasive biota, presence of weir, presence of concrete liner, absence of riparian zone, erosion, substantial nutrient influx, presence of chemical contamination, pine beetle damage, plantation land cover, presence of burial ground, mowing, presence of roads, presence of buildings, and presence of scrap metal.

Moreover, in some instances, disturbed lands may be more ecologically valuable than wilder lands, depending on the ecosystem service under consideration. For example, some species benefit from disturbance at explosives-contaminated military ranges. These include early successional plant species, kangaroo rats (*Dipodomys merriami*), Sonoran pronghorns (*Antilocapra americana sonoriensis*), and frogs that use impact craters. Other species [e.g., black-capped vireo (*Vireo atricapillus*) and Karner blue butterfly (*Lycaecides melissa samuelis*)] use early successional habitats that persist only in the presence of wildfire (Efroymson et al. 2009 and references within).

Habitat Valuation Metrics. Attributes of lands and waters that make them good habitat for multiple species or rare species have been used to estimate habitat value. As early as the 1970s, land areas were prioritized for conservation using one or more of five typical value metrics: quantity of habitat, biodiversity supported, naturalness, rarity, and threat of human interference (Margules and Usher 1981). Although economic factors have always been considered in conservation decisions, habitat benefits are typically described, but not monetized.

Habitat Quantity. Area is a measure of relative habitat value for sites within a single ecosystem. A larger, contiguous habitat patch or stream reach is generally more valuable to a species than a smaller one of the same habitat quality. Rates of species loss are dependent on land or water body area (Margules and Usher 1981). However, area is not a reasonable habitat value metric for comparisons across ecosystem types.

Rarity of Species and Communities. Another determinant of habitat value is rarity, or the lack of substitute habitats. A rare vegetation community is arguably more valuable than a common association, especially if organisms are closely adapted to that vegetation association. The presence of rare species increases the existence value of a community (Rossi and Kuitunen, 1996). Moreover, rare plant or bird species are often indicative of rare vegetation associations (SAMAB 1996). An important dimension of rarity is the region, land area, or stream reach within which a species or biotic community is rare.

Biodiversity Supported. Species diversity or taxa richness are direct measures of use of a site by organisms. Biodiversity is also related to the functional value of ecosystems (Hooper et al. 2005). Some ecologists view biodiversity as insurance against major functional changes in an ecosystem because higher diversity ensures redundancy in ecosystem function among individual species (Doherty et al. 2000). Habitat structural complexity has been found to increase biodiversity by many researchers (Crowder and Cooper 1982, Downes et al. 1998, Benton et al. 2003, Johnson et al. 2003), but not by all (e.g., Doherty et al. 2000). Quantitative methods for assessing habitat structural complexity are much less common in terrestrial systems (Newsome and Catling 1979) and lacustrine systems than in streams (Barbour et al. 1999). Kapustka et al. (2004) modified a model developed by Short (1984) to estimate potential for biodiversity and ecological recovery of habitat. They predicted wildlife species richness for locations surrounding a contaminated copper mine site, based on vertical and horizontal diversity of vegetation cover types.

Habitat valuation schemes based on biodiversity can be refined to account for the fact that species are not valued equally by society. One measure of naturalness and an important determinant of habitat value is the presence, abundance, or land area covered by nonnative and especially invasive species (Burger et al. 2004). The diversity of nonnative species has been used as an indicator of reduced habitat value for native species (Efroymson et al. 2008a). The susceptibility to invasion by exotic species is strongly influenced by species composition, as well as disturbance by stressors such as roads, noise, chemical contaminants, and so on. Invasive exotic plant species are typically assumed to have lower habitat value than their less-invasive counterparts, because some invasive species have the potential to increase their abundances so rapidly that they can dominate the landscape.

Threats to Habitat. Some valuation schemes assume that threatened systems are more valuable for conservation (Margules and Usher 1981). For example, US EPA Region 7 has developed tools for identifying critical terrestrial ecosystems (Missouri Resource Assessment Partnership 2004). In addition to species richness, low number or intensity of stressors, high percentage of public ownership, and connectivity, value in these ecological assessments is based on absence of threats. Threats include land demand, agriculture, and toxic releases.

Case Study. Habitat value metrics representing some of these environmental attributes were recently applied to environmental remediation decisions for chemical contaminants. We conducted a study that was intended to identify metrics of habitat value that might supplement formal EcoRA of contaminants to help decision-makers prioritize wildland and non-wildland sites for remediation (Efroymson et al. 2008a, 2008b). Methods were developed to summarize dimensions of habitat value for several aquatic and terrestrial contaminated sites at the East Tennessee Technology Park (ETTP) on the US Department of Energy (DOE) Oak Ridge Reservation in Oak Ridge, TN, USA. Many locations on Department of Defense (DOD) and DOE reservations where security buffers have been in place for decades have high habitat value (Mann et al. 1996). In this study, an industrialized area with low ecological habitat value and chemical concentrations associated with high ecological risk (but low human health risk) might have a lower priority for remediation than a more natural area with lower ecological risk, but high habitat value. Similarly, the baseline habitat value would provide evidence concerning the potential harm that might be caused by remedial technologies (Whicker et al. 2004, Efroymson et al. 2004).

For this habitat valuation study at ETTP, we developed three broad categories of valuation metrics: onsite use by groups of organisms, value added to onsite use value from spatial context, and rarity (Efroymson et al. 2008a). Use value was measured by taxa richness, a direct measure of number of species that inhabit an area; complexity of habitat structure, an indirect measure of potential number of species that may use the area; and land use designation, a measure of the length of time that the area would be available for use (Table 9.1). Value derived from spatial context was measured by similarity or complementarities of neighboring habitat patches and presence of habitat corridors. Value derived from rarity was measured by the presence of rare species or communities.

Metrics that were more specific to groups of organisms in contaminated streams, ponds, and terrestrial ecosystems, as well those that applied to the east Tennessee region, were selected as examples of the general metrics. Examples of use of value metrics were taxa richness of fish, number of sensitive benthic invertebrate species, riparian wetland coverage relative to Southern Appalachian regional average, and taxa richness of edge-associated breeding birds (Efroymson et al. 2008a). Examples of metrics of rarity were the presence of a rare vegetation community as well as the

Table 9.1. Metrics for Valuing Habitat at Six Contaminated Sites[a]

Type of Value	Metric	Explanation
	Value from Site Alone	
Use	Taxa richness	Direct measure of number of species that inhabit area.
	Number of sensitive species	Subset of diversity and number of species that use area. Absence provides indication of level of degradation of area.
	Complexity of habitat structure	Indirect measure of potential number of species that may use area.
	Presence of special wildlife habitat services	Presence of bird rookeries, bat maternity roosts, male display areas, vernal pools, or other wildlife breeding areas that indicate greater use and importance compared to similar areas without features.
	Habitat suitability relationship for broad taxa	Relationships provide information on whether particular vegetation associations or other environmental quality variables are highly suitable or not suitable for particular broad taxa.
	Number of invasive or nonnative species	Nonnative species decrease use by native species. Invasive species also decrease use by native species, and footprint increases with time, if unchecked (therefore, area-weighted use value for native species decreases with time).
	Land cover designation	If the majority of land area is paved or covered with buildings, habitat value is low because of lack of vegetation, minimal habitat structure, and fragmentation.
	Land use designation	If land used is designated as industrial area, habitat use value may not continue for as long as it would if area were conserved.
	Offsite Value Added	
Rarity	Presence of rare species (state and federally listed)	Current value of habitat is high if rare species use it.
	Presence of rare community with respect to the Oak Ridge Reservation, the region, Ridge and Valley ecoregion, or Southern Appalachians	Rare community implies little redundancy or substitutability for habitat services, along with potentially high demand for site.

Table 9.1. (*Continued*)

Type of Value	Metric	Explanation
Use from spatial context	Presence of similar, adjacent habitat patch	Use value of habitat patch increases with area, because some species need minimal patch areas for home ranges, territories, or viable populations. In addition, size of habitat patch correlated with diversity.
	Presence of ecological corridor	Presence of migration and other movement corridors indicates that community of site in question adds use value to surrounding habitat and that surrounding communities add use value to habitat on site.
	Adjacency to complementary land or water	Arrangement of communities can add value to organisms that enjoy services of each (e.g., terrestrial zones around wetlands and riparian habitats).
	Adjacency to conservation land use area	Habitat value of site adjacent to reserve would probably persist longer than habitat value of other sites.

[a]The major components of value are use, rarity, and use value added from spatial context.
Source: Modified from Efroymson et al. (2008a).

presence of listed species, such as fish and bats (Efroymson et al. 2008a). Examples of metrics for value derived from spatial context were adjacency to a conservation area or part of an ecological corridor linking forests from the Cumberland Plateau to the Smoky Mountains (Efroymson et al. 2008a). For each of these metrics, cutoff values for high, medium, and low habitat value were recommended in the study, based on distributions of organisms and landscape features, as well as habitat use information.

Habitat Equivalency Analysis. Habitat Equivalency Analysis (HEA) is a non-monetary valuation method used to determine locations and land or water areas that provide equivalent ecological services. The method is typically used in Natural Resource Damage Assessment applications or other ecological restoration analyses (NOAA 2000). The HEA might be applied to assign ecological value to alternative wildlands being considered for preservation to compensate for injured ecosystems. A HEA could also be used to evaluate restoration efforts that recreate wildlands from injured resources.

In HEA, ecosystem functions are assumed to be proportional to monetary value; that is, people derive utility from ecological entities correlated with their ecological function(s) (Roach and Wade 2006, Dunford et al. 2004). Thus, resource equivalencies are usually expressed in units of service-acre-years. The relationship between ecological function and economic utility is most likely to apply to relatively small (marginal) changes in habitat services in which changes in scarcity of injured habitat are insignificant (Dunford et al. 2004).

Although ecological restoration decisions commonly rely on HEA, the analysis becomes difficult when the services provided by prospective compensatory resources are not of the same type as those that have been lost. The value of apples may be compared with the value of oranges by gauging human preferences, but the ecological service relationships that HEA draws from are less helpful for comparisons of unrelated ecological entities. For example, the DOE transferred Black Oak Ridge forest land to the state of Tennessee to offset the losses of aquatic resources from chemical contamination in Watts Barr Reservoir from the DOE Oak Ridge Reservation. This exchange of forest for fish and benthic invertebrates could not have been justified by HEA or by comparing ecological relationships, because the forest and fish did not belong to the same ecosystem.

A weakness of HEA is that it assumes that ecosystem function (and therefore ecological value) is proportional to land or water area. Kremen and Ostfeld (2005) recommend that mitigation banks to compensate for damage to wetlands, as well as other applications of HEA, allow factors such as shape of land area, location, connectivity, and species composition to contribute to the relative ecological value of a parcel of land. Landscape Equivalency Analysis is a modification of HEA that incorporates the habitat connectivity value of a particular habitat patch and the tradeoffs between connectivity and area (Bruggeman et al. 2005). In this method, the habitat value of a wildland patch derives from its marginal contribution to metapopulation (group of interacting, spatially separated populations) persistence or the marginal decline in habitat service flows that result from removal of the patch. We believe that habitat connectivity represents an important future direction for habitat valuation (see below).

ENVIRONMENTAL BENEFITS INDICATORS. Environmental benefits indicators (EBI) are being used as nonmonetary measures of ecological value. They take advantage of the increased availability of spatial data and growing literature of ecological indicators. Boyd and others (Boyd 2004, Boyd and Wainger 2002) have pioneered some of these ideas, arguing for the affordability and ease of use of indicators intended to represent some of the same dimensions of ecological value as the habitat valuation metrics described earlier, as well as relative human demand (Table 9.2).

Researchers have used similar types of indicators to represent benefits of ecological services, such as providing habitat, regulating water, and assimilating wastes on military installations (Richard Pinkham, Booz Allen Hamilton, personal communication, September 2006). Pilot tests of these indicators and environmental benefit indices (demand index, scarcity index, risk index) have been conducted to assess the ecosystem service value of providing habitat at Vandenberg Air Force Base and Fort Lewis Army Base (R. Pinkham, personal communication, September 2006). A combined habitat index shows hotspots for habitat value.

Dale and Polasky (2007) discussed the potential use of environmental benefits indicators in measuring ecosystem services from agriculture. Examples of ecological services pertinent to wildlands include pollination, soil retention, nutrient cycling, and maintenance of biodiversity. They argue that useful EBIs must be linked to and predictive of the production of ecosystem services.

Table 9.2. Example Attributes of Value and Related Indicators

Value Attribute	Example Indicator
Demand	Proximity to population
Scarcity, substitutability	Local prevalence
	Abundance of population, ecosystem, land-cover type providing identical service
Complementary inputs	Landscape characteristic or infrastructure allowing access to recreation
Low probability or magnitude of future risks	Measure of stressor such as invasive species, low elevation (vulnerability to flood), etc.

Source: Modified from information in Boyd and Wainger (2002).

Multimetric Environmental-Benefits Indices. Natural systems are inherently multidimensional. Valuation joins the ranks of scientific efforts to project the many dimensions that define ecological systems into one dimension. Measures or indicators of environmental benefits are sometimes aggregated into multimetric indices. Many indices add the component EBI values, often weighting the factors differently. The reductionism of indices is most reasonable if the relationship between environmental variables is well understood [e.g., the relationship between vegetation structure and wildlife habitat and species richness in the habitat model of Kapustka et al. (2004)]. One of the fundamental underpinnings of economic valuation is that different components of value are independent and additive and that the total value of a system or scenario does not either include doubly counted component values or exclude component values. An example of double counting would be adding the contributory value of a prey item (i.e., the value it has as a result of contributing biomass to a valued predator) to the value of the predator.

Multimetric indices are commonly used among aquatic toxicologists and aquatic ecologists to estimate and compare status and trends of ecosystems (Bruins and Heberling 2005). One common multimetric index used in rivers is the index of biotic integrity (Karr 1981), which measures the deviation of a stream invertebrate community from that in a group of pristine reference streams. A challenge for using the index is finding reference streams of approximately the same size and in the same geographic region. An example of a multimetric index that comes closer to measuring ecological value is the index of "ecosystem ecological significance," which is calculated by the US EPA Region 5 Critical Ecosystem Assessment Model (CrEAM). CrEAM is a geographic information system (GIS)-based tool that incorporates ecological diversity, ecological sustainability, rare species, and land cover into one multimetric index of ecosystem value (White and Maurice 2004). More specific habitat quality indices are also available, such as the 64 benthic habitat quality indices summarized in Diaz et al. (2004).

The US Department of Agriculture has developed an EBI to rank offers to enroll lands in the Conservation Reserve Program. Although they are not strictly wildlands, these lands are taken out of agricultural production temporarily or permanently, and participants must show ecological benefits, such as reduced erosion or restoration of vegetation cover for wildlife habitat (USDA 2004). The USDA EBI is the sum of several weighted factors and subfactors. Up to 100 points (of 395 possible points for environmental benefits, exclusive of costs) may be assigned to the "wildlife habitat cover benefits" factor, the only factor that represents ecological benefits.

Within the "wildlife" habitat cover benefits factor, the "cover" subfactor measures management options and seeding mixes that provide habitat for wildlife species of national, regional, state, or local significance (USDA 2004). The "wildlife enhancements" subfactor measures the provision of water to wildlife as well as the degree of conversion of land from a monoculture of vegetation to native species. The "wildlife priority zones" subfactor adds points if the land may contribute to the restoration of habitat of threatened or endangered species or other important or declining species (USDA 2004). However, the tracts of land are not formally examined in their spatial context (e.g., whether they are part of an existing wildlife corridor). Additional environmental benefits in the index relate to water quality, prevention of wind erosion, air quality, and carbon sequestration (USDA 2004).

Banzhaf and Boyd (2005) described how an ecological services index might be developed to summarize beneficial environmental services through time. The index would be based on a comprehensive list of ecological services weighted by proxies for willingness to pay (e.g., human population measure), location-specific quality factors (e.g., proximity of wetlands to polluted runoff), substitution factors (availability of close substitutes), and complementarity factors (i.e., availability of adjacent assets that increase the value of the ecological service) (Banzhaf and Boyd 2005).

Although environmental benefits indices are easily used, their assumptions are not easily understood. Indices can have several disadvantages for valuing ecological stocks and services, such as habitat services. First, if managers or stakeholders have not fully expressed their relative preference for different ecosystem services, then a multimetric index is not useful for estimating ecological value (Efroymson et al. 2008a). Moreover, different weightings of the various indicators might be appropriate for different potential users of environmental benefits indices; a single index is not very useful. Furthermore, indicators developed at one spatial scale may be not be useful to a decision that targets a different spatial scale (Efroymson et al. 2008a, 2008b). Some of Suter's (1993) criticisms of ecosystem health indices also apply to the aggregation of variables into a multimetric index of environmental benefit. Several of his arguments against the use of indices include:

- *Ambiguity.* If the value of an index is low, one cannot tell how many components were low.
- *Arbitrariness of Combining Functions.* An index may be very sensitive to the methods used to calculate it.
- *Arbitrariness of Variance.* The variance of an index does not have a clear relationship to a biological response.
- *Unreality.* Indices do not measure actual biophysical properties.
- *Disconnection from Testing.* Indices cannot be tested in the laboratory or verified in the field.

Complex Models of Value

Complex models of ecological value tend to be used in situations where decision-makers want ecological and economic factors to be integrated, ecological and economic data availability is high, relationships between ecology and economics are understood in a mechanistic way, and adequate funding is available. Although state-of-the-art ecological models produce highly uncertain results, the data to support these models are becoming more readily available, and it is not clear that they are any less predictive than complex economic models.

We identify three classes of complex models. These include (1) integrated models of ecology and economics, (2) models supporting habitat-based replacement costs, and (3) multivariate analysis and optimization.

Integrated Models. "Full ecological-economic models may be the gold standard for establishing the full range of ecosystem service possibilities and management options" (Farber et al. 2006). Integrated ecological–economic systems fit the characteristics of complex systems described in Costanza et al. (1996): strong and usually nonlinear interactions, feedback loops that make cause indistinguishable from effect, lags in time from cause to effect, distance between cause and effect, thresholds, and hierarchical behavior (failure of small-scale results to easily predict large-scale behavior). Costanza et al. (1996) argue that "reductionist thinking fails in its quest to understand complex systems." Thus, previously described simple indicators do not capture all of the dynamics of ecological–economic systems that must be understood in order to inform particular decisions about wildlands. Such dynamics can be simulated, however. Understanding the dynamic behavior of ecological–economic systems and the interdependencies of human and ecological processes has been attempted at the regional scale using ecological–economic models. These have been used to evaluate tradeoffs among policies related to land-use change, development, and ecological value (Costanza et al. 1996). For example, Costanza et al. (2002) developed and demonstrated an integrated ecological economics model for the Patuxent River watershed in Maryland. The goal of these models was "to test alternative scenarios of land-use patterns and management" (Costanza et al. 2002). Simulations incorporated topography, hydrology, nutrient dynamics, and vegetation dynamics with changes in land use.

Habitat-Based Replacement Cost. The Habitat-Based Replacement Cost Method (HRC), a method derived from HEA, generates the habitat restoration (and its cost) needed to offset the losses of a specific number of organisms (Allen et al. 2005, Strange et al. 2004). This method for transferring value from organisms to habitat has been used in the context of replacement of fish lost by impingement and entrainment by power plants. The challenge in HRC is to estimate fish survival, growth, density, movement, and other determinants of productivity in various habitat areas. If HRC is estimated through the use of population models, this method is appropriately included as a complex valuation model.

In the context of HRC, we consider the cost of river habitat required to raise sturgeon—the largest freshwater fish in North America. Maintaining the river as

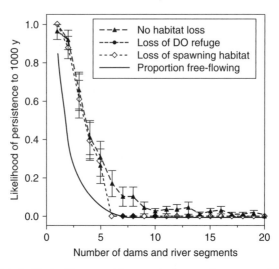

Figure 9.2. Simulated effect of increased fragmentation on the average likelihood of persistence, P_{1000}, for isolated white sturgeon populations. Results are shown for simulations with no loss of habitat and for two habitat-loss scenarios. Circles indicate the average of P_{1000} over populations, and error bars show the standard error in P_{1000} among replicate simulations, averaged over populations. DO is dissolved oxygen. Source: Jager et al. (2001), reprinted with permission.

habitat for sturgeon places constraints on other uses of the river. For example, short river segments appear to be less suitable as white sturgeon (*Acipenser transmontanus*) habitat because they do not provide free-flowing areas used for spawning and for refuge from low oxygen levels in reservoirs (Jager et al. 2002). A population viability analysis model predicted an increase in the likelihood of persistence for white sturgeon populations as a function of the length of river habitat available (Fig. 9.2). Thus, preserving a spawning population may preclude the option of placing dams close together, which reduces the amount of hydropower that can be generated from the same parcel of water. The actual value of this energy depends on the specific characteristics of the site and the local value of alternative fuel sources.

The value of wild rivers may be estimated in part from the difference between the value of wild and hatchery fish. The cost of hatchery operation underestimates total replacement value of fish, because owners assume only the minimum costs by keeping fish until it is no longer cost-effective to do so, and they rely on a continued supply of wild broodstock to persist in the river. The number of adult fish that can practically be kept in a hatchery is low [e.g., 5–15 sturgeon broodstock in Logan et al. (1995)] because it is expensive to house and maintain large enough tanks to accommodate older and considerably larger fish. In addition, the cost of feed increases with sturgeon age due to decreased feed conversion efficiency. Survivorship of various life stages of fish, which can be factored into population models, also addresses differences in value between wild and hatchery fish. For example, in the wild, female white sturgeon produce 5600 eggs/kg compared with 3200 eggs from domestic broodstock, and egg survival increases from 18% to 41% (Logan et al. 1995). However, post-hatch survival of age-zero juveniles is lower in the river than in the hatchery (Jager 2005). Thus, the cost of operating hatcheries to replace reproduction is subsidized by the continued persistence of a wild spawning population and preservation of adequate spawning habitat in rivers.

Multivariate Analysis and Optimization. Ecological valuation brings us a step closer to making optimal decisions that combine ecological and nonecological objectives. This is because multiobjective optimization is facilitated by using a single currency to quantify different objectives. Valuation has been previously used in an optimization context. For example, Field et al. (2004) used decision theory to maximize the ecological value of an endangered koala species. Various mathematical algorithms have been developed to optimize natural reserve design and reserve site selection (Church et al. 2000), two important applied problems that require the valuation of wildlands. These focused on one type of ecological objective—maximizing the number of species represented. Root et al. (2003) refined this objective by weighting species by proxies of extinction risk from organizations such as the World Conservation Union and the US Fish and Wildlife Service.

A number of optimization approaches have recommended or included additional ecological objectives. Ferraro (2004) criticized the use of a single characteristic (e.g., genetic diversity measures, habitat suitability indices, number of species) to represent environmental amenities that are desired at least cost. He provided an alternative optimization approach to allocate funds for conservation cost-effectively by combining multiple biophysical and economic dimensions that contribute to value, using a distance function that can be estimated using nonparametric methods. Church et al. (2000) argued that the "quality" of species representation is just as important to include as number of species in optimizations for reserve site selection—that is, habitat value, adequate population size, presence of critical resources, and presence or absence of nonnative competitors.

Moreover, in an examination of the optimal use of conservation funds by Wu and Boggess (1999), the marginal benefits of additional expenditures on wildlands preservation depended on cumulative benefits and correlations among benefits. Correlations arise because many environmental benefits are produced by the same conservation or natural resource management actions. For example, ceasing crop production may produce enhanced wildlife habitat and decreased groundwater pollution (Wu and Boggess 1999). Thresholds in ecological parameters translate into important thresholds in value that influence on the optimal spatial allocation of conservation funds (Wu and Boggess 1999, Johst et al. 2002, Wu and Skelton-Groth 2002). Such thresholds allowed Wu and Skelton-Groth to determine the optimal allocation of riparian conservation funds to restore salmon populations in Pacific Northwest. Where physical variables, such as stream condition or stream temperatures, were used to allocate conservation funds, the management alternatives did not always provide the greatest benefit to salmonids.

Ideally, wildlands are protected from human influences, but in many cases these lands (or waters) are also used for resource extraction, and the goal of optimization becomes minimizing impacts of resource use on the value of wildlands. Optimization of ecological value has been applied to other applied environmental problems such as timber harvest and reservoir operations. Hof and Bevers (1998) offered numerous examples of spatial resource management decisions aided by spatial optimization, including harvest schedules, containment of pests by optimally treating areas of forest, and harvesting to minimize water quality impacts. In one study, they maximized the long-term diversity of species in a forest, measured by the joint viabilities of multiple species (Hof and Bevers 1998). In general, studies have attempted to optimize land use with regard to either ecological objectives (species preservation; Haight 1995) or human-use objectives (timber production, Nalle et al. 2004). However, ecological optimizations that consider both ecological and economic objectives together are rare.

Not all applications of ecological valuation truly maximize ecological objectives. For example, a recent review characterized the state of the art in reservoir operation toward ecological sustainability (Jager and Smith 2008). The majority of studies, and all that were implemented in practice, used legally mandated restrictions (e.g., minimum flows) as constraints on efforts to maximize other values, such as the amount of hydropower or revenue generated. Consequently, the value of water was not optimized, because the analyses assumed that a fixed amount of instream flow would be best—neglecting the considerable value, as measured by willingness to pay, of higher instream flows (Loomis 1998).

Four approaches to measuring ecological value as a function of flow were considered in reservoir optimizations: (1) the effect of flow regime on water quality in the upstream reservoir, downstream tailwater, or downstream estuary; (2) the effect of flow regime on fish habitat; (3) the deviation of flow regime from a natural flow regime; and (4) the effect of flow regime on simulated fish population viability. At least two model-based approaches have been used to optimize flow regimes, one emphasizing fish population responses to flow and the other emphasizing water allocation aspects of the problem. In an example of the fish modeling approach, Jager and Rose (2003) identified flow regimes to maximize salmon recruitment. In an example of a water-allocation approach, Sale et al. (1982) included more-realistic restrictions on water availability, while treating adequate fish habitat as a constraint.

Some argue that wildlands have the highest value if they not only provide good habitat and associated existence value but also facilitate human access (e.g., with trails or navigable waters) and therefore provide some human use value. However, evidence that willingness to pay for preservation far exceeds other components of ecological value (e.g., Loomis 1998) suggests that access may not be an important part of value. In addition, roads are strongly correlated with human disturbance and consequent loss of ecological value as wildlands for ecosystems ranging from lakes to forests.

FUTURE DIRECTIONS

We believe that three main directions in wildlands valuation share great promise for advancing the science: (1) developing theories and methods for representing temporal variation in ecological value, (2) developing theories and methods for understanding how spatial context influences ecological value, and (3) developing theories and methods for representing ecological relationships in ecological value.

Incorporating the Future in Wildlands Valuation

The future plays a different role in ecological valuation from its role in valuation of nonecological services and commodities. Time is traditionally considered in valuation through discounting—that is, representing the fact that goods and services that are anticipated in the future have lower value than the same goods and services today (Ludwig et al. 2005). Ecological thresholds can be reached beyond which related goods and services will cease to be available. For example, harvest of a fish population today can result in its economic collapse in the future. This outcome likely reduces discounted use value for future users and nullifies existence value. It has been shown that making environmental management decisions based on conventional statistics (low Type I error rate) leads to suboptimal results, because the risk of reaching an ecological threshold is not taken into account (Field et al. 2004).

Quantifying the risk of future extirpation should be a priority for valuation of populations that are rare. Rarity influences value in two major ways. With respect to use value, scarcity leads to increased marginal value of an individual and decreased total value of the population. Rarity also inflates the existence value of ecological entities, because long-term persistence is threatened. Both future use value and existence value are lost when extirpation/extinction thresholds are reached.

A simple approach to assign value based on extirpation risk is to quantify rarity. Value is sometimes assigned to rarity based on semiquantitative indicators (e.g., Efroymson et al. 2008a). A more quantitative and complex approach is to use population models to estimate future risk of extirpation via population viability analysis (PVA). PVA models have only occasionally been used as tools in ecosystem valuation (see HRC discussion above). One use of PVA models is to identify extirpation/extinction thresholds such as the minimum viable population size or the minimum area of suitable habitat (MASH) for a particular species. These thresholds may be important for estimating existence value of a population or the value of a service that is uniquely provided by that population. PVA models can estimate MASH by linking habitat quality and quantity to population processes such as survival and reproduction. The effects of temporal variation on extirpation/extinction risk are simulated by representing (1) environmental stochasticity (year-to-year variation in weather or other environmental variables that influence individual survival or reproduction), (2) demographic stochasticity (chance of extirpation due to small population size), and (3) catastrophes. The use of PVA models has been identified as a priority for advancing the science of ecosystem valuation (US EPA 2006).

Whereas populations face a risk of extirpation, other ecological entities face different risks of irrevocable loss. For example, functioning ecosystems can be destroyed or altered by unnatural and permanent disturbances (e.g., processes of residential or industrial development), particularly when no sources of reintroduction or restorative processes are operating.

Even when extirpation or functional thresholds are remote and the risk of irrevocable loss is zero, changes in ecological value over time can be important. For example, in rivers below dams, both the economic value of hydropower and the ecological value of flow to fishes vary seasonally. If one were trying to design an optimal flow regime to permit sustainable coexistence of salmon populations and hydropower generation, it would be important to consider two things. The first is that salmon require higher flows during spawning migration and outmigration than during other times of the year. The second is that hydropower is more valuable during certain times of the day and week (Jager and Smith 2008). Changes in rarity of species and their habitats are also important components of ecological value.

In the future, we will be challenged to predict the dynamic changes in wildland ecosystems and their value. Changes in rarity of species and their habitats are important components of ecological value that can be very difficult to predict. We anticipate that ecological recovery and succession will be simulated better in the future and that their predictions will help to quantify ecological service value. This will be especially important with trends associated with climatic change. Species niches may change dramatically in the future, with some increasing in suitable area and others disappearing entirely (Hoffman et al. 2005, Best et al. 2007).

Although predicting dynamic changes in wildlands over time is a challenge, perhaps the biggest challenge of all will be to describe changes in human preferences. The vagaries of human preference have a dynamic influence on value, but one that we often neglect. Combined models that forecast changes in human preferences in response to ecological futures can be used to estimate future changes in the value of ecological entities.

Incorporating Spatial Context in Wildlands Valuation

Some aspects of wildland value, like those associated with habitat connectivity and species rarity, come not from qualities intrinsic to individual patches of habitat, but from characteristics of their surrounding landscapes. These contributions arise from the physical placement of the wildland patch and its spatial relationship and juxtaposition with the other patches in the surrounding matrix. Changes to the landscape matrix and to other wildland patches in the constellation can have cascading effects on the value of other wildland patches, even those far from the change. The fact that the ecological value of a wildland site, such as species existence value or value for hunting, derives not only from the site itself, but also from its contextual location, is ignored by EBIs (e.g., USDA 2004).

We anticipate that ecological value will be refined in the future through more complete consideration of the complementarity of ecological services in adjacent lands and waters. Many examples demonstrate how the ecological services of adjacent communities add value to plant and animal habitat (Table 9.1). Lakes and rivers provide critical sources of drinking water for terrestrial organisms. Wetlands increase the habitat value of adjacent land parcels and water bodies by removing toxicants, reducing sediment loads, transforming nutrients, and providing specific habitat needs (e.g., breeding habitat for amphibians) (King et al. 2000, Rosensteel and Awl 1995). Different life stages may require different habitats in close juxtaposition. For example, floodplains provide slow, shallow river habitats that serve as nursery areas and refuge from predators for fishes (Welcomme 1979). Similarly, wooded riparian zones provide maternity roost sites for bats that forage above adjacent ponds. Another illustration of adjacent and complementary ecological services relates to pollination. Kremen et al. (2004) developed a relationship between (a) the proportion of upland natural habitat within several kilometers of an agricultural site and (b) the magnitude and reliability of crop pollination services performed by native bees.

Although the importance of landscape juxtaposition is increasingly recognized in measures of habitat suitability, it is rarely included in ecosystem valuation. Geographic information systems (GIS) are useful to measure distances between areas with particular land-cover or land-use classifications.

Corridors and Connectivity. Movement corridors improve the habitat quality or suitability of adjacent land areas and water bodies. Connectivity increases habitat value of metapopulations because populations in local patches are more likely to be rescued from chance extirpation by immigration from other, connected patches. The presence of habitat corridors has been shown to be correlated with increased native plant species richness in connected patches (Damschen et al. 2006). However, connectivity can also encourage the encroachment of weedy and invasive species, competitors, predators, parasites, and diseases.

The next challenge will be to quantify connectivity and its influence on habitat quality and, ultimately, its contribution to perceived value of a wildland to humans. Many approaches have been used to detect and quantify connectivity among patches within a landscape. Researchers at the Savannah River Site in South Carolina, USA, have taken a direct experimental approach to quantifying connectivity effects by (a) cutting voids in a pine forest to create negative "patches" connected by negative "corridors" and (b) studying the resulting impacts on seeds, plants, rodents, butterflies, and birds (Tewksbury et al. 2002, Haddad et al. 2003, Damschen et al. 2006). Morphometric image analysis, involving sequential dilation and erosion of patches and matrix, has been used to determine the degree of direct and indirect landscape connectivity (Vogt et al. 2007). Even electrical circuit theory has been used to simulate metapopulation connectivity via estimates of impedance and current flow through the habitat patches and surrounding matrix (McRae 2006, McRae et al. 2005). Individual-based models using virtual "walkers" as software agents have also been used to simulate movement preferences of a target species to quantify connectivity and to locate potential optimum movement pathways through a landscape (Gustafson and Gardner 1996, Gardner and Gustafson 2004, Hargrove et al. 2005).

GIS-based analysis of Least-Cost Path (LCP), originally developed to help plan roadway construction routes, was among the first analytical techniques to be borrowed for connectivity analysis. Once parameterized for the cost of movement or friction through each habitat type, LCP results in the pathway of lowest cost between two specified patches of habitat. In one application, the Southeastern Ecological Framework, funded by the US EPA, used GIS-based LCP methods to create a network of forest patches and "linkages" across the southeastern United States (Hoctor et al. 2000).

Graph theory represents individual habitat patches as nodes connected by line segment "edges" to form a connected network (Keitt et al. 1997, Urban and Keitt 2001). Edges may represent simple Euclidean distance, or they may reflect more complex costs of movement. The importance of any connecting edge can be calculated by the number of connections emanating from its two nodes. The minimum spanning tree is the shortest set of edges connecting all nodes. This tree, which shows how to connect all habitat patches with minimum cost, solves problems similar to the famous traveling salesman problem. Graph-theoretic approaches quantify connectivity, but do not explicitly map movement corridors geographically on the landscape.

One should distinguish structural habitat corridors (narrow portions of patches of high-quality habitat) from functional habitat corridors (paths between different patches of high-quality habitat that pass through an intervening matrix of lower-quality habitat). Both structural and functional connectivity affect the habitat value of a particular patch to wildlife, no matter where that patch falls in the continuum of habitat quality for a particular species. In the future, these methods for quantifying connectivity could be integrated into measures of habitat value. Habitat value influences the human use and existence values of relevant species.

Percolation Thresholds. Percolation theory (Stauffer 1985) predicts abrupt thresholds of connectivity as the number and quality of individual connections increases. Nonlinear percolation thresholds, which have been observed empirically in many fields, should have similar, dramatic effects on connectivity-based habitat value (Fig. 9.3). As the number and strength of connections increases, a critical percolation threshold is reached, and connections span the landscape. Spanning connections suddenly and abruptly allow even patches that are separated by significant geographic distances to be open to migrating individuals. Wildland valuations based on habitat connectivity should show a similar nonlinear jump in value near this percolation threshold.

There may, however, be an optimal level of connectivity for patches within a particular landscape. The best degree of connectivity should be one that allows for communication among all patches throughout the metapopulation, but no more.

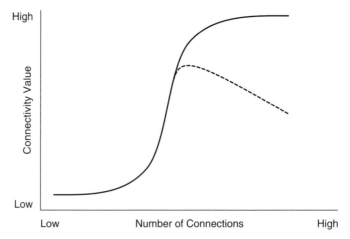

Figure 9.3. Habitat value in relation to landscape connectivity. The dotted line represents situations where connectivity may promote species invasion, disease, or other negative consequences.

Connectivity in excess of this sufficient ideal may make metapopulations too vulnerable to epidemic processes like species invasions, parasitism, disease, and wildfire (Simberloff and Cox 1987, Minor and Urban 2008). If such disturbances can sweep across multiple, connected patches, metapopulations are less likely to find refugia. The connectivity-based value of wildlands could also decrease beyond this optimum connectivity (the dotted line in Fig. 9.3). However, connectivity is both species- and landscape-specific. A landscape feature that serves as a movement corridor for one species can be a barrier to the movement of another. Thus, future research in ecosystem valuation should include methods for optimizing connectivity for multiple species within the same landscape.

Incorporating Spatial Scale into Wildlands Valuation. In the future, ecological valuation also will have to deal more explicitly with notions of spatial scale. Hein et al. (2006) (and references within) have noted that "to date, relatively little elaboration of the scales of ecosystem services has taken place." Thus, research should clarify these spatial scales. Moreover, the relative importance of global value versus national value versus regional value versus local value will have to be negotiated on a case-by-case basis and, more generally, where national or other policy is involved. For example, if wildlands support carbon sequestration (a global value), species or community existence value (variable with scale), and hunting value (primarily regional value), how should these scale-dependent values be weighted? The answer will influence the relative emphasis of ecological valuation research efforts at different scales. One reason that existence value is often higher than other components of ecological value is that estimates are scaled by the number of individuals. Individuals surveyed from distant areas may express preferences for preservation of a given ecosystem or species, but individuals from these same areas may not be counted in the estimated use value for hunting or fishing.

Earlier we described the importance of incorporating influences of the spatial arrangement of the landscape in wildland value. We note that effects of both connectivity and juxtaposition on wildland value are scale-dependent. All maps are finite; consequently, edge effects could cause connectivity effects on habitat value to be underestimated. Likewise, boundaries can cause estimates of how juxtaposition will influence habitat value to be inaccurate. Therefore, it may be important to consider connections with outlying areas in estimates of value of ecosystem components in a smaller area.

Incorporating Ecological Linkages in Wildlands Valuation

The future of ecological valuation will involve more explicit consideration of ecological linkages. A common complaint regarding ecological valuation is that ecological entities are not fully valued, especially in scenarios where monetization is required. Relationships among species and their food, consumers, habitat, limiting nutrients, and functions are only rarely reflected in relative human preferences. Values of populations or services may be extended from one site to another through "benefits transfer." However, until now, benefits transfer methods have rarely taken advantage of ecological relationships to transfer values among related ecological entities, such as habitats and populations, or predators and prey. Transfers of ecological value have previously been extended from predator to prey (Allen and Loomis 2006), ecosystem to ecosystem (compensatory natural resources restoration, NOAA 2000), organism to habitat (Allen et al. 2005), commodity to enabling ecological service (e.g., crop to pollination; Losey and Vaughan 2006), and ecological service on one site to service on another (pollination; Kremen et al. 2004). Ecological benefits transfer may also be used to transfer value from function to structure, population to individual, or population to habitat. We believe that extending monetary values to heretofore unvalued ecological entities through ecological modeling is an important new direction for wildlands valuation.

Integrating the results of ecological models with estimates of monetary value also requires economic research. In addition to developing models of ecological properties that influence value, it is necessary to estimate use and non-use value for different ecological entities. For example, the willingness to pay for a wildlife or plant population of different sizes—that is, those further from versus closer to an extirpation/extinction threshold—may be integrated with PVA results. Likewise, one might estimate willingness to pay for ecosystems that are perceived as more and less wild and ecosystems described as having more or less capacity to recover from disturbance. Efforts are needed to generalize from contingent valuation surveys using meta-analysis and to understand the functional form followed by human values. Development of such general economic models is needed.

Landscapes by Design

In the future, we would like to see spatial optimization used to design efficient, sustainable arrangements of uses and services on the landscape. We envision maximization of ecological value as the objective integrated over a long time horizon. The time horizon is critical, because optimal decisions based on short-term returns inevitably result in poor resource management decisions, as evidenced by numerous overharvested marine fish stocks. Field et al. (2004) demonstrated that management decisions involving rare species based on traditional statistical hypothesis tests resulted in much higher costs than those derived by minimizing long-term management costs. This is because the economic cost of Type II errors (risk of extinction due to a poor decision) is high, and hypothesis tests do not provide a cost-efficient way of deciding whether management intervention is needed.

Another issue is whether to optimize landscapes holistically, permitting mixed arrangements of wildlands with more intensively managed lands. Kareiva et al. (2007) write of the "domestication" of nature, and they suggest that we need to have a willingness to shape such domestication. They assert that we should shun the notion that "wilder is better." Others counter that humans are not capable of understanding ecosystem–human systems well enough for such a utopian vision and that our best bet is to set aside wildlands. From a theoretical standpoint, solutions obtained to problems that permit mixed use will be better than those obtained by separate optimizations of the two types.

We stand to learn a great deal by developing and applying tools that can identify optimal arrangements of alternative land uses that maximize the value of wildlands, possibly along with those of human land uses (e.g., agriculture, rangeland, and urban). Spatial optimization, which allocates human uses and ecosystem services on the landscape, is a tool used in landscape architecture and design (Nassauer et al. 2002, Santelman et al. 2004). Designs may be optimized, tested and evaluated in simulations before they are physically wrought on the landscape (Fernandez et al. 2005). Competing land uses must be evaluated in an even-handed way and must consider all requirements, costs, and benefits (Musacchio and Wu 2004). However, current social and political systems may not allow us to enact, control, enable, and enforce such optimal landscape design solutions (Musacchio et al. 2005). History suggests that governments with the centralized decision-making authority required to implement such regional plans ultimately further political goals rather than scientific strategies for achieving long-term sustainability.

The need to evaluate alternative design schemes will increase as the human population grows and our ecological footprints spread. Landscape construction is a constrained, zero-sum game, because the total available area is fixed. The objective will be to maximize the value of wildlands, and the best designs will harmonize conflicting or competing land uses for optimal value and sustainability. The promise and challenge of wildland valuation will be to provide the tools and functions needed to design better landscapes for our environment and our society.

CONCLUSION

Valuing wildlands is essential to environmental decision-making and landscape design. Without wildland valuation methods, wildlands will be assumed to have no value. Economic valuation methods need to incorporate ecological models to provide reasonable estimates of total value. Limburg et al. (2002) note that "from a purely ecological perspective, valuation begins with identifying the key structures, functions, and interactions of systems, and probing these (via models or experiments) to understand which are important in maintaining their condition, dynamics, and production of ecosystem services." Population dynamics and spatial ecology are disciplines that will come to the forefront of ecosystem valuation. The valuation of wildlands will increasingly incorporate the spatial context of the land and temporal aspects of organisms and their functions, and methods will be selected that are appropriate to the decision context. Research involving extirpation/extinction thresholds and their equivalents at higher levels of ecological organization will achieve prominence in ecosystem valuation. Applications of wildlands valuation will be as diverse as the selection of land areas to conserve, the selection of remediation alternatives, the valuation of benefits of environmental research and development, and the design of multipurpose landscapes.

ACKNOWLEDGMENTS

Research was sponsored by the Laboratory Directed Research and Development Program of Oak Ridge National Laboratory (ORNL), managed by UT-Battelle, LLC for the U. S. Department of Energy under Contract No. DE-AC05-00OR22725. We thank Gbadebo Oladosu of ORNL for reviewing this and the editors/book chapter.

The submitted manuscript has been authored by a contractor of the U.S. Government under contract DE-AC05-00OR22725. Accordingly, the U.S. Government retains a nonexclusive, royalty-free license to publish or reproduce the published form of this contribution, or allow others to do so, for U.S. Government purposes.

REFERENCES

Allen BP, Loomis JB. 2006. Deriving values for the ecological support function of wildlife: An indirect valuation approach. *Ecol Econ* **56**:49–57.

Allen PD II, Chapman DJ, Lane D. 2005. Scaling environmental restoration to offset injury using Habitat Equivalency Analysis. In Bruins RJF, Heberling MT (Eds.), *Economics and Ecological Risk Assessment: Applications to Watershed Management*. CRC Press, Boca Raton, FL, pp. 165–184.

Banzhaf S, Boyd J. 2005. *The Architecture and Measurement of an Ecosystem Services Index*. Discussion paper. RFF DP 05–22. Resources for the Future, Washington, DC.

Barbour MT, Gerritsen J, Snyder BD, Stribling JB. 1999. *Rapid Bioassessment Protocols for Use in Streams and Wadeable Rivers: Periphyton, Benthic Macroinvertebrates and Fish*, second edition. EPA 841-B-99-002. US Environmental Protection Agency, Office of Water, Washington, DC.

Benton TG, Vickery JA, Wilson JD. 2003. Farmland biodiversity: Is habitat heterogeneity the key? *Trends Ecol Evol* **18**:182–188.

Best AS, Johst K, Muenkemueller T, Travis JMJ. 2007. Which species will successfully track climate change? The influence of intraspecific competition and density dependent dispersal on range shifting dynamics. *Oikos* **116**:1531–1539.

Boyd J, Wainger L. 2002. Landscape indicators of ecosystem service benefits. *Am J Agric Econ* **84**:1371–1378.

Boyd J. 2004. What's nature worth? Using indicators to open the black box of ecological valuation. *Resources* **154**:18–22.

Bruggeman DJ, Jones ML, Lupi F, Scribner KT. 2005. Landscape equivalency analysis: Methodology for estimating spatially explicit biodiversity credits. *Environ Manage* **36**:518–534.

Bruins RJF, Heberling MT. 2005. Using multimetric indices to define the integrity of stream biological assemblages and instream habitat. In Bruins RJF, Heberling MT (Eds.), *Economics and Ecological Risk Assessment: Applications to Watershed Management*. CRC Press, Boca Raton, FL, pp. 137–142.

Burger J, Carletta MA, Lowrie K, Miller KT, Greenburg M. 2004. Assessing ecological resources for remediation and future land uses on contaminated lands. *Environ Manage* **34**:1–10.

Cardwell H, Jager HI, Sale MJ. 1996. Designing instream flows to satisfy fish and human water needs. *J Water Res Planning Manage-ASCE* **122**:356–363.

Church R, Gerrard R, Hollander A, Stoms D. 2000. Understanding the tradeoffs between site quality and species presence in reserve site selection. *Forest Sci* **46**:157–167.

Costanza R, Wainger L, Bockstael N. 1996. Integrating spatially explicit ecological and economic models. In Costanza R, Segura O, Martinez-Alier J (Eds.), *Getting Down to Earth: Practical Applications of Ecological Economics*. Island Press, Washington, DC, pp. 249–284.

Costanza R, Voinov A, Boumans R, Maxwell T, Villa F, Wainger L, Voinov H. 2002. Integrated ecological economic modeling of the Patuxent River Watershed, Maryland. *Ecol Monogr* **72**:203–231.

Cromie WJ. 2001. Roads scholar visits most remote spots. Harvard University Gazette. June 14, 2001. http://www.hno.harvard.edu/gazette/2001/06.14/01-roadsscholar.html

Crowder LB, Cooper WE. 1982. Habitat structural complexity and the interaction between bluegills and their prey. *Ecology* **63**:1802–1813.

Dale VH, Polasky S. 2007. Measures of the effects of agricultural practices on ecosystem services. *Ecol Econ* **64**:286–296.

Damschen EI, Haddad NM, Orrock JL, Tewksbery JJ, Levey DJ. 2006. Corridors increase plant species richness at large scales. *Science* **313**:1284–1286.

Diaz RJ, Solan M, Valente RM. 2004. A review of approaches for classifying benthic habitats and evaluating habitat quality. *J Environ Manage* **73**:165–181.

Doherty M, Kearns A, Barnett G, Sarre A, Hochuli D, Gibb H, Dickman C. 2000. *The Interaction Between Habitat Conditions, Ecosystem Processes, and Terrestrial Biodiversity—A Review*. Australia: State of the Environment, Second Technical Paper Series (Biodiversity). Department of the Environment and Heritage, Canberra, Australia.

Downes BJ, Lake PS, Schreiber ESG, Glaister A. 1998. Habitat structure and regulation of local species diversity in a stony, upland stream. *Ecol Monogr* **68**:237–257.

Dunford RW, Ginn TC, Desvousges WH. 2004. The use of habitat equivalency analysis in natural resource damage assessments. *Ecol Econ* **48**:49–70.

Efroymson RA, Morrill VA, Dale VH, Jenkins TF, Giffen NR. 2009. Habitat disturbance at explosives-contaminated ranges. In Sunahara G, Hawari J, Lotufo G, Kuperman R (Eds.), *Ecotoxicology of Explosives and Unexploded Ordnance*. CRC Press, Boca Raton, FL, pp. 253–276.

Efroymson RA, Nicolette JP, Suter GW II. 2004. A framework for Net Environmental Benefit Analysis for remediation or restoration of contaminated sites. *Environ Manage* **34**:315–331.

Efroymson RA, Peterson MJ, Welsh CJ, Druckenbrod DL, Ryon MG, Smith JG, Hargrove WW, Giffen NR, Roy MK, Quarles HD. 2008a. Investigating habitat value to inform contaminant remediation options: Approach. *J Environ Manage* **88**:1436–1451.

Efroymson RA, Peterson MJ, Giffen NR, Ryon MG, Smith JG, Roy MK, Hargrove WW, Welsh CJ, Druckenbrod DL, Quarles HD. 2008b. Investigating habitat value in support of contaminant remediation decisions: Case study. *J Environ Manage* **88**:1452–1470.

Farber S, Costanza R, Childers DL, Erickson J., Gross K, Grove M, Hopkinson CS, Kahn J, Pincetl S, Troy A, Warren P, Wilson M. 2006. Linking ecology and economics for ecosystem management. *Bioscience* **56**:121–133.

Ferarro PJ. 2004. Targeting conservation investments in heterogeneous landscapes: A distance-function approach and application to watershed management. *Am J Agric Econ* **86**:905–918.

Fernandez LE, Brown DG, Marans RW, Nassauer JI. 2005. Characterizing location preferences in an exurban population: Implications for agent-based modeling. *Environ Planning B* **32**:799–820.

Field SA, Tyre AJ, Jonzen N, Rhodes JR, Possingham HP. 2004. Minimizing the cost of environmental management decisions by optimizing statistical thresholds. *Ecol Lett* **7**:669–675.

Gardner RH, Gustafson EJ. 2004. Simulating dispersal of reintroduced species within heterogeneous landscapes. *Ecol Modelling* **171**:339–358.

Gustafson EJ. Gardner RH. 1996. The effect of landscape heterogeneity on the probability of patch colonization. *Ecology* **77**:94–107.

Haddad N, Bowne DR, Cunningham A, Danielson BJ, Levey D, Sargent S, Spira T. 2003. Corridor use by diverse taxa. *Ecology* **84**:609–615.

Haight RG. 1995. Comparing extinction risk and economic cost in wildlife conservation planning. *Ecol Appl* **5**:767–775.

Hargrove WW, Hoffman FM, Efroymson RA. 2005. A practical map-analysis tool for detecting dispersal corridors. *Landscape Ecol* **20**:361–373.

Hargrove WW, Hoffman FM, Efroymson RA. 2005. A practical map-analysis tool for detecting potential dispersal corridors. *Landscape Ecol* **20**:361–373.

Hein L, van Koppen K, De Groot RS, van Ierland EC. 2006. Spatial scales, stakeholders and the valuation of ecosystem services. *Ecol Econ* **57**:209–228.

Hoctor TS, Carr MH, Zwick PD. 2000. Identifying a linked reserve system using a regional landscape approach: The Florida Ecological Network. *Conser Biol* **14**:984–1000.

Hof, JG, Bevers, M. 1998. *Spatial Optimization for Managed Ecosystems*. Columbia University Press, New York, 258 pp.

Hoffman FM, Hargrove WW, Erickson DJ III, Oglesby R. 2005. Using clustered climate regimes to analyze and compare predictions from fully coupled general circulation models. *Earth Interactions* **9**:1–27.

Hooper DU, Chapin FS III, Ewel JJ, Hector A, Inchausti P, Lavorel S, Lawton JH, Lodge DM, Loreau M, Naeem S, Schmid B, Setala H, Symstad AJ, Vandermeer J, Wardle DA. 2005. Effects of biodiversity on ecosystem functioning: A consensus of current knowledge. *Ecol Monogr* **75**:3–35.

Jager HI. 2005. Genetic and demographic implications of aquaculture on white sturgeon (*Acipenser transmontanus*) conservation. *Can J Fish Aquat Sci* **62**:1733–1745.

Jager HI, Bevelhimer MS. 2007. How run-of-river operation affects hydropower generation. *J Environ Manage* **40**:1004–1015.

Jager HI, Rose KA. 2003. Designing optimal flow patterns for fall Chinook salmon in a Central Valley, California river. *N Am J Fisheries Manage* **23**:1–21.

Jager HI, Smith BT. 2008. Sustainable reservoir operation: Can we generate hydropower and preserve ecosystem values? *River Res Appl* **24**:340–352.

Jager HI, Chandler JA, Lepla KB, Van Winkle W. 2001. A theoretical study of river fragmentation by dams and its effects on white sturgeon populations. *Environ Biol Fish* **60**:347–361.

Jager, HI, Van Winkle W, Lepla KB, Chandler JA, Bates P. 2002. Factors controlling white sturgeon recruitment in the Snake River. In *AFS Symposium: Biology, Management, and Protection of Sturgeon*. American Fisheries Society, Bethesda, MD, pp. 127–150.

Johnson MP, Frost NJ, Mosley MWJ, Roberts MF, Hawkins SJ. 2003. The area-independent effects of habitat complexity on biodiversity vary between regions. *Ecol Lett* **6**:126–132.

Johst K, Drechsler M, Wätzold F. 2002. An ecological–economic modelling procedure to design compensation payments for the efficient spatio-temporal allocation of species protection measures. *Ecol Econ* **41**:37–49.

Kapustka LA, Galbraith H, Luxon M, Yokum J, Adams WJ. 2004. Predicting biodiversity potential using a modified Layers of Habitat model. In Kapustka LA, Galbraith H, Luxon M, Biddinger GR (Eds.), *Landscape Ecology and Wildlife Habitat Evaluation: Critical Information for Ecological Risk Assessment, Land-Use Management Activities, and Biodiversity Enhancement Practices*. ASTM STP 1458, ASTM International, West Conshohocken, PA, pp. 107–128.

Kareiva P, Watts S, McDonald R, Boucher T. 2007. Domesticated nature: Shaping landscapes and ecosystems for human welfare. *Science* **316**: 1866.

Karr Jr. 1981. Assessment of biotic integrity using fish communities. *Fisheries* **6**(6): 21–27.

Keitt TH, Urban DL, Milne BT. 1997. Detecting critical. scales in fragmented landscapes. *Conserv. Ecol.* [online] **1**(1): 4. Available from the Internet URL: http://www.consecol.org/vol1/iss1/art4

King DM, Wainger, LA, Bartoldus CC, Wakeley JS. 2000. *Expanding Wetland Assessment Procedures: Linking Indices of Wetland Function with Services and Values*. ERDC/EL TR-00-17. US Army Corps of Engineers Engineer Research and Development Center, Washington, DC.

Kremen C, Ostfeld RS. 2005. A call to ecologists: Measuring, analyzing, and managing ecosystem services. *Frontiers Ecol Environ* **3**:540–548.

Kremen C, Williams NM, Bugg RL, Fay JP, Thorp RW. 2004. The area requirements of an ecosystem service: Crop pollination by native bee communities in California. *Ecol Lett* **7**:1109–1119.

Limburg KE, O'Neill RV, Costanza R, Farber S. 2002. Complex systems and valuation. *Ecol Econ* **41**:409–420.

Logan SH, Johnston WE, Doroshov SI. 1995. Economics of joint production of sturgeon (*Acipenser transmontanus* Richardson) and roe for caviar. Aquaculture **130**:299–316.

Loomis JB. 1998. Estimating the public's values for instream flow: Economic techniques and dollar values. *J Am Water Res Assoc* **34**:1007–1014.

Losey JE, Vaughan M. 2006. The economic value of ecological services provided by insects. *Bioscience* **56**:311–323.

Ludwig D, Brock WA, Carpenter Sr. 2005. Uncertainty in discount models and environmental accounting. *Ecol Soc* **10**: 13 (online). http://www/ecologyandsociety.org/vol10/iss2/art13

Mann LK, Parr PD, Pounds LR, Graham RL. 1996. Protection of biota on nonpark public lands: Examples from the U.S. Department of Energy Oak Ridge Reservation. *Environ Manage* **20**:207–218.

Margules CR, Usher MB. 1981. Criteria used in assessing wildlife conservation potential: A review. *Biol Conserv* **21**:79–109.

McCauley DJ. 2006. Selling out on nature. *Nature* **443**:27–28.

McRae BH, Beier P, Huynh LY, DeWald L, Keim P. 2005. Habitat barriers limit gene flow and illuminate historical. events in a wide ranging carnivore, the American puma. *Molecular Ecol* **14**:1965–1977.

McRae BH. 2006. Isolation by resistance. *Evolution* **60**:1551–1561.

Minor ES, Urban DL. 2008. A graph-theory framework for evaluating landscape connectivity and conservation planning. *Conserv Biol* **22**:297–307.

Missouri Resource Assessment Partnership. 2004. *Development of Critical Ecosystem Models for EPA Region 7. Regional Geographic Initiative (RGI) Report*. Prepared for Holly Mehl and Walt Foster, Environmental Assessment Team, U.S. Environmental Protection Agency, Kansas City, KS.

Musacchio L, Ozdenerol E, Bryant M, Evans T. 2005. Changing landscapes, changing disciplines: Seeking to understand interdisciplinarity in landscape ecological change research. *Landscape Urban Planning* **73**:326–338.

Musacchio L, Wu J (Guest Eds.). 2004. Collaborative research in landscape-scale ecosystem studies: Emerging trends in urban and regional ecology. Special. issue. *Urban Ecosyst* **7**:175–314.

Nalle DJ, Montgomery CA, Arthur JL, Polasky S, Schumaker NH. 2004. Modeling joint production of wildlife and timber. *J Environ Econ Manage* **48**:997–1017.

Nassauer JI, Corry RC, Cruse R. 2002. The landscape in 2025: Alternative future landscape scenarios, a means to consider agricultural policy. *J Soil Water Conserv* **57**(2): 4A–53A.

Newsome AE, Catling PC. 1979. Habitat preferences of mammals inhabiting heathlands of warm temperate coastal, montane and alpine regions of southeastern Australia, In Specht RL (Ed.), *Heathlands and Related Shrublands of the World*, Vol. **9A** of Ecosystems of the World, Elsevier Scientific Publishing Co., Amsterdam, pp. 301–316, as cited in CSIRO 1997.

NOAA. 2000. *Habitat Equivalency Analysis: An Overview*. National Oceanic and Atmospheric Administration, Damage and Restoration Program, Seattle, WA.

Rapport DJ, Turner JE. 1977. Economic models in ecology. *Science* **195**:367–373.

Reijnen R, Foppen R, ter Braak C, et al. 1995. The effects of car traffic on breeding bird populations in woodland. III. Reduction of density in relation to the proximity of main roads. *J Appl Ecol* **33**:187–202.

Roach B, Wade WW. 2006. Policy evaluation of natural resource injuries using habitat equivalency analysis. *Ecol Econ* **58**:421–433.

Root KV, Akçakaya HR, Ginzburg L. 2003. A multispecies approach to ecological valuation and conservation. *Conserv Biol* **17**:196–206.

Rosensteel BA, Awl DJ. 1995. *Wetland Surveys of Selected Areas in the K-25 Site Area of Responsibility*. ORNL/TM-13033. Oak Ridge National Laboratory, Oak Ridge, TN.

Rossi E, Kuitunen M. 1996. Ranking of habitats for the assessment of ecological impact in land use planning. *Biol Conserv* **77**:227–234.

Russell KN, Ikerd H, Droege S. 2005. A potential conservation value of unmowed powerline strips for native bees. *Biol Conserv* **124**:133–148.

Sale MJ, Brill JED, Herricks EE. 1982. An approach to optimizing reservoir operation for downstream aquatic resources. *Water Resources Res* **18**:705–712.

SAMAB (Southern Appalachian Man and the Biosphere). 1996. *The Southern Appalachian Assessment Terrestrial Technical Report*. Report 5 of 5. U.S. Department of Agriculture, Forest Service, Atlanta, GA.

Sanders LD, Walsh RG, Loomis JB. 1990. Toward empirical estimation of the total value of protecting rivers. *Water Resources Res* **26**:1345–1357.

Santelmann MV, White D, Freemark K, Nassauer JI, Eilers JM, Vaché KB, Danielson BJ, Corry RC, Clark ME, Polasky S, Cruse RM, Sifneos J, Rustigian H, Coiner C, Wu J, Debinski D. 2004. Assessing alternative futures for agriculture in Iowa, U.S.A. *Landscape Ecol* **19**:357–374.

Short HL. 1984. *Habitat Suitability Index Models: The Arizona Guild and Layers of Habitat Model*. FWS/OBS-82/10.70, U.S. Fish and Wildlife Service, Fort Collins, CO.

Simberloff D, Cox J. 1987. Consequences and costs of conservation corridors. *Conserv Biol* **1**:63–71.

Stauffer D. 1985. *Percolation Theory*. Taylor and Francis, London, 54 pp.

Strange EM, Allen PD, Beltman D, Lipton J, Mills D. 2004. The habitat-based replacement cost method for assessing monetary damages for fish resource injuries. *Fisheries* **29**(7): 17–24.

Suter GW II, 1993. A critique of ecosystem health concepts and indexes. *Environ Toxicol Chem* **12**:1533–1539.

Tazik DJ, Martin CO. 2002. Threatened and endangered species on U.S. Department of Defense lands in the arid west, USA. *Arid Land Res Manage* **16**:259–276.

Tewksbury JJ, Levey DJ, Haddad NM, Sargent S, Orrock JL, Weldon A, Danielson BJ, Brinkerhoff J, Damschen EI, Townsend P. 2002. Corridors affect plants, animals, and their interactions in fragmented landscapes. *Proc Natl Acad Sci USA* **99**:12923–12926.

Urban DL. Keitt TH. 2001. Landscape connectedness: A graph theoretic perspective. *Ecology* **82**:1205–1218.

USDA. 2004. *FSA Handbook. Agricultural Resource Conservation Program. Short Reference*. 2-CRP. Revision 4. US Department of Agriculture Farm Service Agency, Washington, DC.

US EPA. 2006. *Ecological Benefits Assessment Strategic Plan*. EPA-240-R-06-001. Office of the Administrator, U.S. Environmental Protection Agency, Washington, DC www.epa.gov/economics/

Vogt P, Riitters K, Estreguil C, Kozak J, Wade T, Wickham J. 2007. Mapping spatial. Patterns with morphological. Image processing. *Landscape Ecol* **22**:171–177.

Welcomme RL. 1979. *Fisheries Ecology of Floodplain River*. Longman, New York, 416 pp.

Whicker FW, Hinton T G, MacDonnell MM, Pinder JE III, Habegger LJ. 2004. Avoiding destructive remediation at DOE sites. *Science* **303**:1615–1616.

White ML, Maurice C. 2004. CrEAM: A method to predict ecological significance at the landscape scale. Unpublished manuscript submitted to the EPA Science Advisory Board.

WRI (World Resources Institute). 2005. *Millennium Ecosystem Assessment: Living Beyond our Means—Natural Assets and Human Well-Being*. Washington, DC. http://population.wri.org/mabeyondmeans-pub-4115.html

Wu J, Boggess WG. 1999. The optimal allocation of conservation funds. *J Environ Econ Manage* **38**:302–321.

Wu J, Skelton-Groth K. 2002. Targeting conservation efforts in the presence of threshold effects and ecosystem linkages. *Ecol Econ* **42**:313–331.

10

PREDICTING CLIMATE CHANGE RISKS TO RIPARIAN ECOSYSTEMS IN ARID WATERSHEDS: THE UPPER SAN PEDRO AS A CASE STUDY

Hector Galbraith, Mark D. Dixon, Juliet C. Stromberg, and Jeff T. Price

Riparian areas function as keystone elements of the landscape, having a functional importance that far exceeds their proportional area. Ecosystem services provided by riparian areas include their roles as buffers controlling lateral movements of pollutants or sediments between aquatic and terrestrial environments, corridors for facilitating longitudinal movement of organisms or materials across the landscape, and highly productive habitats that are often hotspots for biodiversity (Naiman et al. 1993, Naiman and Decamps 1997). In arid regions of the southwestern United States, riparian habitats are particularly important for sustaining regional biodiversity, with a large proportion of species dependent on riparian systems (Patten 1998). Southwestern riparian systems have an important influence on continental diversity of neotropical migrant birds, providing critical migratory corridors and stopover habitats through an otherwise arid region (Skagen et al. 1998). Finally, as ecotones between terrestrial and aquatic ecosystems, riparian zones may be highly sensitive indicators or integrators of environmental change in the watersheds within which they occur (DeCamps 1993).

Watersheds and riparian systems of semi-arid to arid regions, such as the southwestern United States, should be particularly sensitive to environmental changes that influence hydrologic processes. Water is a limiting resource in the Southwest, both for natural ecosystems and for humans, presenting an important challenge for balancing economic development and the conservation of riparian and aquatic ecosystems.

Environmental Risk and Management from a Landscape Perspective, edited by Kapustka and Landis
Copyright © 2010 John Wiley & Sons, Inc.

The majority of riparian and river systems in the desert Southwest have already been degraded by a variety of anthropogenic stressors, including flow diversions and dams, groundwater depletion, land use change, urbanization, and overgrazing by livestock (Tellman et al. 1997, Patten 1998). Climate change is another stressor on these already stressed riparian systems. Climate change may directly affect the riparian ecosystems or may interact with other stressors in complex ways, potentially exacerbating or ameliorating their effects.

ECOLOGICAL IMPORTANCE OF THE UPPER SAN PEDRO RIVER

The source of the San Pedro River is near the town of Cananea in the State of Sonora, Mexico, from where it flows 240 km north through Arizona, to its confluence with the Gila River near Winkelman, Arizona (Arias Rojo et al. 1999). The total drainage area is about 1900 km^2 at the international border and about 12,000 km^2 at its confluence with the Gila River (Stromberg 1998). For most of its course, the San Pedro is a low-elevation, low-gradient (0.002–0.005 m/km) alluvial stream, with elevation ranging from 1300 m at the Mexican border to 586 m at the confluence of the Gila River, a distance of 198 km (Huckleberry 1996). The section of the San Pedro that is within the United States (Fig. 10.1) is one of the few low-elevation rivers of its size in the desert Southwest that contains significant reaches of perennial flow and is not regulated by dams.

Throughout its course, the San Pedro flows through an ecological matrix composed mainly of desert or semi-desert, with arid grasslands, Chihuahuan desert, and mesquite scrub being the most prevalent community types. Within this drier matrix, perennial or nearly perennial flows and shallow groundwater support lush riparian vegetation communities along the upper San Pedro (Fig. 10.2), including regionally threatened or rare vegetation types like Fremont cottonwood–Goodding's willow (*Populus fremontii–Salix gooddingii*) gallery forest, riverine marsh or cienega, velvet mesquite (*Prosopis velutina*) woodland (bosque), and sacaton (*Sporobolus wrightii*) grassland. Maintenance of the high vegetative diversity and the cottonwood–willow forests, in particular, are viewed as critical for sustaining the high avian biodiversity of the riparian corridor.

The presence of this riparian corridor within an arid landscape matrix provides critical habitat for migratory and breeding birds, as well as a high diversity of mammals, reptiles, amphibians, butterflies, and other animals (Arias Rojo et al. 1999). For its size, the area has one of the highest vertebrate diversities found anywhere in the United States. Almost 390 bird species have been recorded there, of which 250 are migrants that winter in Central or South America and depend on the San Pedro as a staging post on their journeys to and from their breeding areas in the United States and Canada. Between one and four million songbirds use the area as a migratory corridor each year. Without a riparian north–south habitat corridor provided by the San Pedro, these migratory journeys might not be possible. The San Pedro also provides (a) breeding habitats for some Central American bird species that reach their northernmost outposts in the area and (b) a wintering habitat for species breeding farther north.

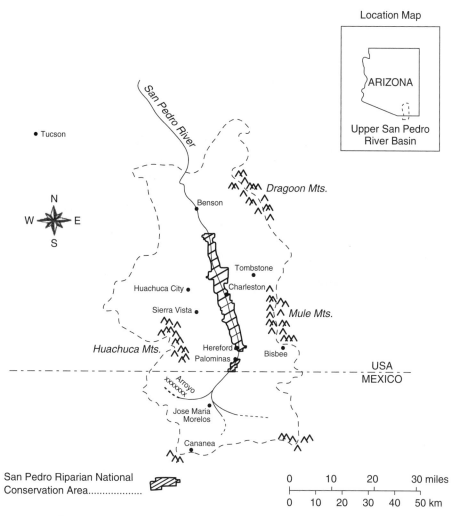

Figure 10.1. Map showing the location of the San Pedro riparian area.

The Nature Conservancy has recognized the high ecological value of the San Pedro through its designation as one of the "Last Great Places." Approximately 58,000 ha (Steinitz et al. 2003) are protected along a 50-km reach of the river within the San Pedro Riparian Conservation Area (SPRNCA), administered by the US Department of the Interior Bureau of Land Management (BLM).

The high biodiversity of the San Pedro River riparian habitats provide a variety of important ecosystem services. For example, the SPRNCA is nationally and internationally recognized for its birdwatching opportunities, one of the fastest growing recreational activities in the United States. Within the United States, Arizona is a

Figure 10.2. A typical view of the San Pedro Riparian National Conservation Area showing the north-south running ribbon of lush hydric vegetation embedded in a more desert-like matrix.

major birding destination with most birdwatching occurring in the southeastern portion of the state, including the SRNCA. By 1997, the annual number of visitors to the SPRNCA had grown to an estimated 100,000 visitors, a high proportion of them being birders and nature viewers.

Previous studies suggest that the San Pedro riparian ecosystem is already being impacted adversely by human use of the underlying groundwater aquifers. Groundwater withdrawals for human consumption jeopardize the riparian habitat by lowering regional and potentially local floodplain water tables. The amount of water pumped from the aquifer supporting the San Pedro riparian ecosystem has increased by an order of magnitude in the last 50–60 years. From small amounts of pumping ($<25,000$ m^3/day) in 1940, the amount of water pumped from the aquifer has increased to $>200,000$ m^3/day over the period 1976–1985, a level of intensity that continued at least into 1997 (Goode and Maddock 2000), and probably continues today.

In this chapter we report the results of modeling studies performed on the riparian ecosystem along the upper San Pedro River in southern Arizona and project how riparian ecosystem structure and function might be affected by changes in climate. Specifically, we model the potential effects of several plausible climate change scenarios on the structure and composition of the riparian vegetation community, and we project how such change may affect the system's ability to continue supporting a high diversity bird community.

MODELING APPROACHES

We used two modeling approaches in this study. The first exercise was to model climate change scenarios. The second involved linking these scenarios to hydrologic, geomorphic, and vegetation responses.

Climate Change Scenarios

We used a 52-year daily time series of historic weather data (1951–2002) from the National Weather Service station at Tombstone (station ID 028619, 31.7° latitude, −110.05° longitude) to create four climate scenarios for the period 2003–2102. The daily 52-year record was cycled through twice to generate the 100-year scenario, with simulation years 2003 and 2054 initialized with the adjusted 1951 daily data. Historic temperature trends from 1951 to 2003 were not removed from the time series. Use of daily historic data preserved the important seasonal and year-to-year patterns of climatic variation (e.g., influences of ENSO and PDO) that characterize the climate of the Southwest. All of the climate scenarios were transient, beginning with the same conditions in 2003 and progressively diverging over the 100-year simulation period. We chose to use transient, rather than fixed, climate change scenarios to more realistically represent gradual, cumulative changes in climate over the next century.

The scenarios were chosen to represent a reasonable set of potential climate trajectories, given the range of projections for the region derived from climate models for the southwestern United States regional assessment (SRAG 2000). The scenarios are as follows:

1. *No Climate Change:* (1951–2002 daily temperature and precipitation repeated)
2. *Warm:* progressive temperature warming over 100 years, with a 4°C increase in maximum daily temperature and a 6°C increase in minimum daily temperature by 2102
3. *Warm Dry:* progressive temperature warming as in #2 and a progressive decline in winter (nonmonsoonal: October 1 to May 31) daily precipitation of 50% by 2102
4. *Warm Wet:* progressive temperature warming as above, with a progressive increase in winter daily precipitation of 50% by 2102.

Changes in temperature and precipitation, relative to historic values, were applied linearly over the 100-year period. Hence, for the warming scenarios (scenarios 2–4), daily minimum temperatures were increased 0.06°C per year and daily maximum temperatures 0.04°C per year over 2003–2102. Similarly, for the precipitation increase (scenario 4), winter precipitation was increased by 0.5% per year from 2003–2102. Changes in precipitation were applied only to days in the historic record that had measurable precipitation. Hence, precipitation totals for individual days were adjusted upward or downward, without changing the frequency of rain events. For the wetter scenario, this effectively increased the magnitude of extreme events, which is one expectation of climatic change (Easterling et al. 2000, Houghton et al. 2001).

Changes in temperature were similar to those projected for the region by the Canadian Climate Centre (CCC), Hadley 2, and NCAR Regional models (SRAG 2000). We assumed that daily minimum temperatures would increase more than maximum daily temperatures, consistent with IPCC projections and recent observations (Houghton et al. 2001). Our consideration of both wetter and drier scenarios reflected differences among climate models in projected winter precipitation changes for the Southwest (SRAG 2000). The CCC and Hadley 2 models suggest a strong increase (perhaps a doubling or more) in winter precipitation, while the NCAR regional model suggests a decrease. None of the models projected a significant change in summer (monsoonal) precipitation, so historic daily values for June 1 to September 30 were retained in the scenarios.

Hydrologic, Geomorphic, and Vegetation Modeling

Following is an overview of these modeling efforts at three study sites. Each study site was evaluated to examine the hydrologic, geomorphic, and vegetation responses, which in turn were used to assess vulnerability of bird species and ultimately the response to overall bird diversity.

Overview. In addition to the direct physiological effects of temperature increases on vegetation, we assumed that important effects of climate change on riparian ecosystems would be mediated by influences on river flow regimes, disturbance (e.g., fire, flood), and geomorphic dynamics (river channel migration). Changes in alluvial groundwater levels would also have an important influence on riparian vegetation. Although we did not simulate groundwater dynamics in responses to climate change or regional groundwater pumping, we did model vegetation dynamics across sites that spanned a gradient in groundwater depth and surface flow intermittency. Comparisons of vegetation dynamics on these sites may yield insights to the potential interactive effects of climate and groundwater change.

Study Sites. The three sites chosen for simulation runs all occur within the SPRNCA and roughly correspond to the three classes of the Riparian Condition Index developed by Stromberg et al. (2006). The perennial flow type (Kolbe site) can be characterized as a hydrologically "gaining" reach, in which shallow groundwater supports baseflow throughout the year; while the intermittent flow types represent "losing" reaches in which baseflows cease during seasons in which the groundwater level falls below the river thalweg. Wet intermittent reaches (Palominas site) are those that have surface flow during the majority of months during a normal year (but are not perennial), while dry intermittent reaches (Contention site) flow less frequently, typically less than 60% of the time.

HYDROLOGY. We simulated the potential effects of the four climate scenarios on daily stream flow at the Charleston USGS gage, using a watershed runoff model, the Soil Water Assessment Tool (SWAT; Arnold et al. 1994) within the Automated

Geospatial Watershed Assessment Tool (AGWA) interface (Hernandez et al. 2000, 2003; Kepner et al. 2004) in ArcView. These simulated stream flows were then used as inputs to the geomorphic and vegetation models (see below). We derived climatic inputs for runoff model from five weather stations (including Tombstone) distributed throughout the upper San Pedro basin, and we adjusted daily precipitation values to reflect the four climate scenarios. Temperature data were derived only from the Tombstone station. Other inputs to the SWAT model included soils, basin topography, and land cover datasets from the upper San Pedro basin. Hydrologic parameter settings were based on previous calibration of the model to historic annual stream flow values in the upper San Pedro basin (Hernandez et al. 2000, 2003).

GEOMORPHOLOGY. We modeled the potential effects of stream flow changes on river channel migration using the program MEANDER (Larsen and Greco 2002). MEANDER simulates lateral channel migration as a function of channel hydraulics, annual stream power, and spatial heterogeneity in bank erodibility, calibrated against past changes in channel location and configuration. Differences in the projected location of the channel midpoint along each study transect at the end of each 5-year interval were used to represent channel location and patch turnover on each study-site cross section, with shifts in channel position assumed to erode portions of existing patches on the outside of the bend and form point bars available for colonization of new vegetation on the inside of the bend. During each 5-year interval, the model initiated new vegetation plots in the year with the highest daily discharge exceeding 170 m^3/s. This peak daily flow threshold was exceeded in about 10% of the years during 1951–2002, and it reflects the assumption that major pioneer forest colonization events are associated with the 10-year return flood (Stromberg et al. 1993, Stromberg 1998).

VEGETATION. Using STELLA II® Dynamic Simulation software (ISEE Systems, Lebanon, NH), we developed a simulation model to project the transient effects of climate change on riparian vegetation composition and structure. The model simulates the recruitment, growth, and mortality of 10 species or functional groups of southwestern riparian plants: Fremont cottonwood, Goodding's willow, saltcedar (*Tamarix ramosissima*), velvet mesquite, velvet ash (*Fraxinus velutina*), a hydromesic shrub group (e.g., *Baccharis salicifolia*), a xeric riparian shrub group (e.g., *Ericameria nauseosa, Hymenoclea monogyra*), herbaceous annuals, wetland herbaceous perennials (e.g., *Typha, Scirpus, Eleocharis*), and mesic perennial grasses (e.g., *Sporobolus wrightii*)—as determined by life history characteristics and changes in environmental drivers (temperature, precipitation, flooding, groundwater depth, fire), at the scale of individual sampling plots (about 10 × 10 m). Greater details of model development are given in Dixon et al. (2009).

We simulated vegetation change at plots distributed along a cross-floodplain transect at each of the three study sites for the four climate scenarios. Four replicate model runs were performed at each plot, under each climate scenario. Only plots in the post-entrenchment floodplain/terrace were modeled; plots on the high terrace were excluded. Initial conditions for each model run consisted of the vegetation, elevation above the riverbed, and seasonal depth to groundwater for each plot, based on field sampling in 2001 and 2002. Vegetation patch types were defined along each transect based on dominant species, age, and physiognomy. Each plot was taken to represent the conditions in the entire patch in which it was found, and its contribution to the vegetation of the entire transect was weighted by the width of the patch relative to the width of the entire floodplain.

BIRD VULNERABILITY AND BIODIVERSITY MODELING. Two main avian modeling approaches were used. In the first, we developed Habitat Suitability Index (HSI) models to examine in detail the potential effects of vegetation change under the four climate scenarios on the habitat quality along the SPRNCA for five species likely to be suitable indicators of climate change impacts on the riparian vegetation communities. Yellow warbler (*Dendroica petechia*), Wilson's warbler (*Wilsonia pusilla*), and yellow-billed cuckoo (*Coccyzus americanus*) are all characteristic of the SPRNCA and depend largely on the existence of the cottonwood/willow gallery forest for their habitats. In contrast, Botteri's sparrow (*Aimophila botterii*) is a resident species that is characteristic of the Chihuahuan desert and mesquite shrubland/sacaton grassland matrix within which the SPRNCA riparian habitat is embedded. Last, the southwestern willow flycatcher (*Empidonax traillii extimus*) is an intermediate species in that while it breeds in riparian habitat, it does not depend on native hydrophytic cottonwood/willow gallery forest, but can breed in shrub-dominated vegetation, even when the dominant species is the mesic invasive saltcedar. The HSI models developed for these species were field-tested in the SPRNCA in areas of known population density and found to be reasonably accurate predictors of habitat suitability. They are described in greater detail in Galbraith et al. (2004).

Whereas the first (HSI) approach focused on a few select species and examined the influence of both drier and wetter climates, the second avian modeling approach explored further how a warmer, drier climate and/or further alluvial groundwater depletion might affect the overall bird diversity in the SPRNCA. Based on our knowledge of the relationships between depth to groundwater, the structure and composition of vegetation communities, and wildlife habitat utilization, it seemed reasonable to assume that some of the most fundamental potential impacts of a warmer, drier climate on vegetation communities in the SPRNCA could include the fragmentation of the existing riparian and wetland communities, and their replacement by communities more typical of the mesic or xeric matrix within which the SPRNCA is set (these initial assumptions were verified by the results of the vegetation modeling performed in this study).

This second modeling approach was to develop a predictive framework, based on the habitat preferences of 87 bird species that inhabit the SPRNCA. These preferences were used to categorize their relative vulnerabilities to the vegetation changes identified above and project potential impacts to their population status within the study area. We focused on the 87 species of birds classified by Kreuper (1997) as "abundant," "common," or "fairly common" within the SPRNCA. Because of their relatively abundant status, the species that we have evaluated will comprise the majority of the birds using the SPRNCA at any particular time and will provide a representative indication of impacts to diversity. This approach does not, however, address the possibility that species that are presently not found in the SPRNCA but that are characteristic of the river further south in Mexico may take advantage of the warming climate and move

Table 10.1. Scoring System for Four Habitat Variables Used in Avian Vulnerability Analysis

Score	Dominance of Riparian Species in Vegetation Community	Dependence on Nonfragmented, Extensive Stands of Riparian Vegetation	Dependence on Wetland Habitat	Dependence on Open Water
5	Found only in riparian habitat	NA	Found only in wetland habitat	Restricted to waterbodies
4	Found mainly in riparian habitat	NA	Found mainly in wetland habitat but also occurs in mesic or xeric habitats	Found mainly in association with waterbodies
3	Occurs in either riparian or more mesic or xeric habitats	Found only in such stands	Occurs in either wetland or more mesic or xeric habitats	No obvious preference for waterbodies
2	Most typical of mesic or xeric habitats, but may also occur in riparian	Found mainly in such stands but may also occur in more fragmented habitats	Most typical of mesic or xeric habitats but may also occur in wetlands	More typical of drier habitats
1	Restricted to mesic or xeric habitats	Found either in fragmented or nonfragmented habitats	Restricted to mesic or xeric habitats	Restricted to drier habitats

NA, not applicable.

north. However, there are only a very small number of species to which this might apply, and their contribution to future biodiversity is likely to be small.

Based on the potential changes in vegetation communities and the known habitat relationships of the 87 bird species, four variables that are most likely to predict future population changes were identified (Table 10.1). Each variable captures the species' need for one or more of three limiting life-history requirements: foraging habitat, cover, and/or nest sites. For each of these variables a scoring system was developed (Table 10.1). The scores for any species are indices of its likely vulnerability to the projected habitat changes.

RESULTS

The modeling results are discussed first in terms of projected changes to hydrology and channel migration. Second, the projected responses of vegetation are presented.

Table 10.2. Cumulative Horizontal Displacement (in meters) in Simulated Channel Position, Relative to 2002, Along Each Transect, for Four Climate Scenarios[a]

Site	Scenario	Cumulative Displacement (m)			
		2027	2052	2077	2099
Kolbe	No change	0.8	0.4	−3.3	−9.8
	Warm	0.8	0.6	−2.1	−5.9
	Warm dry	0.7	1.0	0.5	0.2
	Warm wet	0.9	−1.5	−11.9	−43.4
Palominas	No change	2.6	6.6	9.8	13.0
	Warm	2.5	6.0	9.2	12.8
	Warm dry	2.4	4.2	6.2	8.0
	Warm wet	2.7	8.8	13.8	24.1
Contention	No change	16.9	35.9	49.0	59.3
	Warm	16.7	33.7	45.3	54.3
	Warm dry	15.7	25.6	34.7	36.6
	Warm wet	17.8	43.9	60.5	77.6

[a]Channel migration was simulated using the program MEANDER (Larsen and Greco 2002). Positive values indicate eastward movement of the channel, negative values indicate westward movement.

These are then used to describe projected changes to habitat characteristics of selected species followed by analysis of impacts to bird diversity.

Hydrology and Channel Migration

Climate scenario had a strong influence on simulated flow regimes (Fig. S10.3). Warming alone (the warm scenario) had a relatively weak effect, slightly reducing the magnitude of annual peak flows by the end of the simulation. Changes in winter precipitation had a much stronger effect. Under the warm dry scenario, declines in the size of winter precipitation events led to a near cessation in winter floods, with no fall/winter floods exceeded 170 m³/s (the threshold for vegetation recruitment used in the model). Conversely, under the warm wet scenario, a progressive 50% increase in the size of winter precipitation events resulted in a large increase in the number of years with fall/winter floods exceeding 170 m³/s, with such floods becoming progressively larger and more frequent. By the end of the simulation, peak flood magnitudes were about three times larger under the warm wet scenario than under the no climate change.

Rates of channel migration tracked differences in streamflow among scenarios, but were influenced by individual site differences (Fig. S10.4, Table 10.2). Relative to the no-change scenario, slightly lower migration rates occurred under the scenario with warming and no precipitation change (warm scenario), much higher rates occurred under the warm wet scenario, and much lower rates occurred under the warm dry scenario.

Simulated Vegetation Change

A common pattern in simulated vegetation change across most scenarios and study sites was an increase in cover of mesquite shrubland/woodland, and sometimes sacaton grassland, and a decrease in cottonwood/willow patch types from 2003 to 2102 (Fig. S10.5). Across sites, most patches dominated by cottonwood/willow or saltcedar in 2003 transitioned to other patch types (often mesquite or sacaton) by 2102, as the pioneer stands senesced. Coverage by pioneer riparian patch types in 2102 was closely linked to cumulative channel migration rates, with species composition (i.e., cottonwood/willow vs. saltcedar) influenced by groundwater hydrology. Saltcedar recruitment occurred on all three sites, but was proportionally much higher on the driest site (Contention), consistent with the initial dominance of saltcedar there. Successful recruitment of cottonwood and willow occurred only on the perennial (Kolbe) and wet intermittent (Palominas) sites. However, migration rates under even the warm wet scenario were insufficient to maintain the initially high cottonwood/willow coverage in the floodplain at Palominas. At Kolbe, recruitment of new cottonwood/willow patches was fourfold greater under the warm wet scenario than under the no-climate change scenario and may be sufficient to maintain high coverage of pioneer patch types in the landscape, if continued beyond 2102 into the next century.

Effects on Bird Community: HSI Results for Five Study Species

Even without incorporating climate change into future conditions (the no change scenario), marked changes in habitat quality are projected for the yellow-billed cuckoo and the Botteri's sparrow. This is because even without further climate change, we project that the existing cottonwood/willow forest will senesce and begin to contract about the middle of this century. This will result in a decrease in habitat for the cuckoo, but an increase in Botteri's sparrow habitat as the forest is replaced with grasslands and shrublands.

Our model results (Fig. S10.6) show that different bird species will have different vulnerabilities to climate change; results for the warmer and wetter scenario were not included because this climate future did not produce vegetation changes that were marked enough to result in significant HSI changes. In general, those species that are most dependent on the continuation of the existing hydric riparian forest and wetland habitats (yellow-billed cuckoo, Wilson's and yellow warblers) have the greatest vulnerabilities to climate change-induced effects. In contrast, Botteri's sparrow is relatively invulnerable to the climate change scenarios investigated. Given its preference for more xeric habitats, it may well benefit as mesquite savannah and sacaton grasslands replace the riparian gallery forest. Willow flycatcher may also benefit from the climate change-induced vegetation shifts as mesic shrubs replace the gallery forest in the SPRNCA.

Effects on Bird Diversity in SPRNCA

Total vulnerability scores for the 87 bird species were grouped into four vulnerability categories: Highly Vulnerable, Vulnerable, Less Vulnerable, and Least Vulnerable

(Table 10.3). Twenty-six percent of the 87 species can be categorized as either highly vulnerable or vulnerable. These are the species most likely to be adversely affected by future warming and drying in the SPRNCA due to climate change and groundwater extractions. The 25% of species in the less vulnerable categories can be viewed as species that are relatively insensitive to such changes because they are not closely tied to the existence of riparian or wetland vegetation communities, rather than more mesic or xeric community types, and might not be greatly adversely affected under a shift from continuous riparian forests and wetlands to (for example) increased cover of mesquite woodlands. More than half of the 43 species in the category least vulnerable are adapted to the desert or arid grassland environments of southern Arizona. Such species might be expected to benefit from the conversion of riparian forest and wetland habitats within the SPRNCA to more mesic or xeric environments.

DISCUSSION

Our vegetation modeling results indicate that geomorphic legacies and successional trajectories may play a more important role than climate change in influencing vegetation changes on the San Pedro over the next 100 years. Across all scenarios, simulated coverage of cottonwood and willow forests along the three upper San Pedro sites declined from 2003 to 2102, with these patches largely replaced with mesquite woodland and, to a lesser extent, sacaton grassland. The main reason for this is that most cottonwood and willow patches along the San Pedro were established in response to transient geomorphic processes that are no longer active today. The channel incision and widening events of the late 19th and early 20th centuries were followed by a period of channel narrowing and floodplain reconstruction, during which much of the present-day forest established (Hereford 1993, Dixon and Stromberg, unpublished). With channels having narrowed considerably since the 1950s, opportunities for new cottonwood and willow recruitment are limited by rates of channel migration or localized channel widening, with both processes having slowed over the last 20–30 years (Dixon and Stromberg, unpublished, Stromberg 1998). In the model, riparian forest cover declined over time as the relatively short-lived cottonwoods and willows established during this period of channel narrowing senesced. Simulated coverage of cottonwood/willow forest by 2102 depended largely on new recruitment, which was linked to the creation of moist, mineral seedbeds by winter–spring floods and channel migration (Stromberg et al. 1993; Stromberg 2001, 1998).

Hence, based on the model results, the main effects of climatic change would be to either exacerbate or partially ameliorate this long-term trajectory of cottonwood/willow decline and the corresponding increase in later successional mesquite and sacaton. Drier winters would likely result in less frequent winter floods, lower rates of channel migration, and lower rates of recruitment by cottonwood and willow. Wetter winters would result in larger and more frequent winter floods, higher channel migration rates, and higher recruitment rates by cottonwood and willow. Recruitment and coverage of saltcedar, also a pioneer species, may be strongly influenced by both rates of flooding and channel migration and site-level groundwater conditions. Coverage by mesquite

Table 10.3. Vulnerability Categories Based on Total Framework Scores[a]

Total Score 14–18 Highly Vulnerable	Total Score 10–13 Vulnerable	Total Score 6–9 Less Vulnerable	Total Score < 6 Least Vulnerable
4 (5% of total)	18 (21% of total)	22 (25% of total)	31 (49% of total)
Gray hawk	Great blue heron	Sharp-shinned hawk	Turkey vulture
Belted kingfisher	Virginia rail	Cooper's hawk	Northern harrier
Common yellowthroat	Sora rail	Red-tailed hawk	**Scaled quail**
Yellow warbler	Spotted sandpiper	Western screech owl	**Gambel's quail**
	Wilson's snipe	Great horned owl	American kestrel
	Yellow-billed cuckoo	Gila woodpecker	**White-winged dove**
	Western wood pewee	Ladder-backed woodpecker	Mourning dove
	Black phoebe	Northern flicker	**Common ground dove**
	Tree swallow	Pacific-slope flycatcher	**Greater roadrunner**
	Warbling vireo	Dusky flycatcher	**Lesser nighthawk**
	Wilson's warbler	Violet-green swallow	**Black-chinned hummingbird**
	Yellow-breasted chat	Northern rough-winged swallow	Gray flycatcher
	Summer tanager	Cliff swallow	Say's phoebe
	Lincoln's sparrow	Ruby-crowned kinglet	Vermilion flycatcher
	Red-headed blackbird	Bell's vireo	**Ash-throated flycatcher**
	Yellow-headed blackbird	Yellow-rumped warbler	**Cassin's kingbird**
	Brewer's blackbird	Black-headed grosbeak	**Western kingbird**
	Hooded oriole	Blue grosbeak	**Chihuahuan raven**
		Song sparrow	**Verdin**
		White-crowned sparrow	**Cactus wren**
		Brown-headed cowbird	Bewick's wren
		Bullock's oriole	Northern mockingbird
			Curve-billed thrasher
			Loggerhead shrike
			Lucy's warbler
			MacGillivray's warbler
			Orange-crowned warbler
			Western tanager
			Green-tailed towhee
			Canyon towhee
			Abert's towhee

[a]Species in bold are desert or arid grassland species in southern Arizona.

shrubland/woodland and sacaton grassland may increase over the next 100 years, with fire frequency likely influencing the relative balance between the two.

These projections assume a continuation of present-day geomorphic processes, without reoccurrence of a catastrophic channel incision/widening event in the next century. Improvements in upland land use—and with them, a less "flashy" watershed response to precipitation events (Hereford 1993) over the last 40–50 years—suggest that the occurrence of such a catastrophic flood in the near future is unlikely. However, large increases in the size or variability of precipitation events could increase the vulnerability of the system to channel entrenchment from a catastrophic flood event. Such a major incision and widening event could reset the system to conditions that prevailed in the first half of the twentieth century, before channel recovery processes led to channel narrowing and formation of the vegetated riparian corridor we see today.

Our bird HSI modeling results track those of the vegetation model in that under the no-change, warmer, and warmer and drier scenarios the quality of habitat for the riparian specialist species (yellow-billed cuckoo, yellow and Wilson's warblers) will be reduced. This may not indicate that all of the habitat for these species will be eliminated, but it is likely that the carrying capacity of the area for these species will be reduced. Under these scenarios, the quality of habitat for the Botteri's sparrow (an arid grassland species) is projected to improve within the current riparian area as cottonwood and willow forest is replaced with mesquite and grasses. Willow flycatchers, which typically occur in riparian areas, but are not limited to cottonwood and willow forest, may not suffer a loss of habitat quality. Unlike yellow-billed cuckoo, neither Botteri's sparrow nor willow flycatcher are dependent on large unbroken tracts of habitat and are not projected to pay the price that such specialists will pay if the vegetation communities transition from their current condition to a future more fragmented state.

A complicating factor in our analyses is that we were not able to include potential future changes in aquifer depletion. Current extraction rates may already be adversely affecting the riparian ecosystem (Goode and Maddock 2000), and it is reasonable to assume that the net results of climate change and future groundwater extraction may tend toward a warming and drying of the watershed. We developed our avian community vulnerability model to anticipate the effects that such a warming and drying trend may have on avian diversity. The results of this analysis show that under a warming and drying trend, up to approximately 26% of the bird species that currently are common within the SPRNCA may be highly vulnerable or vulnerable to the resulting changes in the vegetation communities (replacement of hydric species with more xeric species). Seventy-four percent of the species that currently make up the avian community are not so vulnerable, and some may actually benefit from a warming and drying trend, however, these are species that are currently widespread within the watershed (e.g., Botteri's sparrow, verdin, *Auriparus flaviceps*) and in other desert areas of the southeast. Many of the 26% that are likely to lose habitat are rarer and have much more restricted distributions within North America. Thus, the trend will be for overall avian diversity to decline because at least some of these species are replaced with common species from the surrounding arid matrix.

ACKNOWLEDGMENTS

We wish to thank Alex Fremier and Eric Larsen (University of California at Davis) for providing MEANDER output for the analysis of geomorphic change under a changing climate and Mariano Hernandez (USDA Agricultural Research Service in Tucson) for assistance with SWAT parameters. This research was supported by an EPA climate change grant through the American Bird Conservancy, with subcontracts to Arizona State University and Galbraith Environmental Sciences, LLC.

REFERENCES

American Bird Conservancy (ABC). 2005. *Potential Future Climate Change Effects on Ecological Resources and Biodiversity in the San Pedro Riparian National Conservation Area, Arizona*. American Bird Conservancy, Plains, VA.

Arias Rojo HA, Bredehoeft J, Lacewell R, Price J, Stromberg J, Thomas GA. 1999. *Sustaining and Enhancing Riparian Migratory Bird Habitat on the Upper San Pedro River*. Report to the Commission for Environmental Cooperation.

Arnold JG, Williams JR, Srinivasan R, King KW, Griggs RH. 1994. *SWAT: Soil Water Assessment Tool*. USDA, Agricultural Research Service, Grassland, Soil and Water Research Laboratory, Temple, TX.

DeCamps H. 1993. River margins and environmental change. *Ecol Appl* **3**:441–445.

Dixon, M. D., J. C. Stromberg, J. Price, H. Galbraith, A. Fremier, and E. Larsen. 2009. Climate change and riparian vegetation response. In Stromberg J, Tellman B. (Eds.), *Ecology and Conservation of Desert Riparian Ecosystems: The San Pedro River Example*. University of Arizona Press, Tucson, AZ, Chapter 8.

Easterling DR, Meehl GA, Parmesan C, Changnon SA, Karl TR, Mearns LO. 2000. Climate extremes: Observations, modeling, and impacts. *Science* **289**:2068–2074.

Galbraith H, Price J, Dixon M, J Stromberg. 2004. Development of HSI models to evaluate risks to riparian wildlife habitat from climate change and urban sprawl. In Kapustka LA, Galbraith H, Luxon M, Biddinger GR (Eds.), *Landscape Ecology and Wildlife Habitat Evaluation: Critical Information for Ecological Risk Assessment, Land-Use Management Activities, and Biodiversity Enhancement Practices*. ASTM STP 1458, American Society for Testing and Materials International, West Conshohocken, PA, pp. 148–168.

Goode T, Maddock T III. 2000. *Simulation of Groundwater Conditions in the Upper San Pedro Basin for the Evaluation of Alternative Futures*. University of Arizona, Department of Hydrology and Water Resources, HWR No. 00–020, 113 pp.

Hernandez M, Miller SN, Goodrich DC, Goff BF, Kepner WG, Edmonds CM, Jones KB. 2000. Modeling runoff response to land cover and rainfall spatial variability in semi-arid watersheds. *Environ Monitoring Assess* **64**:285–298.

Hernandez M, Kepner WG, Semmens DJ, Ebert DW, Goodrich DC, Miller SN. 2003. Integrating a landscape/hydrologic analysis for watershed assessment. The First Interagency Conference on Research in the Watersheds, October 27–30, 2003, Benson, AZ.

Hereford R. 1993. Entrenchment and widening of the Upper San Pedro, Arizona [Special paper]. *Geol Soc Am* **282**, 46 pp.

Houghton JT, Ding Y, Griggs DJ, Noguer M, van der Linden PT, Dai X, Maskell K, Johnson CA. 2001. *Climate Change 2001: The Scientific Basis*. Cambridge University Press, Cambridge, UK, 944 pp.

Huckleberry G. 1996. Historical channel changes on the San Pedro River, southeastern Arizona. *Ariz Geol Surv Open File Report* 96–15.

Kepner WG, Semmens DJ, Bassett SD, Mouat DA, Goodrich DC. 2004. Scenario analysis for the San Pedro River, analyzing hydrological consequences of a future environment. *Environ Monitoring Assess* **94**:115–117.

Kreuper DJ. 1997. Annotated checklist to the birds of the Upper San Pedro River Valley. Unpublished report to the Bureau of Land Management, Sierra Vista, Arizona.

Larsen EW, Greco SE. 2002. Modeling channel management impacts on river migration: A case study of Woodson Bridge State Recreation Area, Sacramento River, California. *Environ Manage* **30**:209–224.

Naiman RJ, Decamps H. 1997. The ecology of interfaces: Riparian zones. *Annu Rev Ecol Syst* **28**:621–658.

Naiman RJ, DeCamps H, Pollock M. 1993. The role of riparian corridors in maintaining regional biodiversity. *Ecol Appl* **3**:209–212.

Patten DT. 1998. Riparian ecosystems of semi-arid North America: Diversity and human impacts. *Wetlands* **18**:498–512.

Skagen SK, Melcher CP, Howe WH, Knopf FL. 1998. Comparative use of riparian corridors and oases by migrating birds in southeast Arizona. *Conserv Biol* **12**:896–909.

SRAG (Southwest Regional Assessment Group). 2000. *Preparing for a changing climate: the potential consequences of climate variability and change.* http://www.ispe.arizona.edu/research/swassess/pdf/complete.pdf (accessed July 25, 2009).

Steinitz C, Arias H, Bassett S, Flaxman M, Goode M, Maddock T, Mouat D, Peiser R, Shearer A. 2003. *Alternative Futures for Changing Landscapes, the Upper San Pedro Basin in Arizona and Sonora.* Island Press, Washington, DC.

Stromberg JC. 1998. Dynamics of Fremont cottonwood (*Populus fremontii*) and saltcedar (*Tamarix chinensis*) populations along the San Pedro River, Arizona. *J Arid Environ* **40**:133–155.

Stromberg JC. 2001. Restoration of riparian vegetation in the southwestern United States: importance of flow regimes and fluvial dynamism. *J Arid Environ* **49**:17–34.

Stromberg JC, Richter BD, Patten DT, Wolden LG. 1993. Response of a Sonoran riparian forest to a 10-year return flood. *Gr Basin Nat* **53**:118–130.

Stromberg JC, Lite SJ, Rychener TJ, Levick L, Dixon MD, Watts JW. 2006. Status of the riparian ecosystem in the upper San Pedro River, Arizona: Application of an assessment model. *Environ Monitoring Assess* **115**:145–173.

Tellman B, Yarde R, Wallace MG. 1997. *Arizona's Changing Rivers: How People Have Affected the Rivers.* University of Arizona, Water Resources Research Center.

11

INVASIVE SPECIES AND ENVIRONMENTAL RISK ASSESSMENT

Greg Linder and Edward Little

> Now days we live in a very explosive world, and while we may not know where or when the next outburst will be, we might hope to find ways of stopping it or at any rate damping down its force. ... Ecological explosions differ from some of the rest [of the explosions] by not making such a loud noise and in taking longer to happen.
>
> Charles S. Elton, *The Ecology of Invasions by Animals and Plants* (1958)

> The lack of real contact between mathematics and biology is a tragedy, a scandal, or a challenge; it is hard to decide which.
>
> Gian-Carlo Rota, *Discrete Thoughts* (1986)

Invasive species have emerged as recurring concerns of natural resource managers. While Elton (1958) clearly anticipated current issues related to invasive species, his insight foreshadowed the critical role that invasive species play in contemporary natural resource management (e.g., Breithaupt 2003, Drake et al. 1989, Kapustka and Linder 2007, Pimentel et al. 1999, Simberloff 1991, Wein 2002). Over the past 10–20 years, efforts have increased with respect to evaluating risks associated with species invasions, and a wide range of techniques and analytical tools have been developed and deployed (Table 11.1). Similar to the evaluation of environmental risks associated with chemical stressors, the approaches for evaluating risks associated with invasive species have incorporated a variety of tools to assess invasion risks. Our

Table 11.1. Examples of Recent Literature Focused on Risk Assessments or Components of Risk Assessments for Invasive Species

General	Andersen et al. (2004a, 2004b)	Bartell and Nair (2004)
	Campbell and Kreisch (2003)	Clark et al. (2001b, 2001)
	Clark et al. (2003)	Colautti et al. (2006)
	Crall et al. (2006)	Drake (2004), Drake and Lodge (2004)
	Drake et al. (2006)	Ferenc and Foran (2000)
	Finnoff et al. (2005)	Foran and Ferenc (1999)
	Hewitt and Huxel (2002)	Hiddink et al. (2007)
	Hoffman and Hammonds (1994)	Keller et al. (2007)
	Kolar and Lodge (2001)	Kolar and Lodge (2002)
	Kot et.al. (1996)	Landis (2004)
	Levin (1989)	Leung et al. (2004)
	Lodge (1993)	Lodge et al. (2006)
	Maguire (2004)	Marvier et al. (2004)
	Mattson and Angermeier (2007)	NISC (2001)
	Neubert and Parker (2004)	OSTP (2001)
	Olson and Roy (2005)	Rodgers (2001)
	Schnase et al. (2002)	Simberloff (1991)
	Simberloff (1985)	Stahl et al. (2001)
	Stohlgren and Schnase (2006)	Suter (2007)
	US EPA (2003)	US EPA (2000)
	US EPA (1998)	Vermeij (2005)
	Von Holle and Simberloff (2005)	Williamson (1989, 1996)
	Williamson and Fitter (1996)	Wittenberg and Cock (2001)
Aquatic Species	Bartholomew et al. (2005)	Bossenbroek et al. (2006)
	Buchan and Padilla (1999)	Carlton (1993)
	Johnson et al. (2001)	Johnson and Carlton (1996)
	Johnson and Padilla (1996)	Li (1981)
	Light and Marchetti (2007)	MacIsaac et al. (2001)
	Marchetti et al. (2004)	Moyle and Light (1996)
	Ricciardi and Rasmussen (1998)	Stepien et al. (2005)
	Swanson (2004a, 2004b)	USGS (2005)
Terrestrial Species	Goodwin et al. (1998)	Higgins et al. (2003)
	Higgins et al. (1999)	Jeschke and Strayer (2005)
	NAS (2002)	Rejmánek (2000)
	Rejmánek and Richardson (1996)	Sax et al. (2002)
	With (2004)	

focus in this chapter is not to enter the fray of expert opinion regarding invasive species or to resolve which risk methods are "best," but rather to consider the processes and tools for evaluating invasive-species risks in an environmental context. Our overview of invasive species in the EnRA process begins with a brief review of how invasive species are integral to existing EnRA processes (Adler 1996), followed by a description of a generalized process for conducting integrated risk assessment-risk management investigations that are broadly applicable to a range of invasive species of plants, animals, and microbes (Agresti 2002). The current practice largely reflects invasive species as components of a multiple stressor risk management framework. Our discussion will close with an overview of risk management tools such as hazard assessment-critical control point (HACCP) evaluation (Andersen et al. 2004a) and a brief consideration of research and management needs that must be addressed to more fully involve invasive species in the EnRA process (Andersen et al. 2004b).

INCORPORATING INVASIVE SPECIES INTO THE EXISTING ENRA PROCESS

In name, the evaluation of hazards and risks of environmental stressors—biological, chemical, and physical—has a relatively short history (see Chapter 2), yet the concepts of hazard evaluation and risk assessment have long been in practice, including evaluations of multiple stressors (Ferenc and Foran 2000; Foran and Ferenc 1999; Swanson 2004a, 2004b). Although hazards and risks posed by chemicals in the environment initiated the practice of risk assessment to a wide range of users, much of the practice advanced today is out of necessity relatively general in context, providing guidance rather than prescriptions for implementation (NAS 1983, 1993; US EPA 1998, 2003). These generic approaches are process driven, and their implementation often involves application of a wide range of analytical tools determined, in part, by the nature of the risks being considered. A number of implementations have resulted from the risk assessment paradigm initially advanced by the National Academy of Science–National Research Council (NAS 1983), especially the assessment of ecological risks associated with chemical stressors (Suter 2007, US EPA 1998) and, more generally, multiple stressors across a range of spatiotemporal contexts (Ferenc and Foran 2000; Foran and Ferenc 1999; Suter 2007; Swanson 2004a, 2004b; US EPA 2003). Many resource management practices crafted specifically for invasive species have been in place for relatively long periods (e.g., US Department of Agriculture activities focused on weeds, insect and vertebrate pests, and disease agents (USDA 2007, Westbrooks 1998); http://www.invasivespeciesinfo.gov/last accessed 14 July 2009). Invasive species are threads of a multiple-stressors fabric within the context of environmental risks. However, the growing awareness of invasive species to natural resource managers and the public, along with the emerging recognition of biological stressors as ecological hazards (USGS 2005), clearly warrants greater emphasis of invasive species in the evaluation of environmental risks. For example, early contributions by Li (1981) and Levin (1989) that focused on risks associated with invasive species foreshadowed more recent literature developed within contemporary frameworks characterized by Landis (2004) and USGS (2005), then subsequently implemented as exemplified by Colnar and Landis (2007) and generalized by Stohlgren and Schnase (2006).

From an ecological perspective, EnRA provides the context ideally suited for evaluating risks linked to multiple stressors, including biological stressors such as invasive species. The evaluation of hazards and associated risks linked to invasive species can be completed through a systematic process focused on adverse effects that may occur consequent to exposure to one or more invasive species or, alternatively, as a combination of stressors such as an invasive species (for example, a disease agent) and chemical stressors. The tools of EcoRA may be implemented prospectively to predict the likelihood of future adverse effects or retrospectively to evaluate the likelihood that observed effects have been caused by past species invasions. Additionally, the current practice of EcoRA can be applied to many "what-if" analyses related to forecasts of changes associated with various mitigation options (for example, what collateral issues may develop from the selection of management tools to control or eradicate an invasive species) or to evaluate risks associated with natural hazards or other events (for example, road construction may inadvertently increase pathways for dispersal of invasive species).

The evaluation of risks associated with invasive species can be addressed by considering the three phases of risk assessment: problem formulation, analysis, and risk characterization (Suter 2007; US EPA 1998, 2000, 2003). Although the framework for EcoRA has undergone relatively subtle changes since its inception in the late 1980s and early 1990s, the process as applied to environmental issues has largely morphed along lines adaptive to management practices in the field. Problem formulation sets the stage for the evaluation of risks by bounding the risk assessment through critical definition of the questions being considered in the process. Derivative to the formulation of questions, a plan for analyzing and characterizing risk is then developed. Regardless of the type of stressor or the hazards being considered, problem formulation identifies and integrates existing data and information on sources, stressors, and effects within the bounds of the system at risk and the receptors characteristic of that system. Primary products of problem formulation include characterization of assessment endpoints and conceptual models, as well as implementation plans for analysis.

Analysis focuses on exposure to invasive species and effects potentially linked to exposure. Frequently a desktop analysis would be implemented, wherein the strengths and limitations of existing data and information would be considered. Characterizing the exposure and effects for the system and receptors at risk subsequently develops as an outcome of risk characterization, the third phase of the risk assessment process.

During risk characterization, the analysis of exposure and stressor-response are integrated, yielding an estimation of risks, which is detailed with respect to assumptions underlying the risk estimation process, scientific uncertainties, and strengths and limitations of the analysis. Risk characterization includes a complete description of outcomes of risk estimation, including interpretation of adverse effects and attendant consequences. The lines of evidence and the strength of that evidence which inform the risk characterization are detailed along with a characterization of uncertainties associated with the risks. This is particularly important for justifying risk management decisions.

Problem formulation, analysis, and risk characterization may follow a simple linear implementation, although as with most risk assessment processes, the ideal practice follows an iterative course. For example, investigative discovery may bring new or additional data or information to problem formulation, and updated derivative products such as revised or supplemental conceptual models may enter analysis, ultimately yielding a recharacterization of risks. Risk assessment ideally occurs interactively with risk management, where the former evaluates the likelihood of adverse effects and the latter places those risk estimates into a larger context involving social, legal, and political, or economic factors that shape the bounds of acceptable risk.

Depending on the contingencies that drive the EnRA process and the role that invasive species play in that particular implementation, the analysis of risks may be categorized as being hazard-based or as being risk-based to inform resource management decisions. The distinction between these ends of the technical-analysis spectrum is largely data- and information-dependent. A hazard-based process generally reflects limited empirical data interpreted by "experts" within the context of risk, whereas a risk-based analysis is more data-rich and amenable to empirical modeling. In practice, a mix of hazard-based and risk-based tools are in play, particularly when the analysis of multiple stressors reflects the increasingly role of invasive species in the menu of resource management concerns.

To consider invasive species and their role in environmental risk assessment, key terms are defined following the terminology of the National Invasive Species Council (NISC) as specified in Executive Order 13112 (Office of the President 1999) where an "introduction" means the intentional or unintentional escape, release, dissemination, or placement of a species into an ecosystem as a result of human activity. A "native species" is one that has historically occurred or currently occurs in a specific region other than as a result of an introduction, while an "alien species" is any species, including propagules such as seeds, eggs, spores, or other biological material capable of propagating that species that is not native to that ecosystem. An "invasive species" then follows as an alien species whose introduction causes or is likely to cause economic or environmental harm or harm to human health. Hence, the distinction between alien species and invasive species is dependent on a resource valuation and reflects adverse effects or impacts associated with the presence of the alien species in its expanded species distribution. Both monetary and nonmonetary values of a resource adversely affected by the introduction of the invading species influence the distinction between an alien species and an invasive species. Equally critical to defining terms directly linked to invasive species are distinctions between the terms "hazard" and "risk," particularly given the varying usage of these key terms throughout resource management.

WHAT ARE HAZARDS AND RISKS, AND HOW DO INVASIVE SPECIES FIT INTO THE RISK PARADIGM?

Regardless of whether we are focused on chemical, physical, or biological stressors, hazards are events or processes linked to adverse effects associated with these stressors and result in loss of system integrity. Within an ecological context and depending on the level of biological organization—individual, population, community, ecosystem—loss of system integrity may be characterized as decreased survival, local extirpation or extinction, and habitat degradation. Hazards may be natural or anthropogenic in origin. Frequently, natural hazards and anthropogenic hazards are linked in their expression—for example, by the location of natural gas pipelines through zones of high seismic activity. Regardless of hazards being natural, anthropogenic, or a mix of natural and anthropogenic processes, risk is simply defined as the chance of hazard occurring. Most often, risk considers the likelihood of occurrence and the magnitude of the adverse effects linked to the hazard, if the event occurs. Exposure captures the frequency and length of time a system is subjected to a hazard (for example, if dispersal of invasive species occurs and eradication efforts are unsuccessful, exposure may be continuous). Measures of severity also characterize exposure, wherein expected consequences of an event are considered in terms of degree of adverse impacts such as injury, property damage, or other metrics that gage a system's impaired structure and function. Risk may be expressed in terms of the probability of hazard as a numeric value bound by 0 and 1, or as a categorical estimate wherein ordinal assignments of risks are characterized. Attendant to risks are consequences, most often characterized by frequency of occurrence [such as a numeric or categorical estimate of time) and severity (such as a numeric or categorical estimate of adverse impact; see Agresti (2002) for background on categorical analysis]. A risk decision hinges on definition of acceptable risk associated with a management action intended to offset risks linked to particular hazards, ranging from no management action to investments of time and resources to attain a level of acceptable risk (Holling 1978, Stahl et al. 2001, Walters 2001). Risk management, then, is the process of identifying and controlling hazards, where controls are intended to eliminate hazards or, more often, reduce their risk and attendant consequences. For example, risk managers may consider incursions of alien species as "chronic outcomes" of repeated dispersal events with consequences observed at increasing frequency, eventually yielding a successful species invasion.

Risk analysis focused on invasive species begins with the identification of candidate species and pathways. The process is commonly built around a review of scientific and resource management literature, expert opinion, and various qualitative and quantitative analytical tools. When empirical data are not sufficient to conduct a quantitative, probabilistic evaluation, environmental risks captured by invasive species may be variously characterized as "high," "moderate," or "low." Alternatively, risk may be quantitatively characterized as a point estimate with upper and lower bounds (e.g., probabilities characterized by a median bound by 25th percentile and 75th percentile values), in part, as a function of the data available to support the analysis of risks. Categorical risk analysis is often the primary tool for evaluating events such as species invasions, since empirical data may be insufficient to inform a quantitative, probabilistic risk analysis. The paucity of empirical data necessary to support a quantitative analysis reliant on spatial and time-series analysis, however, need not make the categorical analysis of risk any less rigorous than that informed by quantitative estimates supported by robust existing data.

Regardless of whether risks are considered quantitatively, based on widely available, high-quality empirical data, or whether risks are considered categorically because empirical data are limited, there are many different kinds of risk that must be considered when completing an environmental risk assessment, particularly when invasive species are integral to an analysis of systems at risk to various biological, chemical, and physical stressors.

Competing Risks

In field settings, resource managers primarily face problems that stem from competing risks. While the literature dealing with competing risks is ample with respect to conceptual background, for natural resource managers the issue simplifies to multiple hazards linked to chemical, physical, or biological stressors that might adversely affect a system at various levels of biological organization. The analysis of competing risks—regardless of the focus being on biological and ecological systems, or on engineering systems integrated into managed landscapes—revolves about specification of "system failure" or loss of that system's integrity. Systems typical of natural resources are generally challenged by a range of biological, chemical, and physical stressors in the field, as well as potential interactions among these stressors. Each of these stressors may potentially be characterized as a hazard capable of yielding a loss in system integrity because a physical hazard such as wildfire may interactively compete with hazards associated with invasive species to alter landscapes across various temporal and spatial scales [see, for example, D'Antonio and Vitousek (1992)]. Similarly, anthropogenic interactions entangled with chemical stressors (Deines et al. 2005) or socioeconomic issues potentially influence the likelihood of species invasions, particularly within the context of competing views of natural resource management [see, for example, Erickson (2005) and Maki and Galatowitsch (2004)].

Aggregate Risks

Aggregate risks share attributes of competing risks and may be regarded as hazard-specific competing risks. Simply defined, aggregate risks are those risks associated with single stressors that are potentially linked with receptors through many possible pathways or routes of exposure. For example, aggregate risks would be focused on a single-species invasion potentially achieved through multiple routes; for example, propagules of a species of invasive plant may be windblown or may be carried via human-mediated processes ("hitchhike") to previously unoccupied habitats in an expansion of species distribution. From a competing risk perspective, the invasion process might consider these competing pathways as independent events in a chain of events capable of achieving the end-state, an expanded species distribution. For example, a species invasion may be viewed as a series of events that initially entail the chain of events characteristic of dispersal, colonization, and establishment. Each competing chain of events, such as windblown dispersal versus human-aided dispersal of plant seeds, may be seen as competing processes, each potentially capable of achieving the end state, a successful invasion. Establishing a sustainable population in previously unoccupied habitat may represent the initial failure sufficient to cause loss of system integrity, or the initial failure might represent establishment of a beach head in previously unoccupied habitats from which expansion of species distribution is continued. Although evaluations of ecosystems at risk to species invasions may seek to reduce the system's interconnectedness and complexity to equivalent simple systems, even those models resulting from wishful application of Ockham's razor may be characterized by marked interdependence and interconnectedness and reflect operational structures that make simple network flows difficult to manage (Brandes and Erlebach 2005, Nelson 1995, Paine et al. 1998).

Cumulative Adverse Effects and Cumulative Risks

When evaluating chemical and physical stressors as part of an EnRA, cumulative risk is narrowly defined as that risk associated with exposure to all chemical and physical stressors having a common mechanism of toxicity (US EPA 2003). For example, for chemical hazards, cumulative risk assessments combine the aggregate risks from multiple exposure pathways and routes for all substances that have a common mechanism of toxicity. Within a multiple stressors context, extension of the term to include biological stressors would follow a similar thread wherein cumulative risks would be captured by cumulative adverse affects associated with a loss of system integrity yielding similar, if not identical, end-states. Cumulative ecological risks are simply the total risk linked to multiple stressors. For example, multiple species of invasive plants might be linked to cumulative adverse affects associated with loss of community structure and function. Cumulative risks, then, would be characterized as those that are manifested by end states whose loss of integrity stemmed from multiple stressor exposures. For biological stressors, the mechanisms linked to those adverse effects would display common end states across a landscape. Management actions might benefit from recognizing that similar end states may result from cumulative adverse effects, but different "modes of action" may mediate the observed loss of system integrity. For example, phytotoxic effects associated with chemical exposure may yield end states (e.g., altered vegetation communities) similar to those effects linked to exposure to biological or physical stressors, such as epidemic plant disease outbreaks or changes in soil water, respectively.

Residual Risks

Residual risk is that risk remaining after control measures have been identified and selected for managing hazards. To manage residual risks, a system's multiple stressors and their aggregate and cumulative risks must be well-characterized. Similarly, the uncertainties associated with residual risks must be understood, since management of residual risks generally responds to uncertainties and those risks that are considered acceptable and manageable. To address these residual risks, resource management often focuses on the system at risk through a "top-down" holistic approach and seeks to resolve problems by identifying proximal causes to maintain system integrity. In contrast, much of the analysis focused on risks associated with multiple stressors (including invasive species), and much of the regulatory focus guiding risk management activities related to natural resources relies on "bottom-up" approaches. A bottom-up approach is generally characterized by efforts to identify whether suspected causes are critical to resolving resource management problems. Subsequent corrective actions made to eliminate those risks based on apparent causality may not resolve the management problem and potentially result in conflicting solutions relative to the management of other stressors. For multiple risks potentially requiring conflicting solutions, a bottom-up approach may contribute to management conflicts when priorities are set among risks. Paradoxically, identifying such conflicts may help optimize management actions and increases the likely of successfully resolving system conflicts.

Landscape and Regional Risk

Landscape and regional risk assessment have increasingly been incorporated into the environmental decision-making process. While the methods and tools applied in such analyses vary depending on the scope and spatial scale of the problem, decision-making may benefit from the melding of risk disciplines focused on evaluating multiple stressors and their effects on biological and ecological structure and function. Much of the impetus for developing these larger spatial scale processes resulted from increasing applications of the environmental risk assessment process, which provides for a systematic approach to evaluating undesired effects linked to exposures to environmental stressors, including chemicals, land-use changes, altered hydrology, invasive species, genetically modified organisms, and climate change. Landscape perspective approaches to evaluating multiple stressors, including invasive species, could benefit from opting into an EcoRA framework, since the analysis would reflect explicit consideration of scale and spatial organization during problem formation (Adler 1996), account for spatial heterogeneity in exposure characterization (Agresti 2002), and consider extrapolation from small-scale studies focused on mechanisms of dispersal to broad-scale effects capturing adverse effects linked to dispersal events yielding species invasions (Andersen et al. 2004a; see, for example, Andow et al 1993, Bartell and Nair 2004, Brown 1989, Buchan and Padilla 1999, Bullock et al. 2002, Cantrel and Cosner 2003, Colbert et al. 2001, Hengeveld 1989, Pascual 2005, Shigesada and Kawasaki 1997). By opting to the multiple stressors mind-set for the analysis and characterization of ecological risks, we can optimize risk management decisions wherein competing risks linked to each stressor type are considered within the context of acceptable risk. When invasive species are fully integrated into the landscape perspective and regional risk assessment process, then selection of appropriate assessment endpoints and measures of effects could be shaped within context of spatial scale and attendant uncertainties. From a risk management perspective, opting into a nested, spatiotemporal hierarchy for evaluating risks would also provide visualization techniques (most commonly maps) for risk communication while informing environmental decision-makers within a spatiotemporal context.

Vulnerability Analysis

Vulnerability is defined as a measure of a system's likelihood to experience adverse effects consequent to exposures to environmental perturbations or stress (see, for example, D'Antonio et al. 1999, US EPA 2003). Landscape or regional vulnerability analysis is comparable to chemical stressor evaluations of individual species or their populations, and it may be illustrated by the role that landscape heterogeneity may play in identifying habitats at greater risk to invasion (see, for example, Condesco and Meentemeyer 2007). Exposures to multiple stressors, however, are characterized by many potentially interrelated factors, which may be simplified if any one type of hazard such as invasive upland plants is the primary concern in the risk evaluation. Sensitivities to environmental stressors and hazards such as invasive species may predispose some organisms as "receptors" or "target species" regarded as representatives of guilds of similarly exposed biota or ecosystems upon which those species are critically linked.

In parallel to risk analysis, vulnerability analysis focuses on the quality of a resource at risk, including the frequency of occurrence and intensity of the hazards. Regardless of the spatial scale, for many environmental stressors a fundamental limitation to analysis of risk is lack of data, be that for individual stressors or for multiple stressors jointly through time and space to characterize the system at risk. Yet, in the likely event that data are not sufficient for a fully implemented quantitative analysis, reliance on categorical tools to evaluate hazards can assure that risk management practices can be developed with a better understanding of the uncertainties associated with sparsely available empirical data. For example, for invasive species, categorical analysis may rely predominantly on the analysis of life history data available for biota suspected of becoming an invasive species problem (see Table 11.2). Successful invasions reflect the life history attributes of the invading species and the habitat vulnerabilities characteristic of the unoccupied area potentially at risk for invasion. Similarly, at-risk areas are characterized by vulnerabilities commonly shared with previously invaded areas.

Shared Attributes of Invasive Species. Invasive species successfully established in previously unoccupied landscapes exert adverse effects on challenged systems, regardless of the geographic location. For example, the occurrence of invasive species in those systems will likely stem from direct or indirect gains in competitive advantage over indigenous species subsequently displaced, yielding disruption in community structure and function in habitats previously not occupied by the invasive species. These adverse effects may range from being relatively limited, direct interactions that result in reduced populations of displaced target species or groups of closely related species to widespread effects manifest by alterations in community structure (for example, brown treesnakes, *Boiga irregularis*, were linked to adverse impacts on native avian and herpetofauna communities of Guam). When adverse effects of invasive species exist singly or in combination with other environmental stressors such as land-use practices or chemicals released to the environment, loss of native species is a recurring effect. Population declines may follow the time course of the

Table 11.2. Examples of Literature Focused on the Role of Life-History Attributes and Other Species-Specific Data Needs that Are Critical to Characterization of Invasive Species

Goodwin et al. (1998)	Kolar and Lodge (2001, 2002)
Marchetti et al. (2004)	Moyle and Light (1996)
Rejmánek and Richardson (1996)	Rejmánek (2000)
Stohlgren and Schnase (2006)	

invasion process, wherein initial conflicts between targeted native species and invasive species result in decreased populations of natives through direct competition and predation. Effects are also manifest by habitat modification, especially for invasive species that exert a dominant effect on habitat structure and function. These adverse effects are commonly indirect in their action. Mechanisms that increase the invasiveness of a species range from genetic capacities enabling hybridization with native species to those life-history attributes ensuring a species' capacity to modify previously unoccupied communities (hence, rendering habitats amenable to their continued colonization).

In general, attributes of highly successful invasive species may be considered as:

- having high fecundity and reproductive rates common to "pioneer species" that characteristically have a relatively young-age at first reproduction, and may have relatively long reproductive life, if not bearing great numbers of propagules with each generation,
- having high dispersal rates, which may be enhanced through abiotic (such as wind and water) or phoretic mechanisms (such as "hitchhiking"),
- being successful as a "single parent"—for example, parthenogenic species, species with limited parental investment, or asexual species,
- having vegetative or clonal reproduction as a common life-history attribute,
- presenting high phenotypic plasticity,
- presenting a wide physiological tolerance to environmental stressors, including life history traits to assure passage through relatively long periods (in lifetimes) of dormancy, encystment, or similar adaptations,
- presenting a large native range, characteristically linked to having relatively wide latitudinal or altitudinal range in its native setting,
- being characterized as a habitat generalist, and
- being omnivorous in food habit (for potentially invasive animals).

Perhaps the most telling attribute that characterizes a species as being an invasive threat is whether the species has a history of being invasive elsewhere. Historical performance is a key predictor of a species' future performance: If the species has proven invasive in the past, it will very likely be invasive in the future when opportunity arises. When coupled with similarities in a species' preferred climate and habitat, risk of invasion increases. Additionally, a species' native distribution may not necessarily reflect limits in its potential range of habitat conditions. In part, the limits of native distribution may reflect interactions that go beyond physical habitat constraints and tap into the interspecies and intraspecies factors (e.g., competitors, predators) that potentially restrict a species' distribution. Simply stated, release from these biological constraints on distribution in a species' native range may not be fully appreciated until a chance dispersal event provides opportunity for a species to reach potentially amenable habitats and become invasive, expanding its distribution to habitats other than those characteristic of its native area.

Species attributes that reflect mechanisms of species' dispersal may also be keys to determining whether an organism is a candidate invasive species. Life history characteristics related to reproduction (e.g., "can the organism reproduce asexually or is the organism parthenogenic?") and tolerance to physical habitat variables such as temperature and osmotic challenge (e.g., "can an aquatic organism tolerate exposure to wide ranges in ambient salinity, can a terrestrial organism tolerate wide ranges in rainfall and humidity?") may be critical to establishing sustainable populations in previously unoccupied areas. Invasive species tend to be highly malleable physiologically or display predisposing life history attributes that might be adaptive and contribute to a species becoming invasive—for example, for species dependent on vectors for dispersal, life history attributes that assure their being reliant on a wide range of vectors or, alternatively, being dependent on a few vectors but capable of relying on multiple mechanisms for dispersing to areas beyond immediate sources. Dispersal may also be variously influenced by stochasticity, including jump events within stratified diffusion processes that may result in long-distance dispersal events typical of a "Markov jump" [see, e.g., Bartlett (1995, 1960), Boyce et al. (2005), Breuer (2003), Bullock (2002), Durret (1996), Gamerman (1997), Nathoo and Dean (2007), Nelson (1995), Neubert and Parker (2004), and Paine (1998) for mathematical background and potential applications].

Attributes of Systems Vulnerable to Invasion. Invasive species accidentally or intentionally introduced to a previously unoccupied landscape generally are more likely to be successful as invaders if land masses are small and isolated, as would be the case for islands or highly fragmented landscapes. History also suggests that invasions are more likely to be successful in receiving areas characterized by high endemism, again a condition typical of islands or disturbance communities. In general, communities most vulnerable to invasion are characterized by having:

- climates similar to those of the invading species source area,
- attributes characterized as "early-succession stage" communities or disturbance communities,
- low species diversity within the native species currently occupying the target area,
- a relatively low abundance, if not absence, of predators and parasites that might limit success of invasive species, if the species was successful in reaching the previously unoccupied area,
- relatively "simple" predator–prey systems characterized by food webs having few interconnections,
- few species that would directly compete with candidate invasive species (for example, systems at risk lack ecological equivalents), and
- a previously unoccupied area that presents a history of past invasions (e.g., co-occurring biological stressors may increase likelihood of invasion by enhancing the capacity to transport invasive species as hitchhikers, or increased number of pathways linked to corridors of human transportation and migratory animals may increase the likelihood of invasion).

Community response to invasion varies and depends on species composition and the extent to which invasion succeeds. Outcomes will range from complete extirpation of members of the community to relatively minor impacts directly linked to invasion. Often, the extent to which invasive species dominate the landscape depends on the location's previous history of disturbance; highly disturbed habitats present greater vulnerability for invasion than do habitats that are relatively undisturbed. Habitat fragmentation and increased human activity linked to activities, such as construction of engineering structures in the "built environment," foster increased risks of establishment of invasive species, and reduced habitat heterogeneity may facilitate more rapid rates of dispersal and expansion of species distribution.

If empirical data were readily available for a range of candidate species, invasion success associated with these species would likely be relatively low, although such an estimate would be highly dependent on the system at risk and the life-history attributes of specific candidate invasive species. The likelihood of a successful invasion increases when species likely to be displaced occur at low density and the invading species encounters limited resistance—for example, when competitors and disease agents of the invasive species are few, if any, in the area invaded. In contrast to displaced species that generally display low viability under conditions that reduce population size, invasive species tend to be less vulnerable to endogamy. Displaced species also tend to be more likely to display adverse effects when confronted with an invasive species, if their populations display marked oscillations and their life history is characterized by limited variation and trophic specialization, or the species likely to be displaced has a relatively limited food choice. Successful invasions depend on population-level responses of both the species entering the previously unoccupied landscape and the target species most likely displaced by the invader. Hence, species' attributes that would characterize invasive species should also focus on the extirpation process.

Generalized Extirpation Process. Population viability is a problem common to both the invader and the species likely to be displaced consequent to invasion. Within the context of population biology and conservation biology, two guiding principles influence population viability analysis: (Adler and Ziglio 1996) any finite population will eventually become extirpated and (Agresti 2002) population size cannot be predicted with absolute certainty; it can only be specified as probabilities of particular outcomes. Population viability analysis concerns a relatively simple, yet challenging, question: "How large must a population be for it to have a reasonable chance of survival for a reasonably long period of time?" The term *viability* considers the persistence of the population during a specified period of time, with an emphasis on characterizing population levels that would be considered self-sustaining. Both founding populations of invasive species and remnant populations of species on the decline potentially share a common problem: If their population reaches some minimum size, it may no longer be able to sustain itself, if its population numbers go below some threshold which leads to extirpation or extinction. For a native species confronted by a challenge such as a species invasion, its pre-invasion sustainability does not necessarily imply continued population numbers and avoidance of future declines, given the multiple post-invasive threats the species encounters. Threats to population persistence are systematic, and analysis of life history should identify the life-history stages that are most critical to risk-management efforts to control species invasions [see for example, Beissinger and McCullough (2002) for detailed overview of population viability analysis].

As a general rule, population viability most often becomes a limiting factor for insular species or species occupying habitat islands in a fragmented landscape. In these settings, demographic and population genetic problems associated with reduced populations are more likely to be observed. And, if changes to their environment co-occur (e.g., habitat alteration, releases of chemical stressors and increased predation or competition from invasive species or disease agents), then multiple stressors may exasperate exposure, yielding adverse effects characterized by extirpation or extinction. Reduced population viability reflects an integrated response oftentimes initiated by a limited number of events, including species invasion. Stochastic events also threaten the persistence of small populations regardless of their status as founding populations of an invasive species or the waning numbers of a species in the process of being displaced by a species whose arrival heralds the establishment of a species in previously unoccupied territory. Various "flavors" of stochasticity may influence population viability:

- Demographic stochasticity
- Environmental stochasticity
- Natural catastrophes
- Genetic stochasticity

Demographic stochasticity must not be undervalued in resource management plans, especially in those early stages of species invasion or in those latter stages of species extirpation or extinction when numbers dramatically affect outcomes of model projections or, more importantly, outcomes in the field. Increasing environmental variability unavoidably affects population dynamics, especially when outcomes from deterministic analyses may underestimate forecasts of future populations. Incorporating stochastic variation in structured population models will better frame management strategies, since demographic stochasticity will influence population growth rate, persistence, and resilience and, ultimately, community composition, species interactions, distributions and harvesting [see, e.g., Beissinger and McCullough (2002), Boyce et al. (2005), Drake et al. (2006), Kuparinen et al. (2006), Levins (1969), Ogut and Bishop (2007)].

Environmental stochasticity becomes less problematic when population numbers are high and widely distributed, in which case numbers and their spatial pattern buffer a system and minimize adverse effects linked to natural catastrophes or large-scale anthropogenic events that might play out on regional scales, including unplanned releases of toxic chemicals, regional landscape changes linked to sociopolitical events, or locally catastrophic events (e.g., volcanic activity, wildfires). However, exposure of any system through time and space may mean that the magnitude of any hazard may be sufficient and no single population can be large enough to buffer against adverse effects linked to natural catastrophes that represent low-probability–high-consequence events (e.g., asteroid or comet impacts). When population size sufficient to buffer systems from commonly observed environmental events that are less than catastrophic in scale tends to be large, genetic stochasticity is not generally encountered unless small populations become isolated and are susceptible to endogamy and genetic effects linked to stochastic processes often associated with isolated, small populations (e.g., founder effects, Sewall Wright effects). Attributes of a species' life history are critical to the evaluation of a population's viability, with the most critical stages of an organism's life cycle likely yielding the greatest impact on population dynamics. For example, attributes of an organism's life history that limit population size, population growth rate, or species distribution are generally most critical in projecting whether a founding population will become established and invasion ensured. Ultimately, reduced population viability is likely the manifestation of multiple interacting factors yielding population declines bounded by extinction or local extirpation [see, e.g., Beissinger and McCullough (2002), Mooney and Drake (1986), Sakai et al. (2001)].

Categorical analysis has commonly been applied as a tool in the evaluation of risks associated with invasive species, particularly in identifying candidate invasive species on the basis of life-history attributes. Categorical analysis has a long history of application to a wide range of issues encountered in risk analysis, including data typical of problems focused on invasive species.

Categorical Analysis

The analysis of risks linked to invasive species has historically fallen short of relying on quantitative approaches regardless of data available for the analysis. All too frequently, data are deemed insufficient to warrant more than a narrative analysis of risks with poorly characterized uncertainties. Frequently, a sole reliance on "best professional judgment" yields clouded technical analysis for informing risk management decisions and subsequently crafting risk management plans [see, e.g., Burgman (2005), Jensen and Bourgeron (2001), Walters (2001), Wittenberg and Cock (2001)].

Categorical analysis relies on objects being grouped into categories based on some qualitative trait (Agresti 2002). Depending on the number of categories potentially applied to a variable, categorical data are classified as being nominal, ordinal, or binary (dichotomous). Nominal data are not ordered categories (such as flower colors), while ordinal data are categorical data in which order is important; for example, developmental stages of some invertebrates are an ordered set referred to as eggs, larvae, juveniles, and adults or pathological states such as morbidity may be scored as none, mild, moderate, and severe. Binary or dichotomous data are categorical data that occur as one of two possible states; that is, there are only two independent categories; for example, species occurrence data are "present" or "absent" (or more appropriately, "found" or "not found"). Binary data can either be nominal or ordinal.

Given the relatively sparse empirical data available for evaluating risks associated with many candidate invasive species, categorical tools frequently are applied to existing information and field data, where risks may be ordinal data; for example, ranked as "very low," "low," "moderate," and "high," and "very high." Historically, categorical methods have been a commonly applied tool in the evaluation of invasion risks. Similarly, an increasing reliance on Delphi methods has enhanced the risk analysis process for invasive species despite acknowledged short comings of the method. Delphi methods are highly dependent on the selection and implementation of the Delphi group [see, for example, Adler (1996) and Surowiecki (2004)], which in part may be offset by bringing multiple tools to the analysis. Multiple tools should be incorporated into the analysis and characterization of risks. For example, a categorical analysis dependent on the collective knowledge of a wide range of experienced field biologists could be completed and would be amenable to implementations of a Delphi method. These used in conjunction with a quantitative, spatial tool would forecast potential species distributions. Together, these tools would tend to minimize the shortcomings of either tool as a stand-alone method. As such, risks and uncertainties characterized consequent to these analyses should be more adequately developed and should better inform risk management decisions. Equally important, derivatives of categorical analysis are amenable to integration (especially for multiple stressor applications) and visualization using graphical tools.

For example, radar graphs, sometimes referred to as star or spider graphs, may be used as a graphic tool for comparison between and among datasets, especially those characterizing risks associated with multiple stressors—chemical, physical, and biological—or multiple representatives within a given stressor type (for example, multiple biological stressors). Depending on the identified hazards, biological, physical, and chemical stressors linked to those hazards may be graphically summarized using a simple multiple axis graph depicting preliminary outcomes of exploratory data analysis wherein quartile plots for numeric or categorical data may be evaluated. From a synthesis and integration perspective, risks may be simply characterized as simple polygons (the pentagon in Fig. 11.1) with each vertex a quartile score for a given stressor or hazard. Each stressor's magnitude would simply reflect its individual score by quartile ranking, and the magnitude of all stressors would be captured in the polygon's area; the greater the area, the greater the hazard. Applying this graphic analysis to risk would simply mean incorporating a measure of probability into each axis of the multiple-axis plot.

Categorical assignments of risk may also be developed as derivatives of simple probability calculations generated through an elementary stochastic simulation (see, for example, USGS 2005). For example, field observations of exposure and effects data may yield rough estimates of Bernoulli trials where observed dispersal events may be considered relative to number of possible dispersal events. When empirical data are limited, general distribution properties [for example, assumptions of normality as limiting distribution for Poisson processes; see Bedford and Cooke (2001), Olofsson (2007) for background] may be applied to initial characterizations of the multiple-step invasion process. Each of these data-dependent assumptions, however, influences uncertainties associated with subsequent interpretations of simple probability estimates of risk derived from the analysis.

From a synthesis and analysis perspective, characterization of categorical risks and graphic presentation of outcomes of that analysis may be depicted in a variety of ways.

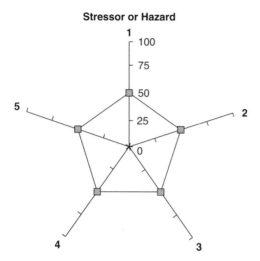

Figure 11.1. Illustration of a radar graph where five stressors or environmental hazards are plotted as a function of their quartile assignments following a multiple-participant-scoring activity used in the risk analysis process.

Frequently, graphic presentations will incorporate temporal and spatial components of risks that are all too often lost in simple deterministic implementations of the risk analysis process—for example, illustrations of a snapshot of risks associated with biological, chemical, and physical hazards at two locations in time (Fig. 11.2). Here, the three-axis spider graphic views categorical risks for each hazard as percentile outcomes characterized from exploratory data analysis completed on ordinal data, where rank scores for each stressor plotted for Location A at t_0 and Location B at t_0 indicate that biological hazards such as presence of invasive species and chemical hazards such as control agents targeted on invasive species are greater at Location A than at Location B. In contrast, physical hazards reflecting habitat manipulation as a control measure in response to invasive species appear to be greater at Location B than at Location A—again, as captured in a snapshot of risks at t_0.

For the categorical evaluation of risks, alternative methods of scoring are amenable to the risk assessment process, particularly one with a wide range of stakeholder perspectives. While alternative methods are numerous, especially in the sample survey literature [see, for example, Groves et al. (2004)], one may illustrate alternatives that account for reducing epistemic uncertainty reflected by having numerous stakeholders participate in an "expert panel" scoring process wherein Delphi methods are employed. Applying Delphi methods, risk analysts implement a systematic, interactive forecasting tool reliant on independent input based on expert opinion, experience, and intuition, thus allowing use of available, but limited, information.

The track record of Delphi methods is mixed, but its strengths generally offset weaknesses inherent to the tool (Adler and Ziglio 1996). As a predictor of future events, outputs from the Delphi method may be incorrect, although poor performance may reflect poor application of the method rather than a weakness of the method

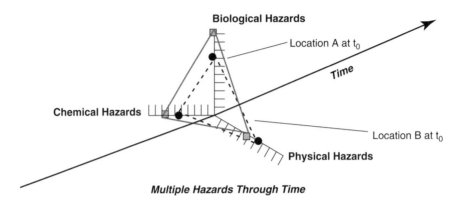

Figure 11.2. Adverse effects linked to chemical, biological, and physical stressors may change through time as illustrated by triaxial plots viewed as snapshots of a system at risk as a function of time.

itself. Also, application of the Delphi method in data-limited problem areas may yield forecasts associated with high degrees of uncertainty. A high degree of error may be expected even with assembly of the "best" of expert panels (Adler and Ziglio 1996, Surowiecki 2004). Another weakness of the Delphi method is that future developments are not always predicted correctly by developing an iterative consensus of experts, and "unconventional thinking" of "nonexpert outsiders" may be as likely to yield a good forecast of future events. Delphi method has been a widely accepted forecasting tool and has been used successfully for forecasting technical outcomes when data and information are sparse.

Categorical analysis is a valuable tool for evaluating risks associated with multiple hazards, including biological hazards such as invasive species. Yet, the tool alone provides incomplete answers to questions related to risks, an observation quickly realized when the spatiotemporal character of the invasion process is considered. Spatial and time-series analysis tools are available for the evaluation of risks associated with invasive species when data are sufficient and problems are well-characterized in problem formulation.

Ecological Forecasting and Spatiotemporal Analysis of Invasion Risks

Ecological forecasts consider the effects and potential impacts of biological, chemical, and physical stressors at various levels of biological organization, most commonly examined on watershed and regional spatial scales. By focusing on this spatial scale, not only will forecasts feature resources at a level of management action, but also resolution on smaller-scale endpoints may follow. Adverse effects on populations and communities may be readily linked to stressors whose signatures are more easily characterized at larger spatial scales. Ecological forecasts potentially inform resource managers regarding an area's current status with respect to presence of invasive species and system vulnerabilities; hence, resource management plans may better reflect risks and uncertainties captured in a snapshot of the resource at a particular time.

Refining the conceptual framework advanced in US EPA (1998, 2000, 2003) or Suter (2007), ecological forecasts potentially reflect a spatial time series that captures multiple hazards as those vary in time and space relative to land, water, and resource use. Depending on the system, stressor exposure and response components influence the signature of risks linked to particular environmental hazards (Fig. 11.3). Exposure and response may occur as short-term or long-term events, and for multiple hazards these events more than likely are asynchronous, yet interacting. Spatial scales for exposure and response vary over ranges from local to regional to global. As various authors suggest (see, for example, Clark et al. 2001a, 2001CENR 2001, Henebry and Merchant 2002, Henebry and Goodin 2002, NOAA 2001, Schnase et al. 2002), ecological forecasts reflect responses to various, and commonly interacting, agents of change linked to use of land, water, and natural resources (Adler 1996), extreme natural events, (Agresti 2002), climate change (Andersen et al. 2004a), and magnitude of biological, chemical, and physical stressors as those occur and interact through time and space (Andersen et al. 2004b). This oversimplified conceptual model suggests that one of the greatest forecast challenges will be predicting cumulative

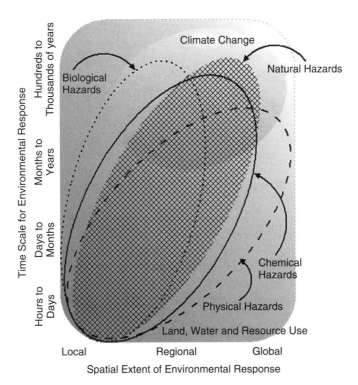

Figure 11.3. Conceptual model of environmental hazards and their relationships in time and space. [Modified from CENR (2001).]

impacts of multiple stressors and the effects linked to their interactions. While inter-active effects complicate efforts to minimize risks and to anticipate problems arising from uncertainty, it is not uncommon for multiple causes being linked to ecological change. For example, extreme natural events such as wildfire may enable species dis-persal and initial establishment, but the success of that invasion may be predicated by altered precipitation and temperature patterns associated with climate change. Sim-ilarly, implementation of conservation tillage in agricultural landscapes may reduce surface water loss and erosion, but may be unavoidably linked to changes in agrichem-ical use and changes in groundwater quality. Indeed, forecasting system signatures of multiple stressor exposure will be a challenge shared by natural resource managers and risk analysts.

Anthropogenic effects on natural systems have been casually and anecdotally evaluated throughout history, but systematic monitoring, often outpaced by develop-ment, has only occurred over the past few decades. Indeed the absence of empirical data sufficient to characterize baseline and to differentiate effects from natural hazards and those linked with human activities confounds interpretation of current observa-tions relative to prior conditions. The analysis of risks associated with invasive species reflects this same history when human-facilitated mechanisms of species invasions are discussed.

Spatiotemporal analysis of invasion risks, particularly as described by forecasts of potential species distribution, relies on a variety of tools. There are numerous tools that project spatial point data—for example, simple dot maps—to projections of species distribution as continuous patterns across a landscape. Alternatively, projections of spatial point data may assume disjunct collections of continuous distributions that result from point data that occur as clusters on a simple dot map. Though simple dot maps are examples of one of our oldest tools for evaluating species distributions, contemporary tools such as geographic information system (GIS) applications reflect the development of computer-based analysis of spatial processes and their applications to biological and ecological questions. Additionally, there are several mathematical, statistical, and computational algorithms for forecasting species distributions, including Genetic Algorithm for Rule-set Prediction (GARP), general additive models (GAM), generalized linear models (GLM), and BIOCLIM [see Scott et al. (2002) of overview]. For example, as one tool available to the risk analyst, GARP is finding increased application across a range emerging diseases and invasive species issues [see Haupt and Haupt (2002), Stockwell and Peters (1999), Stockwell and Noble (1992) for general background on genetic algorithms and Kluza and Jendek (2005), Peterson et al. (2002), Peterson (2006), and Peterson et al. (2004), USGS (2005) for examples of GARP's application; see also Table 11.1]. There are also new knowledge areas such as cellular automata, fuzzy logic, neural nets, and cognitive agents that could be used to generate such forecasting algorithms, but given their relatively underexploited use in biological and ecological forecasting, these tools may be delegated to roles in comparative analysis until sufficiently developed and validated for wider application by natural resource managers.

Regardless of the analytical tools applied to any analysis of risks, no risk assessment is complete unless uncertainties linked to these risks are sufficiently characterized. Most often, screening-level and reconnaissance-level evaluations of risks incorporate an uncertainty analysis into the synthesis and integration of the technical findings, since policy and management decisions potentially critical to the resource at risk strive to reduce uncertainty in the face of policy-dependent definitions of acceptable risks.

Uncertainty

As Brewer and Gross (2003) observe, the ability to forecast consequences linked to system-wide change faces a number of challenges, prominent among those being an accounting for uncertainly in models of ecological and physical processes [see, for example, Helton (1994), Hoffman and Hammonds (1994), and Puccia and Levins (1985) for complex systems overview]. These challenges confront risk analysts and natural resource managers who work within a network of multiple stressors and multiple hazards where invasive species have become central characters. Two general types of uncertainty: Aleatory uncertainty (also referred to as random uncertainty or stochastic uncertainty) and epistemic uncertainty affect the characterization of risks, especially within the context of their roles in influencing risk management. Aleatory uncertainty deals with the randomness (or predictability) of an event, while epistemic uncertainty reflects our "state of knowledge." Hence, epistemic uncertainty is also referred to as subjective uncertainty or parameter uncertainty. Aleatory uncertainty would be illustrated by a forecast of species invasion (or system failure), but we cannot predict exactly when that failure will occur, even if large quantities of empirical data were available. In contrast, epistemic uncertainty includes parameter-specific uncertainty and model-specific uncertainty, which reflects our inability to fully characterize a model of a system that represents higher levels of development than those detailed by basic events in a process. In any process, these basic events are characterized by lower-level outcomes, each potentially characterized by processes subject to failure or loss of integrity.

The concept of uncertainty captures complexity that is often inadequately appreciated. Currently, when faced with analysis of complex adaptive systems such as ecosystems, the evaluation of model, parameter, and aleatory uncertainty is often, out of necessity, based on expert opinion; hence, a Delphi method may be applied (see Adler and Ziglio 1996, Agresti 2002). Some types of uncertainty are more easily quantified than others, although a complete quantitative treatment of all types of uncertainty is oftentimes not achievable. Uncertainty arising through error, bias, and imprecise measurement, along with uncertainty arising through inherent variation in natural parameters, can be addressed through sampling in the field or in data-mining efforts, wherein data quality and quantity are specified to ensure these sources of uncertainty are characterized. Uncertainty that arises through lack of knowledge or scientific ignorance reflects uncertainty related to state of knowledge which may indicate research needs. In contrast, characterizing aleatory uncertainty is more straightforward, although no less strenuous, and depends on risk management for directing technical analysis, which most often relies on statistical and numerical simulation methods to better characterize aleatory uncertainty.

In order to move the technical analysis of risk and uncertainty associated with invasive species into the risk management arena focused on natural resources, various categorical and graphical tools have been developed. Each looks with different perspectives at criticality as a way of prioritizing risks.

Criticality Analysis and Companion Risk Management Tools

Criticality analysis is applied in the risk assessment and risk management process to rank decisions. If not in name, then in practice, criticality analysis is a key element in risk management, particularly when decisions are made in the face of uncertainty in its various guises. Some species may be considered greater invasion risks than other species because of their past record as successful in expanding their distribution, once dispersal has occurred. There are a wide variety of methods used to conduct a criticality analysis which is predicated on risk management needs such as those for managing invasive species. For example, risk managers often opt for Pareto analysis as their tool of choice in developing decisions based on outcomes of a risk analysis. On the other hand, environmental engineers of various sorts work with failure mode, effects, analysis (FMEA) or failure mode, effects, and criticality analysis (FMECA) as aids to decision-making, particularly as their activities may facilitate the invasion process—for example, in ballast water management and its role in mediating dispersal of aquatic nuisance species. Pareto analysis, FMEA, and FMECA are special cases of categorical analysis and present flexible implementations to help develop risk management plans focused natural resources potentially at risk from invasive species.

Pareto analysis may be regarded as a categorical method for addressing a range of risk-based decisions. This analysis considers alternative actions for managing risks. Uncertainty may be incorporated into the Pareto analysis, but regardless of how the analysis is developed, these alternatives are considered relative to anticipated benefits likely to be realized as outcomes of alternative actions. Depending on the kind of benefits, uncertainty may be equally considered with respect to these anticipated benefits. As the reader may anticipate, Pareto analysis is amenable to application of the Delphi method. Ideally, Pareto analysis identifies which of the alternatives are most likely to maximize benefits. Another method commonly used to distinguish between multiple hazards focuses on calculation of a risk priority number (RPN) or criticality index (CI). These tools are primarily embedded in evaluations of complex systems as a result of a categorical analysis where scores are developed for each of the hazards facing the risk manager. For example, a five-category scale (very low, low, moderate, high, and very high) may be used to rate the severity of each effect, the probability (numerical or categorical) of the effect's occurrence, and the detectability of the event once it has occurred. The number of factors captured by a CI may vary from application to application, but the tool would be directly responsive to ecological uncertainty as detailed in Brewer and Gross (2003). A categorical risk table may also be used to weigh the likelihoods and consequences of risks in order to integrate invasion risks into a risk management plan, using, for example, a 5-bin Likert scale or similar metric to evaluate risk (Fig. 11.4). For multiple stressors this simple "frequency of occurrence" by "consequence of occurrence" matrix may suffice for characterizing CI as the

			Likelihood of Occurrence		
		Very High	High	Moderate	Low	Very Low
	Very High	Highest priority	Highest priority	High priority	High priority	Moderate priority
	High	Highest priority	Highest priority	High priority	Moderate priority	Moderate priority
Consequences	Moderate	High priority	High priority	Moderate priority	Low priority	Low priority
	Low	Moderate priority	Moderate priority	Moderate priority	Low priority	Low priority
	Very Low	Moderate priority	Moderate priority	Low priority	Low priority	Lowest priority

Figure 11.4. As derivatives of a risk analysis, graphics may be constructed that reflect a risk management orientation relative to risks considered as likelihood of occurrence and consequences.

product of the two factors. Depending on input data, outcomes for CI may be numeric or ordinal and are commonly summarized in a table, often referred to as a "hazard matrix" (Fig. 11.4). Risk management criteria developed in parallel with a criticality analysis might focus on the significance of consequences as those contribute to the prioritization process summarized in the hazard matrix. Similarly, the frequency of occurrence may be variously weighted as a contributing factor in characterizing criticality for specific events linked to species invasions.

For invasive species concerns targeted by various risk management practices, the hazard assessment-critical control point (HACCP) process may be adopted by resource managers confronting multiple stressor issues in the field [see, e.g., Minnesota Sea Grant/Michigan Sea Grant (2001); see also www.haccp-nrm.org last accessed 14 July 2009]. HACCP in many respects reflects analytical procedures similar to those of Pareto, FMEA, or FEMCA applications [see, for example, Martel et al. (2006)]. When applied to risk management of multiple hazards, such methods are commonly referred to as tools of composite risk management (CRM). HACCP evaluation has become increasingly applied to natural resource management and environmental engineering problems, particularly as hazards may be managed with respect to the safety of processes or release of materials or products to the environment. HACCP evaluation is a simple linear process or a network of linear processes that represent the structure of any event; the hazard analysis (HA) depends on the data quality and data quantity available for the evaluation process, especially because that relates to critical control points (CCPs) characterized in completing HACCP. Control measures target CCPs and serve as limiting factors or control steps in a process that reduce or eliminate the hazards that initiated the HACCP evaluation initially. The main reason for implementing HACCP is to prevent problems associated with a specific process, practice, material, or product. HACCP should be an integral part of engineering or resource management practices used to develop aquatic, wetland, and terrestrial habitats for human use (for example, related to agriculture or construction activities) or to enhance habitats for fish and wildlife.

HACCP is a systematic and preventive approach that addresses biological, chemical, and physical hazards through anticipation and prevention, rather than through end-product inspection and testing or retrospective engineering solutions necessitated because of previous undertakings. The HACCP system is intended for assessing and managing risks and safety concerns associated with a wide range of materials, products, and management practices with an emphasis on a total systems approach to improve environmental quality. HACCP emphasizes control of a process as far upstream in the processing system as possible by utilizing operator control or continuous monitoring techniques, or a combination of both, at critical control points. The HACCP system uses the approach of controlling critical points in any process to reduce or eliminate risks and prevent safety problems from developing. The identification of specific hazards and measures for their control to ensure the safety of a process, material, or product through prevention and reduces the reliance on end-product inspection and testing (as for agrichemicals), remedial measures (such as those related to construction practices), or mitigation measures as part of a control program (for example, quarantine or disinfection for control of invasive species) are integral components of any HACCP system. Any HACCP system should be capable of accommodating change, such as advances in equipment design or developing alternative resource management practices, changes in processing procedures, or technological developments.

As an iterative procedure for assessing hazard and characterizing CCPs, HACCP is a proactive management tool that serves to reduce hazards potentially expressed as adverse biological or environmental effects associated with chemical releases, changes in natural resource or engineering practices and their related impacts, or intentional or unintentional releases of biological stressors such as invasive species. For different processes and situations, HA may be based on substantially different amounts and kinds of biological, chemical, physical, and toxicological data, but the identification of CCPs serving to reduce hazard is key to successful implementation of HACCP, which characteristically is an open-ended process intended to be updated as systems change, for example, as land-use practices change, habitat vulnerabilities to species invasions may change.

Regardless of any method's acronym or its application, the evaluation process is systematic and reflects a generalized approach to any problem wherein events potentially linked to unwanted outcomes are minimized. For risks associated with invasive species, the value of a resource will influence implementation of management strategies supporting prevention and control or mitigation, but HACCP has evolved into a frequently instituted resource management practice, especially for managing risks related to human-aided dispersal of invasive species. As illustrated (Fig. 11.4), risks expressed as products of occurrences and consequences focused on invasive species may be considered independent of other hazards and characterized categorically as being very likely, likely, unlikely, or highly unlikely to occur. In turn, these matrix-summarized products could be assigned a management priority. These measures of occurrence and consequences may be treated as numerical or categorical data, depending on data available to the analysis. Categorical assignments may even have numeric probabilities linked to them, depending upon application-specific contingencies [see, for example, USGS (2005)]. HACCP procedures may identify management tools to reduce risks through changes in practices—for example, increased training and continuing education in field decontamination to reduce unintended transfer of aquatic nuisance species through "bait bucket" transfers. Incorporating risks linked to biological hazards such as invasive species into a broader resource management perspective that considers chemical and physical hazards and their attendant risks, however, may require additional specification for scaling of risks to assure that the aggregate and cumulative risks to the system are considered. Simply, risks of concern to resource managers are not only those characteristic of invasive species; other biological hazards (such as cosmopolitan distributions of disease-causing agents) and chemical and physical hazards (including releases of environmental chemicals and habitat disturbance) are clearly present in the multiple stressor exposures commonly encountered in field settings that natural resource managers tend in their routine operations. And, interactions among these hazards are commonly presenting a joint effect that may not necessarily follow from an analysis focused on component stressors alone.

Risk Management Practice for Invasive Species

Risk management practices for invasive species have been variously developed. In many respects, these practices reflect the practical experience gathered over years of, for example, weed management and other "lessons learned" stemming for classical case studies for a wide range of invasive microorganisms, plants, vertebrates, and invertebrates. Regardless of the biological stressor of immediate concern, the risk management practice for invasive species is characterized by four interrelated activities: prevention, early detection, eradication, and control and mitigation.

Preventing species invasions is the "best" option and the most cost-effective tool available to the risk manager, although the tools used to prevent species invasions will be highly species- and pathway-dependent. Even the best of tools may prove inefficient if multiple pathways are available, and each will bring risks to the decision-making process. Preventive measures may not be equally applied across all pathways, but preventive measures may target multiple invasive species through, for example, common disinfection practices sufficient to prevent dispersal of various aquatic nuisance species. Although risk management concerns may be focused on a single species, past histories of successful invasions such as the New Zealand mudsnail (*Potamopyrgus antipodarum*) and zebra mussel (*Dreissena polymorpha*) have contributed to developing prevention plans that minimize chances of initial incursions of invasive species into previously unoccupied areas. Generic prevention strategies are primarily focused on regulatory tools that enable interception of propagules (for example, through inspections) or prohibit imports of species considered highly likely to become invasion risks, if unintentionally or intentionally released to previously unoccupied areas (for example, "white lists" and "black lists" of regulatory agencies that clearly identify which introduced species are allowed or not allowed entry to a country or state, respectively). Depending on the risk management practices implemented in prevention strategies, quarantine and treatment options are also available for initially prohibiting access of potential vectors to an area at risk or for decontaminating vectors or materials that potentially serve as carriers of nonindigenous species.

When prevention fails, early detection of initial incursions of invasive species becomes crucial to limiting the success of invasion and subsequently enhancing the effectiveness of management countermeasures such as eradication. Early detection ideally enables identification of early events characteristic of dispersal, which may be well-served through designed surveys focused on particularly problematic species or highly vulnerable areas of concern. While single-species and area-specific surveys may be designed to address specific situations, a robust survey design capable of monitoring multiple candidate invasive species sharing similar life history attributes, as exemplified by some aquatic nuisance species displaying similar dispersal mechanisms, may be critical to avoiding, or at least minimizing, potential for loss system integrity consequent to species invasions.

Because detection capabilities critical to prevention and early detection programs are frequently less than optimal, risk management plans must incorporate control and eradication strategies into their programs. Depending on the invasive species of concern, eradication can be successful, although costs associated with eradication programs strongly encourage development of better prevention and early detection tools. A wide range of tools is available for eradication programs, including mechanical or physical measures, chemical treatments, biological controls, and various habitat management practices such as prescribed burning to control some invasive plants.

When eradication is not achieved, effective countermeasures that offset mechanisms that enable continued expansions of invasive species must be incorporated into risk management plans. These countermeasures are reflected in control programs developed to manage invasive species, and most often they strive to reduce the density and abundance in the species' expanded range. Countermeasures are similar, if not identical, to mechanical, physical, chemical, and biological controls that are frequently applied in eradication programs. Each of these mitigation measures, whether applied to control or eradication programs, should be considered within the context of competing risks, for example, some chemical controls may display nontarget effects that are inappropriate for some risk management programs.

Risk Assessment and Invasive Species: National and International Perspectives

Given the scope of the invasive species problems that are gaining visibility in the technical and resource management community, various local, state, and federal initiatives have been developed in the recent past. Similarly, the global character of invasive species problems is illustrated by long-standing risk assessment and risk management programs in countries such as Australia and New Zealand, whose vigilance on invasive species and bio-security are well-recognized. Although brief, this overview illustrates (a) the level of effort practiced in the United States to offset the hazards associated with a wide range of invasive species and (b) the differing perspectives on how risks and consequences of species invasions should be addressed.

In the United States the Lacey Act (16 U.S.C. §701, May 25, 1900) initially authorized the Secretary of the Interior "to adopt measures to aid in restoring game and other birds in parts of the United States where they have become scarce or extinct [sic] and to regulate the introduction of birds and animals in areas where they had not existed." Lacey Act Amendments of 1981 (P.L. 97-79, 95 Stat. 1073, 16 U.S.C. 3371-3378, approved November 16, 1981, and as amended by P.L. 100–653, 102 Stat. 3825, approved November 14, 1988, and P.L. 98–327, 98 Stat. 271, approved June 25, 1984) repealed the Black Bass Act and sections sec43 and 44 of the Lacey Act of 1900 (18 U.S.C. 43–44) and replaced them with a single comprehensive statute. The Lacey Act Amendments prohibit the import, export, transport, buying, or selling of fish, wildlife, and plants taken or possessed in violation of federal, state or tribal law. Interstate or foreign commerce in fish and wildlife taken or possessed in violation of foreign law also is illegal. The Act requires that packages containing fish or wildlife be plainly marked. Enforcement measures include civil and criminal penalties, cancellation of hunting and fishing licenses, and forfeiture. Additionally, the Act establishes marking requirements, making it illegal to import, export, or transport in interstate commerce a container or package containing fish or wildlife unless the container or package is plainly marked, labeled, or tagged in accordance with regulations issued under the Act.

The US Department of Agriculture (USDA) involvement with invasive species issues has a similarly long history (see http://www.aphis.usda.gov/about_aphis/history. shtml last accessed July 14, 2009). Although initially charged with ministering to animal health and disease issues, the present-day Animal and Plant Inspection Service (APHIS) within the USDA provides technical support and contributes regulatory guidance for invasive species issues. APHIS was established in 1972, but it traces its history within USDA back to the formation of its regulatory Veterinary Division (1883), which was subsequently renamed the Bureau of Animal Industry (BAI). BAI was created to "promote livestock disease research, enforce animal import regulations, and regulate the interstate movement of animals," which leads inspection activities at US ports of entry. The Plant Quarantine Act of 1912 and formation of other bureaus within USDA led to BAI's incorporation into USDA's Agricultural Research Service (ARS), and in 1972 APHIS was formed. In 1985 Animal Damage Control (ADC) from US Fish and Wildlife Service was transferred to USDA and became part of APHIS, where it was renamed Wildlife Services. Through plant, wildlife, and biotechnology programs, APHIS has a wide scope of activities focused on invasive species, particularly through prevention programs for the introduction of foreign pests and diseases. For example, Wildlife Services has been a key participant in characterizing risks and for developing risk management plans for the brown treesnake (*Boiga irregularis*), particularly because that species has been identified to present a moderate to high risk for invasion of southeast United States, islands of Hawaii, and other vulnerable habitats in North America.

More recently, and in recognition of the increasing awareness of invasive species and their potential adverse effects and consequences, the National Invasive Species Council (NISC) was created under the auspices of Presidential Order 13112 (1999).

NISC has published their invasive species management plan and related guidance to address concerns related to invasive species [see, for example, Campbell and Kreisch (2003), NISC (2001), and Ruiz and Carlton (2003)]. Similarly, National Biological Information Infrastructure (NBII) is a collaborative program managed by the US Geological Survey that serves as a portal to data and information regarding biological resources, including databases, information, and analytical tools focused on invasive species. The NBII is composed of numerous government agencies, academic institutions, nongovernment organizations, and private industry that are developing standards, tools, and technologies that help find, integrate, and apply biological resources information. For example, the NBII has assembled regional programs and is compiling existing information about invasive species, as part of coordinated international efforts such as the Global Invasives Species Programme (GISP) under a United Nations initiative. These and similar efforts recognize the transboundary issues related to invasive species, especially the enhanced transport of invasive species and their propagules from distant sources to local environs; global transportation networks have unintentionally facilitated dispersal through emigration of potentially invasive species via direct and indirect mechanisms involving global trade and travel, importation of agricultural and horticultural goods, increased aquaculture practices, and commercial enterprises focused on the pet trade.

Risk Assessment, Risk Communication, and Risk Management

Risks can be managed on a species-specific basis and on a larger spatial scale with a primary focus on higher orders of biological organization, including communities within a landscape setting. Much of the current practice of invasive species risk management is focused on single species, so the often unobserved events of dispersal and colonization are only fully appreciated when a previously unrecorded invasive species has become established in an area previously unoccupied. Because of the problems linked to early observation of dispersal and colonization events that foreshadow full-blown species invasions, a wide range of invasive species management practices have been developed with a focus on rapid assessment of the potential spread of the newly established invader, particularly when the species is linked to unacceptable economic losses or unacceptable ecological effects such as extirpation of native species. Whether or not countermeasures to control or eradicate the invader should be considered is oftentimes directly related to risk management decisions based on the species of concern past performance as an invader of other systems. Species-specific evaluations may be completed proactively, a process commonly practiced in horticultural enterprises for exotic species proposed for introduction as landscape plants. The process characteristic of risk assessment and risk management can be a valuable tool for resource managers; however, the numbers of potentially invasive species preclude a universal practice of the species-by-species tool. Inevitably, our lack of knowledge about many species may limit quantitative analysis based on empirical data for a species-by-species approach, and alternative approaches to risk analysis such as categorical evaluation will become a tool of choice. Categorical analysis is also imperfect: There is no single correct answer, but many incorrect answers, and minimizing the number of incorrect answers is the primary goal of any risk analysis. Risk assessment, however, assures that a logical process for collecting and evaluating existing data is implemented, and it also assures that the analysis and synthesis of these data are targeted on critical resource management questions that can inform decision-makers and convey these findings and decisions to the wide range of stakeholders sharing concerns about invasive species.

Risk management deals with what to do about identified risks. Management of identified risks begins by setting the results of the risk assessment process and other analysis against available options through a decision-making process. The objective is to develop a strategy and plan of action. There are usually a number of risks and limited resources to deal with them. For established pests, several management options (ranging from doing nothing to exclusion to eradication to control measures) are available. Physical, biological, and chemical control options each have advantages and disadvantages. Various techniques such as the use of probability theory can be used to support the decision-making process.

Risk communication is the communication of risk assessment results so that they are clearly understood and rational decisions can be made. Risk assessment results must be communicated both to decision-makers and to the public, which must support decisions and the resultant actions. It is important that the process be open and honest and that public input or participation be solicited at appropriate points throughout the process. Public understanding, acceptance, and support are usually essential for effective action against a pest species. Deliberate introduction of potentially risky species should only be done with the informed consent of the public.

Economic and ecological models can be used as part of the assessment, management, and communications process to estimate the potential consequences of the establishment of invasive species (Bruins and Heberling 2005, Sharov 2004, Stahl et al. 2001, USGS 2005). Economic analytical tools can be used to analyze potential consequences of risks associated with species invasions, including (a) methods that estimate the net economic values associated with these consequences and (b) regional economic impact analysis. Methods have been developed to estimate net economic values. However, these methods generally rely on public surveys, which require significant investments in time and budget resources to design and implement. Such methods may also involve highly technical economic behavioral modeling and statistical estimation techniques that may not be amenable to risk communication needs of resource managers working with a wide range of stakeholders. Economic impact analysis can also involve public surveys and sophisticated modeling efforts that do not require as much time and funding to implement.

Of those tools available to the task, habitat equivalency analysis (HEA) and bioeconomic analysis are potentially applicable to consequence analysis. HEA does not estimate economic values, but considers these values in quantifying the consequences of risk management actions. HEA determines the size of ecological restoration projects that provide replacement services with an economic value at least as great as the economic value of the lost ecological services associated with the particular risk under consideration. That is, the size of the restoration project is determined to offset the economic value of lost ecological services. Therefore, the impacts are quantified as the size or cost of the required restoration project. The analytic inputs and results of HEA are directly associated with the potentially affected resources and their services. The results of HEA are easily understood by a broad range of interested parties. As a risk management tool, HEA is a relatively transparent economic approach and describes consequences in terms of the amount of restoration that would be needed to address potential impacts. The analytic inputs and results of HEA are directly associated with the potentially affected ecological resources and their services. Because of that, the results of HEA are easily understood by a broad range of interested parties. From a risk manager's perspective, HEA is applicable to an evaluation of consequences, because the tool is readily available in terms of the time and budget resources required for implementation. Unlike methods relying on public surveys, HEA can be conducted relatively quickly and at a modest cost. This feature has allowed the estimation of potential consequences over the broad geographical range and provides a consistent method to estimate and compare the potential consequences of different components.

Bioeconomic analysis combines biological and economic models to evaluate economic consequences associated with a wide range of resource management issues—for example, related to fisheries and invasive species (Helton 1994, Jiao et al. 2005, Westra et al. 2005). In general, bioeconomic analysis focuses on endogenous risk and explicitly recognizes that risk management decisions and adaptive responses of private entities ultimately affect the likelihood of management outcomes and resulting consequences related to a successful species invasion. As a tool applicable to the risk manager, bioeconomic analysis evaluates the risks posed by invasive species and quantifies the relative merits of different management strategies (for example, allocation of resources between prevention and control). The model identifies the optimal allocation of resources to prevention versus control, acceptable invasion risks, and consequences of invasion to optimal investments of capital and labor. As an optimization approach to invasive species management, bioeconomic analysis would be appropriate for analyzing invasive species management issues, yet the cost and time required to employ that approach may be daunting and data-intensive. Jointly, HEA and bioeconomic analysis illustrate the range of tools available to risk managers as they consider their particular invasive species problems. If data are sufficient and the number of invasive species are limited to a single species of concern—or, perhaps, limited to a set of invasive species that share a common habitat—bioeconomic analysis may be the tool of choice. In contrast, HEA may be more amenable to resource management issues that involve multiple species of concern whose life histories range vary widely in both data quantity and quality. As options available to the risk manager, both tools offer strengths tempered by limitations that ensure that uncertainties captured by either may be diagnosed, if not addressed as data gaps in any iteration of the analysis of consequences.

As with any assessment tool, economic analysis usually requires making assumptions, yet when incorporated with the analysis of risks, decision-makers and the public are better informed regarding the monetary impacts (i.e., benefits and costs) related to invasive species management. Resource valuation influences economic analysis, and agreement on resource values must be gained, if outcomes of the economic analysis are going to be appreciated. Similarly, clearly identifying acceptable risk is critical to achieving successful outcomes in the risk assessment process. Understanding how people accept risk requires an understanding of how preferences are accepted and measured. Varying perceptions of risks are important to keep in mind when dealing with differences in defining acceptable risk. Needless to say, the estimation of risk is largely a technical undertaking and differs from the social context facing risk managers when dealing with issues related to acceptability of risk.

Invasive Species Risk Assessment: Research Needs, Summary, and Conclusions

Invasive species are but one issue contributing to regional and local challenges confronting natural resource managers (Heinz Center 2002). Yet, not unlike other resource issues, we have the technical capabilities to inform resource management decisions. Assessing risks of species invasions requires our understanding of the stepwise process of dispersal, establishment, and spread of potentially invasive species. Needless to say, we have much to learn and the evaluation of risks linked to species invasions remains a field of study "under construction." From a regulatory perspective, only a few countries have implemented formal risk assessment frameworks for evaluating candidate invasive species, and much of the effort expended by resource managers is devoted to control and mitigation of losses linked to invasive species that have already circumvented prevention and early detection countermeasures. Nonetheless, risk analysis contributes to resource management practices for invasive species, since the process focuses on species attributes that are highly correlated with successful introduction, establishment, and spread. The reiterative process of risk analysis helps shape risk management, given the former's reliance on open-ended analyses that may be updated as new information is garnered as part of the adaptive management process. Equally important, within a multiple stressor context, risks are considered as outcomes of nonlinear processes commonly encountered in systems displaying interaction among numerous risk factors, which contributes to discrimination between risks that may be ranked with priority.

The crafting of risk analysis, assessment, and management tools expressly focused on invasive species could benefit from the risk community's experience in EnRA, particularly those efforts focused on evaluating ecological risks. Although not in name, EcoRA has long been practiced. Yet, in the 20–25 years since ecological systems have been formally considered within the context of risk, the EcoRA process has moved from a solitary focus on chemical stressors to one focused on multiple stressors, including invasive species as a group of biological stressors that have increasingly become hazards that require the attention of resource managers. Application of the EcoRA process has expanded since its inception, incorporating increase spatial and temporal scales that enable risk assessments focused on landscapes and regions and the interacting networks of stressors and receptors.

Decisions regarding the management of invasive species should be informed by the risk assessment approach. At present, invasive species risk assessment reflects much of the early developmental angst of ecological risk assessment, with multiple competing "best practices" working with diffuse data sources and multiple species that should have topmost priority. Out of necessity, generic approaches based on general knowledge in the absence of specific knowledge to the contrary provide a starting point for evaluating risks associated with invasive species. This reliance on a relatively generic approach revisits the risk assessment paradigm advanced by the National Academy of Science over 25 years ago. As work with invasive species continues, use of more data-intensive, quantitative tools will help hone our risk assessment process, particularly when guided by sources of uncertainty identified in a hazard-based risk assessment process. While a tension between a hazard-based and a quantitative risk approach is unavoidable, these ends of the risk assessment spectrum capture the extremes, wherein hazard-based approaches reflect limited data, are largely qualitative in character, and are easy to communicate, and the quantitative risk assessment process requires substantial data and may be more difficult to communicate.

When considering the risk assessment process for invasive species as a component of EnRA, many research needs are apparent. For example, research on environmental and ecological conditions, their variability, and role that human impacts have on early records of species distributions need to be addressed in order to adequately characterize baseline in a highly dynamic system. Although invasive species are one of many biological stressors and hazards of concern in the environmental assessment process, the tools available to evaluate risks linked to species invasion are underdeveloped or incompletely applied to the risk assessment process. In the absence of a widely implemented strategy focused on managing invasive species, baseline characterizations and initial surveys of native and alien species (and their impacts) for evaluating mitigation practices will be wanting or only haphazardly developed. Legal and institutional frameworks must be in place to better define prevention and management plans for invasive species.

Current techniques for management of invasive species are at best inadequate, oftentimes impractical or not economical. Resource managers are facing a recurring option of peaceful coexistence for managing invasive species that adversely affect the resources under their charge. Early recommendations from the National Academy of Sciences suggest that inference guidelines be applied in the risk assessment process, since such methods are consistent in their implementation and efficient in their application. Inference guidelines, however, tend to oversimplify any analysis, and for environmental issues such as invasive species, oversimplification may be inappropriate, especially when working within an ecological context.

While a call for more quantitative risk practices focused on invasive species is a recurring comment from a wide range of visionaries, Lodge et al. (2006) most recently presented their overview of the current state and research needs related to invasive species, including recommendations to assure that risk assessments for invasive species develop a trajectory that yields improved forecasts of invasions and their associated costs. Lodge et al. (2006) warned that harmful, nonnative species are dispersing, colonizing, and establishing sustainable populations in terrestrial and aquatic habitats in the United States at an increasing rate. In the absence of an improved science-based national strategy to respond to challenges, they concluded that invasive species will surely cause increasingly adverse impacts to natural and economic resources. Key to these increasing species invasions is the continuing globalization of trade and travel.

While initial efforts to offset these risks and consequences of species invasions should not be dismissed out of hand, current efforts in the United States to manage invasive species lack coordination and is relatively ineffective. Effective means of prevention, eradication, and control of invasive species are clearly indicated by current conditions.

Lodge et al. (2006) recommend that intergovernmental (federal, state, and local agencies) alliances take no fewer than six actions to remedy issues currently apparent in managing invasive species in the United States:

- Use new information and practices to better manage commercial and other pathways to reduce the transport and release of potentially harmful species.
- Adopt new, more quantitative procedures for risk analysis and apply them to every species proposed for importation in the country.
- Use new, more cost-effective diagnostic techniques to increase active surveillance and sharing of information about invasive species, so that responses to new invasions can be more rapid and effective.
- Create new legal authority and provide emergency funding to support rapid responses to emerging invasions.
- Provide funding and incentives for cost-effective programs to slow the spread of existing invasive species, in order to protect still uninvaded ecosystems, social and industrial infrastructure, and human welfare.
- Establish a National Center for Invasive Species Management (under the existing National Invasive Species Council) to coordinate and lead improvements in federal, state, and international policies on invasive species.

These actions are consistent with recommendations advanced earlier by global partners such as GISP, which reinforces the common grounds for invasive species problems throughout the world. Furthermore, while action by governments is critical to developing and implementing invasive species programs, local efforts involving nongovernmental organizations must not be undervalued. The education and involvement of the public are vital to a thriving resource management policy, wherein many actively participate in resource monitoring programs. With a focus on invasive species management programs, an increasing awareness of downside risks associated with invasive species may contribute to increased local involvement in control and eradication efforts. Joint actions by governmental and nongovernmental organizations are instrumental to involving everyone in the environmental management process. For reducing risks linked to invasive species, resource managers benefit from invasive species sharing a common set of biological processes and pathways for introduction, colonization, and establishment of sustainable populations. By recognizing these common species' traits and educating technical and nontechnical members of the community, risks linked to invasive species may be better managed in the future. Invasive species recognize no political borders, yet these common attributes of invasive species afford an opportunity for a science-based, cost-effective approach to protecting systems at local and global scales from the risks associated with these biological stressors.

REFERENCES

Adler M, Ziglio E. 1996. *Gazing into the Oracle: The Delphi Method and Its Application to Social Policy and Public Health*. Jessica Kingsley Publishers, London, 252 pp.

Agresti A. 2002. *Categorical Analysis*, second edition, John Wiley & Sons, Hoboken, NJ, 710 pp.

Andersen MC, Adams H, Hope B, Powell M. 2004a. Risk analysis for invasive species: General framework. *Risk Analysis* **24**:893–900, published online, doi:101111/j0272-200400487x.

Andersen MC, Adams H, Hope B, Powell M. 2004b. Risk assessment for invasive species. *Risk Analysis* **24**:787–793, published online, doi:101111/j0272–43322004,00487x.

Andow DA, Kareiva PM, Levin SA, Okubo A. 1993. Spread of invading organisms: Patterns of spread. In Kim KC, McPheron BA (Eds.), *Evolution of Insect Pests: Patterns of Variation*. John Wiley & Sons, New York, pp. 219–242.

Bartell SM, Nair SK. 2004. Establishment risks for invasive species. *Risk Analysis* **24**:833–845, published online, doi:101111/j0272–4332200400482x.

Bartholomew J, Kerans BL, Hedrick RP, Macdiarmid SC, Winton JR. 2005. A risk assessment based approach for the management of whirling disease. *Rev Fish Sci* **13**:205–230.

Bartlett MS. 1955. *An Introduction to Stochastic Processes with Special Reference to Methods and Applications*. Cambridge at the University Press, Cambridge, UK, 312 pp.

Bartlett MS. 1960. *Stochastic Population Models in Ecology and Epidemiology*. Methuen & Co, London, 90 pp.

Bedford T, Cooke R. 2001. *Probabilistic Risk Analysis*. Cambridge University Press, Cambridge, UK, 393 pp.

Beissinger SR, McCullough DR (Eds.). 2002. *Population Viability Analysis*. The University of Chicago Press, Chicago, 577 pp.

Bossenbroek JM, Johnson LE, Peters B, Lodge DM. 2006. Forecasting the expansion of zebra mussels in the United States. *Conserv Biol* **21**:800–810, published online, doi:101111/j1523-1739200600614x.

Boyce MS, Haridas CV, Lee CT. 2005. The NCEAS Stochastic Demography Working Group, 2006. Demography in an increasingly variable world. *Trends Ecol Evolut* **21**:141–148 Epub Dec 27.

Brandes U, Erlebach T (Eds.). 2005. *Network Analysis*. Springer, New York, 471 pp.

Breithaupt H. 2003. Aliens on the shores. *EMBO Report* **4**:547–550, doi: 101038/sjemborembor877.

Breuer L. 2003. *From Markov Jump Processes to Spatial Queues*. Kluwer Academic Publishers, Norwell, MA, 156 pp.

Brewer CA, Gross LJ. 2003. Training ecologists to think with uncertainty in mind. *Ecology* **84**:1412–1414.

Brown JH. 1989. Patterns, Modes and extents of invasions by vertebrates. In Drake JA, Mooney HA, di Castri F, Groves RH, Kruger FJ, Rejmanek M, Williamson M (Eds.), *Biological Invasions: A Global Perspective*. John Wiley & Sons, New York, pp. 85–109.

Bruins RJF, Heberling MT (Eds.). 2005. *Economics and Ecological Risk Assessment: Applications to Watershed Management*. CRC Press, Boca Raton, FL.

Buchan LAJ, Padilla DK. 1999. Estimating the probability of long-distance overland dispersal of invading aquatic species. *Ecol Applic* **9**:254–265.

Bullock JM, Kenward RE, Hails RS (Eds.). 2002. *Dispersal Ecology*. Published by Blackwell Science, Ltd/Blackwell Publishing, for British Ecological Society, Oxford, UK, 458 pp.

Burgman M. 2005. *Risks and Decisions for Conservation and Environmental Management*. Cambridge University Press, Cambridge, UK, 488 pp.

Campbell F, Kreisch P. 2003. *Invasive Species Pathway Team: Final Report*. National Invasive Species Council, Washington, DC, 25 pp.

Cantrell RS, Cosner C. 2003. *Spatial Ecology via Reaction-Diffusion Equations*. John Wiley & Sons, New York, 411 pp.

Carlton JT. 1993. Dispersal mechanisms of the zebra mussel (*Dreissena polymorpha*). In Nalepa TF, Schloesser DW (Eds.), *Zebra Mussels: Biology, Impacts, and Control*. Lewis Publishers, Boca Raton, FL, pp. 677–697.

CENR (Committee on Environmental and Natural Resources). 2001. *Ecological Forecasting: Agenda for the Future*. Washington, DC.

Clark JS, Carpenter SR, Barber M, Collins S, Dobson A, Foley J, Lodge D, Pascual M, Pielke R, Jr, Pizer W, Pringle C, Reid WV, Rose KA, Sala O, Schlesinger WH, Wall D, Wear D. 2001a. Ecological forecasts: An emerging imperative. *Science* **293**:657–660.

Clark JS, Lewis, M, Horvath L. 2001b. Invasion by extremes: Variation in dispersal and reproduction retards population spread. *Am Natur* **157**:537–554.

Clark J, Horvath L, Lewis M. 2001c. On the estimation of spread rate for a biological population. *Stat Prob Lett* **51**:225–234.

Clark J, Lewis MA, McLachlan J, Hille Ris Lambers J. 2003. Estimating population spread: What can we forecast and how well? *Ecology* **84**:1979–1988.

Colbert J, Danchin E, Dhondt AA, Nichols JD (Eds.). 2001. *Dispersal*. Oxford University Press, Oxford, UK, 452 pp.

Colautti RI, Grigorovich IA, MacIsaac HJ. 2006. Propagule pressure: A null model for biological invasions. *Biol Invasions* **8**:1023–1037, doi:101007/s10530-005-3735-y.

Colnar AM, Landis WG. 2007. Conceptual model development for invasive species and a regional risk assessment case study: The European green crab, *Carcinus maenas*, at Cherry Point, Washington USA. *Hum Ecol Risk Assess* **13**:120–155.

Condesco TE, Meentemeyer RK. 2007. Effects of landscape heterogeneity on the emerging forest disease sudden oak death. *J Ecol* **95**:364–375, doi:101111/j1365-2745200601206x.

Crall AW, Meyerson LA, Stohlgren TJ, Jarnevich CS, Newman GJ, Graham J. 2006. Show me the numbers: What data currently exist for non-native species in the USA? *Frontiers Ecol Environ* **4**:414–418.

D'Antonio CM, Vitousek PM. 1992. Biological invasions by exotic grasses, the grass/fire cycle, and global change. *Annu Rev Ecol System* **23**:63–87.

D'Antonio CM, Dudley T, Mack M. 1999. Disturbance and biological invasions: Direct effects and feedbacks. In Walker L (Ed.), *Ecosystems of Disturbed Ground*. Elsevier, Amsterdam, The Netherlands, pp. 413–452.

Deines AM, Chen V, Landis WG. 2005. Modeling the risks of non-indigenous species introductions using a patch-dynamics approach incorporating contaminant effects as a disturbance. *Risk Analysis* **26**:1637–1651.

Drake JM. 2004. Allee effects and the risk of biological invasions. *Risk Anal* **24**:795–802, published online, doi:101111/j0272-4332200400479x.

Drake JM, Lodge DM. 2004. Allee effects, propagule pressure and the probability of establishment: Risk analysis for biological invasions. *Biol Invasions* **8**:365–375, doi:101007/s10530-004-8122-6.

Drake JA, Mooney HA, di Castri F, Groves RH, Kruger FJ, Rejmanek M, Williamson M (Eds.). 1989. *Biological Invasions: A Global Perspective*. John Wiley & Sons, New York, 525 pp. Available online at http://wwwicsu-scopeorg/downloadpubs/scope37/scope37html, last accessed February 28, 2007.

Drake JM, Drury KLS, Lodge DM, Blukacz A, Yan ND, Dwyer G. 2006. Demographic stochasticity, environmental variability, and windows of invasion risks by *Bythotrephes longimanus* in North America. *Biol Invasions* **8**:843–861.

Durrett R. 1996. *Stochastic Calculus*. CRC Press, Boca Raton, FL, 341 pp.

Elton CS. 1958. *The Ecology of Invasions by Plants and Animals*. The University of Chicago Press, Chicago, 181 pp.

Erickson JA. 2005. The economic roots of aquatic species invasion. *Fisheries* **30**:30–33.

Ferenc SA, Foran JA (Eds.). 2000. *Multiple Stressors in Ecological Risk and Impact Assessment: Approaches to Risk Estimation*. SETAC Press, Pensacola, FL, 264 pp.

Finnoff D, Shogren JF, Leung B, Lodge D. 2005. Risk and nonindigenous species management. *Rev Agric Econ* **27**:475–482, published online, doi:101111/j1467–9353200500247x.

Foran JA, Ferenc SA (Eds.). 1999. *Multiple Stressors in Ecological Risk and Impact Assessment*. SETAC Press, Pensacola, FL, 100 pp.

Gamerman D. 1997. *Markov Chain Monte Carlo*. Chapman & Hall/CRC, Boca Raton, FL, 245 pp.

Goodwin BJ, McAllister AJ, Fahrig L. 1998. Predicting invasiveness of plant species based on biological information. *Conserv Biol* **13**:422–426.

Groves RM, Fowler FJ, Jr, Couper MP, Lepkowski JM, Singer E, Tourangeau R. 2004. *Survey Methodology*. John Wiley & Sons, Hoboken, NJ, 424 pp.

Hannesson R. 1993. *Bioeconomic Analysis of Fisheries*. Blackwell Publishing Professional (Iowa State University Press), Ames, IA, 144 pp.

Haupt RL, Haupt SE. 2002. *Practical Genetic Algorithms*, second edition, Wiley-Interscience, Hoboken, NJ, 253 pp.

Heinz Center. 2002. *The State of the Nation's Ecosystems*. The H. John Heinz Center for Science, Economics, and the Environment, published by Cambridge University Press, Cambridge, UK, 270 pp.

Helton JC. 1994. Treatment of uncertainty in performance assessment for complex systems. *Risk Anal* **14**:483–511.

Henebry GM, Goodin DG. 2002. Landscape trajectory analysis: spatio-temporal dynamics from image time series. *Proc Internl Geosci Remote Sensing Symp, IEEE Internl* **4**:2375–2378 (Posted online 2002-11-07, ISBN: 0–7803–7536-X, INSPEC Accession Number: 7541524.

Henebry GM, Merchant JW. 2002. Geospatial data in time: Limits and prospects for predicting species occurrences. In Scott JM, Heglund PJ, Morrison M. (Eds.), *Predicting Species Occurrences: Issues of Scale and Accuracy*. Island Press, Covello, CA, pp. 291–309.

Hengeveld R. 1989. *The Dynamics of Biological Invasions*. Chapman & Hall, London, UK.

Hewitt CL, Huxel GR. 2002. Invasion success and community resistance in single and multiple species invasion models: Do the models support the conclusions? *Biol Invasions* **4**:262–271.

Hiddink JG, Jennings S, Kaiser J. 2007. Assessing and predicting the relative ecological impacts of disturbance on habitats with different sensitivities. *J Appl Ecol* (Online Early Articles), doi:101111/j1365-2664200701274x.

Higgins SI, Richarson DM, Cowling RM, Trinder-Smith TH. 1999. Predicting the landscape-scale distribution of alien plants and their threat to plant diversity. *Conserv Biol* **13**:303–313.

Higgins SI, Clark JS, Nathan R, Hovestadt T, Schurr F, Fragoso JMV, Aguiar MR, Ribbens E, Lavorel S. 2003. Forecasting plant migration rates: Managing uncertainty for risk assessment. *J Ecol* **91**:341–347.

Hoffman FO, Hammonds JS. 1994. Propagation of uncertainty in risk assessments: The need to distinguish between uncertainty due to lack of knowledge and uncertainty due to variability. *Risk Anal* **14**:707–712.

Holling CS (Ed.). 1978. *Adaptive Environmental Assessment and Management. International Institute for Applied Systems Analysis*. John Wiley & Sons, New York, 377 pp.

Jensen ME, Bourgeron PS (Eds.). 2001. *A Guidebook for Integrated Ecological Assessments*. Springer-Verlag, New York, 536 pp.

Jeschke JM, Strayer DL. 2005. Invasion success of vertebrates in Europe and North America. *Proc Natl Acad Sci USA*, **102**:7198–7202, published online, doi:101073/pnas0501271102.

Jiao Y, Chen Y, Wroblewski J. 2005. An application of the composite risk assessment method in assessing fisheries stock status. *Fish Res* **72**:173–183.

Johnson LE, Carlton JT. 1996. Post-establishment spread in large-scale invasions: dispersal mechanisms of the zebra mussel (*Dreissena polymorpha*). *Ecology* **77**:1686–1690.

Johnson LE, Padilla DK. 1996. Geographic spread of exotic species: Ecological lessons and opportunities from the invasion of the zebra mussel (*Dreissena polymorpha*). *Biol Conserv* **78**:22–33.

Johnson LE, Ricciardi A, Carlton JT. 2001. Overland dispersal of aquatic invasive species: A risk assessment of transient recreational boating. *Ecol Appl* **11**:1789–1799.

Kapustka L, Linder G. 2007. Invasive species: A real but largely ignored threat to environmental security. In Linkov I, Kiker G, Wenning R (Eds.), *Environmental Security in Harbors and Coastal Areas*. Springer, The Netherlands, 2007, pp. 175–188.

Keller RP, Lodge DM, Finnoff DC. 2007. Risk assessment for invasive species produces net bioeconomic benefits. *Proc Natl Acad Sci* **104**:203–207, published online, doi:101073/pnas0605787104.

Kluza DA, Jendek, E. 2005. Potential distribution of the emerald borer (Agrilus planipennis). *Proc XV US Department of Agriculture Interagency Research Forum on Gypsy Moth and Other Invasive Species 2004, GTR-NE-332*. USDA Forest Service, USDA Forest Service, Publications Distribution, Newtown Square, PA, 50 pp.

Kolar CS, Lodge DM. 2001. Progress in invasion biology: Predicting invaders. *Trends Ecol Evolut* **16**:199–204.

Kolar CS, Lodge DM. 2002. Ecological predictions and risk assessments for alien species. *Science* **298**:1233–1236.

Kot M, Lewis MA, van den Driessche P. 1996. Dispersal data and the spread of invading organisms. *Ecology* **77**:2027–2042.

Kuparinen A, Snäll T, Vänskä S, O'Hara RB. 2006. The role of model selection in describing stochastic ecological processes. *Oikos* **116**:1037–1050 (Online Early Articles., doi:101111/j20060030–129915563x).

Landis, WG. 2004. Ecological risk assessment conceptual model formulation for nonindigenous species. *Risk Anal* **24**:847–858, published online, doi:101111/j0272–4332200400483x.

Leung B, Drake JM, Lodge DM. 2004. Predicting invasions: Propagule pressure and the gravity of Allee effects. *Ecology* **85**:1651–1660.

Levin SA. 1989. Analysis of risk for invasions and control programs. In Drake JA, Mooney HA, di Castri F, Groves RH, Kruger FJ, Rejmánek M, Williamson M (Eds.), *Biological Invasions: A Global Perspective*. John Wiley & Sons, New York, 1989, pp. 425–435.

Levins R. 1969. Some demographic and genetic consequences of environmental heterogeneity for biological control, *Bull Entomol Soc Am* **15**:237–240.

Li HW. 1981. Ecological analysis of species introductions into aquatic systems. *Trans Am Fish Soc* **110**:772–782.

Light T, Marchetti MP. 2007. Distinguishing between invasions and habitat changes as drivers of diversity loss among California's freshwater fishes. *Conserv Biol* **21**:434–446(OnlineEarly Articles., doi:101111/j1523-17392006000643x.

Lodge DM. 1993. Species invasions and deletions. In Kareiva PM, Kingsolver JG, Huey RB (Eds.), *Biotic Interactions and Global Change*. Sinauer, Sunderland, MA, pp. 367–387.

Lodge DM, Williams S, MacIsaac HJ, Hayes KR, Leung B, Reichard S, Mack RN, Moyle PB, Smith M, Andow DA, Carlton JT, McMichael A. 2006. Biological invasions: Recommendations for US policy and management. *Ecol Appl* **16**:2035–2054.

MacIsaac HJ, Grigorovich IA, Riccardi A. 2001. Reassessment of species invasions concepts: The Great Lakes basin as a model. *Biol Invasions* **3**:405–416, published online, doi:101023/A:1015854606465.

Maguire LA. 2004. What can decision analysis do for invasive species? *Risk Anal* **24**:859–868, published online, doi:101111/j0272-200400484x.

Maki K, Galatowitsch S. 2004. Movement of invasive aquatic plants into Minnesota (USA) through horticultural trade. *Biol Conserv* **118**:389–396.

Marchetti MP, Moyle PB, Levine R. 2004. Alien fishes in California watersheds: Characteristics of successful and failed invaders. *Ecol Appl* **14**:587–596.

Martel K, Kirmeyer G, Hanson A, Stevens M, Mullenger J, Deere D. 2006. *Application of HACCP for Distribution System Protection*. AWWA Research Foundation, Denver, CO, 148 pp.

Marvier M, Kareiva P, Neubert MG. 2004. Habitat destruction, fragmentation, and disturbance promote invasion by habitat generalists in a multispecies metapopulation. *Risk Anal* **24**:869–878, published online, doi: 101111/j0272-200400485x.

Mattson KM, Angermeier PL. 2007. Integrating human impacts and ecological integrity into a risk-based protocol for conservation planning. *Environ Manage* **39**:125–138.

Minnesota Sea Grant/Michigan Sea Grant. 2001. *Hazard Analysis And Critical Control Point Analysis*. MN SG-F11/MSG-00–400, University of Minnesota, Duluth MN.

Minton MS, Verling E, Miller AW, Ruiz GM. 2005. Reducing propagule supply and coastal invasions via ships: Effects of emerging strategies. *Frontiers Ecol Environ* **3**:304–308.

Mooney HA, Drake JA (Eds.). 1986. *Ecology of the Biological Invasions of North America and Hawai'i*. Springer-Verlag, New York, 321 pp.

Moyle PB, Light T. 1996. Biological invasions of fresh water: Empirical rules and assembly theory. *Biol Conserv* **78**:149–161.

NAS (National Academy of Science). 1983. *Risk Assessment in the Federal Government: Managing the Process*. Committee on the Institutional Means for Assessment of Risk to Public

Health, Commission of Life Sciences, National Research Council, National Academy Press, Washington, DC, 1983, 191 pp.

NAS (National Academy of Science). 1993. *Issues in Risk Assessment*. Committee on Risk Assessment Methodology, Board on Environmental Studies and Toxicology Commission of Life Sciences, National Research Council, National Academy Press, Washington, DC, 356 pp.

NAS (National Academy of Science). 2002. *Predicting Invasions of Nonindigenous Plants and Plant Pests*. National Academy Press, Washington, DC, 194 pp.

Nathoo F, Dean CB. 2007. A mixed mover–stayer model for spatiotemporal two-state processes. *Biometrics* **41**:91–101 (Online Early Articles., doi:101111/j1541-0420–200700752x.

Nelson BL. 1995. *Stochastic Modeling, Analysis & Simulation*. McGraw-Hill, New York, 321 pp.

Neubert MG, Parker IM. 2004. Projecting rates of spread for invasive species. *Risk Anal* **24**:817–831, published online, doi:101111/j0272–332200400481x.

NISC (National Invasive Species Council). 2001. *Meeting the Invasive Species Challenge: The National Invasive Species Management Plan*, 80 pp.

NOAA (National Oceanic and Atmospheric Administration). 2001. *Ecological Forecasting: Expanding NOAA's Assessment and Prediction Capabilities to Support Proactive Ecosystem Management*. NOAA National Ocean Service.

Office of the President. 1999. Executive Order 13112, February 3, 1999 Established Invasive Species Council and specified its duties.

Ogut H, Bishop SC. 2007. A stochastic modeling approach to describing the dynamics of an experimental furunculosis epidemic in Chinook salmon, *Oncorhynchus tshawytscha* (Walbaum). *J Fish Disease* **30**:93–100, published online doi:101111/j1365-2761200700791x.

Olofsson P. 2007. *Probabilities, The Little Numbers that Rule Our Lives*. John Wiley & Sons, Hoboken, NJ, 267 pp.

OSTP (Office of Science and Technology Policy). 2001. *Ecological Forecasting: Agenda for the Future*. Committee on Environment and Natural Resources, Subcommittee on Ecological Systems, Office of Science and Technology Policy, Washington, DC, 8 pp.

Olson LJ, Roy S. 2005. On prevention and control of an uncertain biological invasion. *Rev Agric Econ* **27**:491–497, published online, doi:101111/j1467–9353200500249x.

Paine RT, Tegner MJ, Johnson EA. 1998. Compounded perturbations yield ecological surprises. *Ecosystems* **1**:535–545.

Pascual M. 2005. Computational ecology: From the complex to the simple and back. *PLoS Comput Biol* **1**: e18, published online 2005 July 29, doi:101371/journalpcbi0010018.

Peterson AT, Stockwell DRB, Kluza DA. 2002. Distributional prediction based on ecological niche modeling of primary occurrence data. In Scott JM, Heglund PJ, Morrison ML (Eds.), *Predicting Species Occurrences: Issues of Scale and Accuracy*. Island Press, Washington, DC, pp. 617–623.

Peterson AT, Bauer JT, Mills JN. 2004. Ecological and geographic distribution of filovirus disease. *Emerging Infect Dis* **10**:40–47.

Peterson A. 2006. Ecologic niche modeling and spatial patterns of disease transmission. *Emerging Infect Dis* **12**:1822–1826.

Pimentel D, Lach L, Zunia R, Morison, D. 1999. *Environmental and Economic Costs Associated with NonIndigenous Species in the United States*. Available at http://wwwnewscornelledu/releases/Jan99/species_costshtml last accessed 27 February 2007.

Puccia CJ, Levins 1996, R. 1985. *Qualitative Modeling of Complex Systems*. Harvard University Press, Cambridge, MA, 259 pp.

Rejmánek M. 2000. Invasive plants: approaches and predictions. *Aust Ecol* **25**:497–506.

Rejmánek M, Richardson DD. 1996. What attributes make some plant species more invasive? *Ecology* **77**:1655–1660.

Ricciardi A, Rasmussen JB. 1998. Predicting the identity and impact of future biological invaders: A priority for aquatic resource management. *Can J Aquat Sci* **55**:1759–1765.

Rodgers CJ (Ed.). 2001. *Risk Analysis in Aquatic Animal Health*. World Organization for Animal Health (Office International des Epizooties), Paris, France, 346 pp.

Rota G-C. 1986. *Discrete Thoughts*. In Kac M, Rota G-C, Schwartz JT (Eds.), *Discrete Thoughts, Essays on Mathematics, Science, and Philosophy*. Birkhauser, Boston, pp. 1–3.

Ruiz GM, Carlton JT (Eds.). 2003. *Invasive Species: Vectors and Management Strategies*. Island Press, Washington, DC, 518 pp.

Sakai AK, Allendorf FW, Holt JS, Lodge DM, Molofsky J, With KA, Baughman S, Cabin RJ, Cohen JE, Ellstrand NC, McCauley DE, O'Neill P, Parker IM, Thompson JN, Weller SG. 2001. The population biology of invasive species. *Annu Rev Ecol Syst* **32**:305–332.

Sax DF, Gaines SD, Brown JH. 2002. Species invasions exceed extinctions on islands worldwide: A comparative study of plants and birds. *Am Natur* **160**:766–783.

Schnase JL, Smith JA, Stohlgren TJ, Graves S, Trees C. 2002. Biological invasions: A challenge in ecological forecasting. *Proc Int Geosci Remote Sensing Symp IEEE Int* **1**:122–124.

Scott JM, Heglund PJ, Morrison ML, Haufler JB, Raphael MG, Wall WA, Samson FB (Eds.). 2002. *Predicting Species Occurrences, Issues of Accuracy and Scale*. Island Press, Washington, DC, 868 pp.

Sharov, AA. 2004. Bioeconomics of managing the spread of exotic pest species with barrier zones. *Risk Anal* **24**:879–892, published online, doi:101111/j0272-200400486x.

Shigesada N, Kawasaki K. 1997. *Biological Invasions: Theory and Practice*. Oxford University Press, Oxford UK, 1997, 205 pp.

Simberloff D. 1985. Predicting ecological effects of novel entities: Evidence from higher organisms. In Halverson HO, Pramer D, Rogul M (Eds.), *Engineered Organisms in the Environment: Scientific Issues*. American Society for Microbiology, Washington, DC, pp. 152–161.

Simberloff D. 1991. Keystone species and community effects of biological invasions. In Ginzburg LR (Ed). *Assessing Ecological Risks of Biotechnology*. Butterworth-Heinemann, Boston, pp. 1–19.

Simberloff D. 1996. *Impacts of Introduced Species in the United States, Consequences: The Nature and Implications of Environmental Change 2.2*. Updated November 11, 2004 Available online at http://wwwgcrioorg/CONSEQUENCES/vol2no2/article2html, last accessed February 28, 2007.

Stahl RG, Bachman RA, Barton AL, Clark JR, deFur PL, Ells SJ, Pittinger CA, Slimak MW, Wentsel RS (Eds.). 2001. *Risk Management: Ecological Risk-Based Decision-Making*. SETAC Press, Pensacola, FL, 191 pp.

Stepien CA, Brown JE, Neilson ME, Tumeo MA. 2005. Genetic diversity of invasive species in the Great Lakes versus their Eurasian source populations: Insights for risk analysis. *Risk Anal* **25**:1043–1060.

Stockwell DRB, Noble IR. 1992. Induction of sets of rules from animal distribution data: A robust and informative method of analysis. *Math Comput Simul* **33**:385–390.

Stockwell DRB, Peters DP. 1999. The GARP modelling system: Problems and solutions to automated spatial prediction. *Int J Geogr Inform Syst* **13**:143–158.

Stohlgren TJ, Schnase JL. 2006. Risk analysis for biological hazards: What we need to know about invasive species. *Risk Anal* **26**:163–173, published online, doi:101111/j1539-6924200600707x.

Surowiecki J. 2004. *The Wisdom of Crowds*. Doubleday, New York, 296 pp.

Suter GW II. 2007. *Ecological Risk Assessment*, second edition. CRC Press, Boca Raton, FL, 643 pp.

Swanson SM. 2004a. Multiple stressors: Literature review and gap analysis. *Water Intelligence Online*, IWA Publishing #200408WF00ECO2B, 142 pp.

Swanson, SM. 2004b. Multiple stressors: Risk-based framework and experimental design for cause–effect relationships. *Water Intelligence Online*, IWA Publishing #200408WF00E/ci2A, 145 pp.

USDA (US Department of Agriculture, National Agricultural Library). 2007. *Economic Impacts*. online at http://wwwinvasivespeciesinfogov/economic/usshtml, last accessed February 28, 2007.

US EPA (US Environmental Protection Agency). 1998. *Guidelines for Ecological Risk Assessment (Final)*. Washington DC: USEPA EPA/630/R-95/002F.

US EPA (US Environmental Protection Agency). 2000. *Stressor Identification Guidance Document*. US Environmental Protection Agency, Office of Water, Office of Research and Development, Washington, DC, EPA-822-B-00-025.

US EPA (US Environmental Protection Agency). 2003. *Framework for Cumulative Risk Assessment*. Risk Assessment Forum, US Environmental Protection Agency, Washington, DC, EPA/630/P-02/001F.

USGS (US Geological Survey). 2005. *Risk and Consequence Analysis Focused on Biota Transfers Potentially Associated with Surface Water Diversions Between the Missouri River and Red River Basins*. Written, edited, and compiled by Linder G, Little E, Johnson L, Vishy C. USGS, Columbia Environmental Research Center [CERC], Columbia, Missouri. and Peacock B, Goeddecke H. National Park Service [NPS], Environmental Quality Division, Fort Collins, CO, Volumes 1 and 2, Pagination by section and appendix.

Vermeij GJ. 2005. Invasion as expectation: A historical fact of life. In Sax DF, Stachowicz JJ, Gaines SD (Eds.), *Species Invasions: Insights into Ecology, Evolution, and Biogeography*. Sinauer Associates, Sunderland, pp. 315–339.

Vitousek PM, D'Antonio CM, Loope LL, Westbrooks R. 1996. Biological invasions as global change. *Am Sci* **84**:468–478.

Von Holle B, Simberloff D. 2005. Ecological resistance to biological invasion overwhelmed by propagule pressure. *Ecology* **86**:3212–3218.

Walters C. 2001 *Adaptive Management of Renewable Resources* [originally published 1986]. Republished by The Blackburn Press, Caldwell, NJ, 374 pp.

Wein J. 2002. Predicting species occurrences: Progress, problems, and prospects. In Scott, MJ et al. (Eds.). *Predicting Species Occurrences, Issues of Accuracy and Scale*. Island Press, Washington, DC, pp. 739–749.

Westbrooks R. 1998. *Invasive Plants, Changing the Landscape of America: Fact Book*. Federal Interagency Committee for the Management of Noxious and Exotic Weeds (FICMNEW), Washington, DC, 109 pp.

Westra JV, Zimmerman JKH, Vondracek B. 2005. Bioeconomic analysis of selected conservation practices in soil erosion and freshwater fisheries. *J Am Water Resources Assoc* **41**:309–322.

Williamson M. 1989. Mathematical models of invasion. In Drake JA, Mooney HA, di Castri F, Groves RH, Kruger FJ, Rejmánek M, Williamson M (Eds.), *Biological Invasions: A Global Perspective*. John Wiley & Sons, New York, pp. 329–350.

Williamson M. 1996. Biological Invasions. *Population and Community Biology Series 15*. Chapman & Hall London, 244 pp.

Williamson M, Fitter A. 1996. The characteristics of successful invaders. *Biol Conserv* **78**:163–170.

With KA. 2004. Assessing risks of invasive spread in fragmented landscapes. *Risk Anal* **24**:803–815, published online, doi:101111/j0272–433220040048.

Wittenberg R, Cock MJW (Eds.). 2001. *Invasive Alien Species: A Toolkit of Best Prevention and Management Practices*. Published on behalf of Global Invasive Species Programme, CAB International, Wallingford, Oxon, UK, 228 pp.

12

LANDSCAPE NONINDIGENOUS SPECIES RISK ASSESSMENT: ASIAN OYSTER AND NUN MOTH CASE STUDIES

Wayne G. Landis, Valerie C. Chen, Audrey M. Colnar, Laurel Kaminski, Goro Kushima, and Ananda Seebach

The idea that invasive or nonindigenous species produce a societal cost has become a cultural cliché. Despite the high level of societal concern, the risk assessment process has not been widely used to quantify the risks of invasive species.

The study of invasive species has been approached largely on a case-by-case basis, and field studies often have been too idiosyncratic to be used to derive general hypothesis of invasive species establishment (Vermeij 1996). Laboratory and garden studies are hampered by problems of scale, replication, and control (Wardle 2001, Doak et al. 1998). Hypothesized mechanisms for invasive species establishment and spread abound and most may be separated into two general categories: (1) attributes of the nonindigenous species and (2) attributes of the community into which the invasive species has arrived. Studies considering the former are typically searches for lists of common traits among the various species of invasive species, and exceptions to these lists are common (Mack et al. 2000).

One of the strongest predictors of plant introductions is if the species has established in another location (Kolar and Lodge 2001). In fish invasions of the Great Lakes, the factors that determined establishment were relatively faster growth, toleration of a wider range of temperature and salinity, and a past history of invasiveness (Kolar and Lodge 2002). Quickly spreading fish have the features of slower relative

Environmental Risk and Management from a Landscape Perspective, edited by Kapustka and Landis
Copyright © 2010 John Wiley & Sons, Inc.

growth rate, tolerance of a wide temperature range, and poor survival at higher temperature ranges. Examining past patterns of invasion or home range characteristics can also prove predictive for a variety of species (Kolar 2004). The second category of hypotheses includes vacant niche, enemy escape, disturbance, and species richness or diversity (Mack et al. 2000, Shea and Chesson 2002). However, we may not yet know enough about ecosystem functioning to relate these types of observations to the larger questions of prediction, policy, and management (NSTC 1999).

General patterns of invasion by nonindigenous species have to be inferred from other research dealing with the Great Lakes and a variety of other systems (Ricciardi and Cohen 2007, Ricciardi 2006, Kolar and Lodge 2002, Kolar 2004). Computer models also have been useful in exploring invasive risk (Anderson et al. 2004a, 2004b; Bartell and Nair 2004, Deines et al. 2005). Some general patterns of invasive characteristics have been identified where habitat destruction or alteration is followed by the introduction of invasive generalists (Marvier et al. 2004), although invasiveness does not necessarily predict impact as defined by loss of diversity (Ricciardi and Cohen 2007).

While it is possible only to assess the risk of introduction, in which the endpoint is the introduction to a new environment, the investigator implicitly assumes that the establishment of any invasive species in a new region is an undesired event (Hewitt and Hayes 2002). Another way of defining risk is the likelihood of impacts following introduction. This acknowledges that the undesired impacts must occur for the particular introduction event to be a cause for concern. The Aquatic Nuisance Species Task Force (ANSTF 1996) used this definition of risk to develop the Generic Nonindigenous Aquatic Organisms Risk Analysis Review Process. While the review process identifies important considerations in evaluating risk of introduction and impacts, it is only a "skeleton" process, designed to accommodate a variety of approaches from very subjective to quantitative and, thus, lacks detailed standardized methods.

HIERARCHICAL INVASIVE RISK MODEL (HIRM)

The HIRM (Colnar and Landis 2007) is an outgrowth of a research program to develop conceptual models and eventually a risk assessment strategy to deal with invasive species at regional scales. The initial conceptual model (Landis 2004) is based upon the relative risk model (RRM) for regional scale risk assessment (Landis and Wiegers 1997, 2005). The RRM has been used successfully at 17 different sites across the world (see Chapter 7 for additional details). The next sections outline the basics of the RRM approach and the development of the HIRM.

The RRM is an approach for conducting regional scale risk assessment that inherently incorporates the features of the landscape, the variety of types of stressors, and endpoints. The methods have been extensively described (Landis and Wiegers 1997, 2005; Wiegers et al. 1998, Hart Hayes and Landis 2004, Landis et al. 2004) and are quickly summarized below. The RRM is inherently landscape in approach. Sources of stressors are identified; these may be effluents, patches of invasive species, or a climatic feature. The location of each is specified, usually using a GIS framework.

The stressors resulting from these sources are specified such as amount and type of contaminants, a particular type of invasive, or a change in temperature regimen. The connection between the stressor and the habitats supporting the valued ecological resources is evaluated with the magnitude of exposure evaluated in part by its proximity to the source and the likelihood of pathways existing between these locations. The habitats are also the location of multiple receptors, and the interactions within the habitat may generate effects by direct or indirect pathways. The effects upon the multiple receptors within the habitat have a probability to produce impacts upon the ecological resources being managed. Areas within the landscape with similar sources, stressors, habitats, and endpoints are broken into risk regions, and risk is computed and compared for each.

The risk regions allow a comparison of risks within a landscape. Risk regions are based in part on management priorities, collections of important habitats, types of similar stressors, and other features. In cases where stressors in one risk region transport into another, this feature can be incorporated in the risk calculation. The use of risk regions has also been an important communication tool for decision-makers and stakeholders. Part of the region at higher risk can be identified easily. Management scenarios can be put through the risk model and changes in risk observed (Thomas 2005).

A fundamental feature of the RRM is the use of ranks to order magnitudes of sources, stressors, and amount of habitat and the size of the impact. The use of ranks allows the combination of stressors and other features with different metrics into an overall ranking of relative risk. A variety of approaches are used to set the ranks, concentration response relationships, natural breaks in distributions, and expert judgment when sufficient data do not exist. Filters are also used in the calculations to signify an exposure to a habitat or the existence of a causal pathway to generate an effect (Landis and Wiegers 2005, Hart Hayes and Landis 2004). Uncertainty is described throughout the process, and recently Monte Carlo analysis has been employed to capture the propagation of uncertainty in the modeling process (Hart Hayes and Landis 2004). Sensitivity analysis is performed to illuminate the features driving the output of the risk assessment.

Relative risks among the risk regions and the associated uncertainty can then be calculated and presented as a series of maps and tables [see Wiegers et al. (1998), Moraes et al. (2002), Obery and Landis (2002), and Hart Hayes and Landis (2004) for examples]. Sources and stressors producing the greatest risk can be identified. Endpoints and habitats at risk are mapped and can be presented in tabular form.

To identify and incorporate the multiple scales involved in a risk assessment for invasive species, the features of the Hierarchical Patch Dynamics Paradigm (HPDP) were incorporated into conceptual model development. The HPDP incorporates three levels or scales: local, focal, and regional. The focal scale is defined as the scale at which the phenomenon or process under study characteristically operates (Wu and Loucks 1995, Wu and David 2002) and is the primary scale in which the analyses are conducted. In addition to the focal scale, Wu and Loucks (1995) and Wu and David (2002) recommend considering the two scales adjacent to the focal scale: the local and regional scales. The advantage to using the HPDP as a framework is that interactions between patches of invasive species, the physiological effects of contaminants,

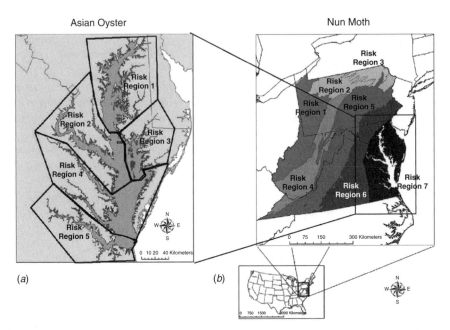

Figure 12.1. Study areas and risk regions in this report. (**A**) Regions of the Chesapeake Bay included for the Asian oyster. (**B**) Regions of the Mid-Atlantic States considered for the nun moth.

and large-scale climate phenomena can be organized and relationships delineated. Modeling by Deines et al. (2005) has demonstrated that the relationship between patch distance, contaminants, and differential fitness are all important considerations in calculating the probability and outcome of a successful invasion. Using an HPDP approach allows incorporation of each of these factors in a spatially explicit fashion. In the next sections we demonstrate the applicability of the HIRM and the incorporation of the HPDP in two case studies.

We are presenting two case studies to illustrate the use of HIRM. The scenarios for the nun moth and Asian oyster case studies (Fig. 12.1) are based on the information available in the summer of 2006. In the case of the Asian oyster, the site was limited to the marine and brackish environment of Chesapeake Bay, the largest estuary in the United States (Fig. 12.1A). The area of interest for the nun moth was for the mid-Atlantic states of the United States (Fig. 12.1B). We present the Asian oyster scenario in the next section.

CASE STUDY 1: ASIAN OYSTER IN CHESAPEAKE BAY, MARYLAND

There are efforts to reconstruct the oyster industry in Chesapeake Bay by introducing the Asian oyster (*Crassostrea ariakensis*) into the region. The goal of this case study was to examine the potential risks of escape of Asian oyster from the areas of culture to other areas of the estuary.

The natural range of the Asian oyster is the coastlines of China, Southern Japan, Taiwan, the Philippines, Thailand, Vietnam, northern Borneo, Malaysia, Pakistan, and India (Tschang and Tse-kong 1956, Rao 1987, Zhou and Allen 2003). The Asian oyster was accidentally introduced to Oregon in the 1970s with *Crassotrea gigas* and *Crassotrea sikamea* (Breese and Malouf 1977). No Asian oyster populations have established on the west coast of the United States presumably because the water temperature is too low (NRC 2004).

The Asian oyster can survive in a temperature range of 14°C to 32°C and salinity range of 7 to 30 g/L (ppt) and settle and support larval growth at about 28°C and 20–30 g/L (ppt) (Cai et al. 1992). There is relatively little information about the ecology of the Asian oyster in the native habitat. In China, they are found to build reefs and have larval settlement on the shady sides of hard surfaces (Cai et al. 1992). In Japan, they are only found on muddy surfaces (Amemiya 1928, Hirase 1930) and in Pakistan, they are found in both muddy and hard surfaces (Patel and Jetani 1991, Ahmed et al. 1987).

The main food supply for the Asian oyster includes phytoplankton and detritus. The Asian oyster will probably have the same predators as *Crassostrea virginica*, the Eastern oyster, which includes sponges, annelids, gastropods, and crabs (NRC 2004). Several diseases have been documented to infect the Asian oyster. A rickettsia-like organism might have caused an 80–90% mortality in the China population since 1992 (Wu and Pan 2000). Some Asian oyster under quarantine in France was infected with Bonamia parasite (Cochennec et al. 1998). The Asian oyster can also be infected by *Perkinsus marinus*, but there were no effects on growth and survival.

Chesapeake Bay

The Chesapeake Bay watershed contains the Chesapeake Bay and is located in parts of New York, Pennsylvania, West Virginia, Delaware, Maryland, Virginia, and the entire District of Columbia. The study area for this risk assessment only included the Chesapeake Bay. The bay is the largest estuary in the United States, has 7100 km (4400 mi) of shoreline, is important for fisheries, shipping, many industries, and provides habitat for various organisms and recreations (US EPA 1996). Recently, oyster and blue crab populations have declined dramatically because of overharvesting, diseases, and degradation of habitat. The bay was once the dominant oyster source in United States, but now it provides only 3% of the total supply. Between 1974 and 2000, Maryland and Virginia had a 65% drop in the number of processing plants, which affected oystermen and processors (NRC 2004).

The study area was divided into subregions mostly by the salinity of the bay and partly by the types of land-use. The salinity was considered because the bay has a large salinity gradient and salinity affects the survival and reproduction of oysters. The upper part of the bay has much lower salinity than the lower part. The land-use types were also considered when dividing the subregions because there is a difference between the eastern and western shore. The eastern shore has more urban development, and the western shore has more agricultural land.

Sources of Asian Oysters

Two possible sources for the Asian oyster are aquaculture and larval current dispersal. There are efforts by groups such as the Virginia Seafood Council to start aquaculture in order to rebuild the oyster industry in the Chesapeake Bay. As of July 2004, about 860,000 triploids have been tested. The triploid oyster is infertile, whereas the diploid oyster is fertile. The Asian oyster can also be introduced through illegal introduction not in compliance with the International Council for the Exploration of the Sea. Accidents such as storms can destroy aquaculture bio-security measures and spread the triploids. The triploids might then be converted to the fertile form.

In the risk assessment for the Asian oyster, two scenarios are considered. Scenario 1 is for spread of the organism only by aquaculture. Scenario 2 includes spread by aquaculture and also by larval dispersal.

Habitats

The Chesapeake Bay consists of subtidal and intertidal habitats. Smith et al. (2001) identified six subtidal benthic habitats: (1) sand, (2) sand and shell, (3) mud, (4) mud and shell, (5) hard bottom and oyster rock.

Assessment Species

Assessment species were chosen based on their importance economically, ecologically, culturally, and suggestions from the US EPA project manager, Daniel Kluza. Five species were chosen: the native Eastern oyster (*Crassostrea virginica*), striped bass (*Morone saxatilis*), blue crab (*Callinectes sapidus*), piping plover (*Charadrius melodus*), and eelgrass (*Zostera marina L.*).

Native Eastern Oyster (**Crassostrea virginica***).* The Eastern oyster is found on the Atlantic Coast from the Gulf of St. Lawrence to the Bay of Campeche in Mexico (Carriker and Gaffney 1996). The average life-span for the oyster is 6–8 years, but they have been found to live for as long as 25 years (NRC 2004). Some predators of the oyster include worms, crabs, oyster drills, starfish, and finfish. Oyster reefs provide habitat for fish such as the striped bass and for invertebrates such as shrimps and blue crabs (Coen and Luckenbach 2000, Bahr and Lanier 1981).

Striped Bass (**Morone saxatilis***).* Striped bass are anadromous fish that spawn once a year. The inhabitant range of striped bass on the Atlantic coast extends from the St. Lawrence River in Canada to the St. Johns River in Florida (Magnin and Beaulieu 1967, McLane 1955). The Mid-Atlantic region is important to the striped bass because it provides spawning grounds; a large amount of recreational fishery activities occur in the region (Fay et al. 1983).

Blue Crab (**Callinectes sapidus***).* The Blue crab is found in the estuaries between Massachusetts Bay and the Eastern coast of South America (Piers 1923,

Scattergood 1960). It is important in the Mid-Atlantic region because it is important in the structure and functions of estuarine communities and it also supports a commercial fishery. It preys on clams and oysters and is preyed by different estuarine and marine animals, including the striped bass (Newcombe 1945, Manooch 1973).

Piping Plover (Charadrius melodus). The piping plover, a small North American migratory shorebird, is a listed endangered species. There are three breeding piping plover populations and they are at the beaches of the Atlantic Coast, at the shorelines of the Great Lakes, and along wetlands and rivers in the Northern Great Plains (Ferland and Haig 2002). Piping plovers prey on larvae and adult macroinvertebrates, mollusks, and crustaceans as well as on other small marine animals (Bent 1929; Shaffer and Laporte 1994, Cuthbert et al. 1999).

Eelgrass (Zostera marina L.). Eelgrass is found on the eastern coast of North America from Nova Scotia to the Carolinas. It is the dominant species of submerged aquatic marine vegetation (SAV) in its range. Some important factors of eelgrass as listed by Thayer et al. (1975) include the following: (1) It supports for epiphytic organisms, (2) leaves produce large amount of organic material that can decompose and be transported to other systems, (3) detritus supports local communities, and (4) the shoots help stabilize sediments.

Conceptual Model

We adopted the conceptual model developed by Colnar and Landis (2007) for the green crab for this study. The model addressed the different scales that can affect the succession of invasion. At the local scale, physical and biological parameters as well as disturbances or alterations can affect the local habitat suitability. The physical parameters include local substrate, local depth, local salinity range, and local temperature range. The biological parameters include local resource competition and local predation. Some disturbances that can occur include land-use change and xenobiotics.

At the regional scale, there can be ocean regime and climate changes that alter the western Atlantic temperature and salinity range. The Gulf Stream, Chesapeake Bay currents, and estuarine-forced water flow make up parts of the physical parameters that affect the habitat suitability at a regional scale.

Some possible undesirable effects of Asian oyster include resource competition, reproductive interference, disease transmission, physical habitat alteration, and destruction/degradation of habitat for other species. Some potential desirable effects of Asian oyster include increased reef habitat and increased prey availability.

Conceptual models used in the RRM have a different organization compared to other risk assessments (Fig. 12.2). The stressor is the Asian oyster, and the relationships between habitats and endpoints are graphed.

Risk Analysis

The analysis phase involves relating exposure and effects to each other (US EPA 1998) and investigating each route to the impact. To analyze the risk of exposure

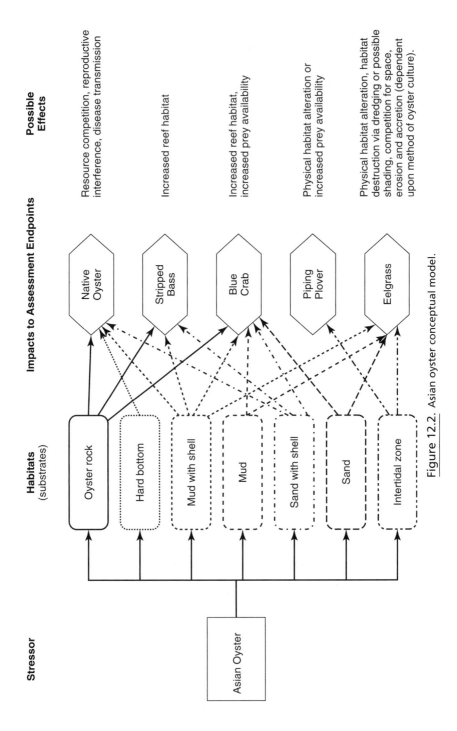

Figure 12.2. Asian oyster conceptual model.

and effects, we used HIRM previously introduced. The following assumptions were considered (Landis and Wiegers 1997, 2005, Wiegers et al. 1998):

1. The type and density of assessment endpoints is related to the available habitat.
2. The sensitivity of receptors to stressors varies between habitat.
3. The severity of effects in subregions of the Chesapeake Bay region depends on relative exposures and the characteristics of the organisms present.

Development of Habitat and Source Ranks. The subtidal habitats were ranked by the areas (km^2) in each subregion, and the ranks ranged from zero to six on a two-point scale. Intertidal habitat was ranked by whether or not the habitat is present and the ranks were either a zero or a six. The ranking categories were determined by using natural breaks in GIS datasets. A rank of zero represents no habitat present, and six represents the greatest relative amount of habitat present.

The source ranks were based solely on presence/absence of each source, with zero representing absence of source (low exposure potential) and six indicating presence of source (high exposure potential). To be consistent with the habitat-ranking scheme, zero and six were again used as the minimum and maximum ranks possible for the sources.

Development of Exposure and Effects Filters. Filters are weighting factors used to determine the relationship between risk components: sources, habitats, and impacts to assessment species (Wiegers et al. 1998). The exposure and effects filters were determined with similar criteria as developed by Colnar and Landis (2007). The filter values ranged between zero to one, with zero indicating an incomplete pathway and one indicating a complete pathway. We developed different decision trees to provide guidelines for filter assignments. When multiple components were present for the filter, all components were multiplied together to give individual filter values.

Two exposure filter components were used to indicate whether or not the source would release the stressor and whether the stressor could occur and persist in the habitat. Exposure filter component A considers whether or not the source will release the stressor, and it is dependent upon the origin of the source, application of treatment or precautionary methods, and the interaction of the source with the aquatic environment. Exposure filter component B considers the habitat suitability and is dependent upon temperature, salinity, substrate type, and predation.

Three effects filter components were used to indicate (a) whether or not the endpoint occurs in and utilizes the habitat, (b) whether there is seasonal overlap in habitat usage between the stressor and the endpoint, and (c) whether effects to the endpoint are possible from interaction with the stressor. Effects filter component A considered whether or not the endpoint occurs in and utilizes the habitat. For blue crab, striped bass, and Eastern oyster, rank categories were determined using natural breaks on landing (pounds) GIS data. There was not specific distribution data for eelgrass and piping plover, so we estimated whether or not the species was present or absent. Effects filter component was determined by whether or not there was seasonal habitat overlap between the stressor and each species. A value of 0 indicates no overlap, and a value of 1 indicates that there is possible overlap. Effects filter component was determined by whether or not there are possible effects to the endpoints by the stressor. A value of 0 indicates no effects possible, a value of 1 indicates possible undesirable effects, and a value of −1 indicates possible desirable effects.

Risk Characterization

Risk characterization is the final phase of a risk assessment and is the process of integrating exposure and effects data to estimate the risk. All ranks were converted into a point system. To generate the exposure and effects filters for each pathway, the filter components of the exposure and effects filters were multiplied. Finally, the source and habitat ranks were integrated with the exposure and effects filter products to generate the risk for each source–habitat–endpoint pathway. The risk scores were then summed to produce the following risk predictions: (1) risk in subregions, (2) risk in habitats, (3) risk to assessment species, and (4) risk from each source.

Uncertainty Analysis. The amount of uncertainty for the ranks of components in the model was classified as low, medium, or high based on the amount and the types of data available. The uncertainty was addressed using Monte Carlo analysis, a probabilistic approach. A discrete statistical distribution was assigned to components where the uncertainty was medium or high. The ranks or filters with low uncertainty retained their original values and were not assigned a distribution.

All source ranks were assigned low uncertainty because we assumed that aquaculture would most likely occur in all risk regions. All habitat ranks were assigned with high uncertainty for different reasons. Sand, sand and shell, mud and shell, hard bottom, and mud habitat for RR1, RR2, and RR3 were assigned high uncertainty because the data was based on 1975–1983 Maryland Department of Natural Resource survey to reassess oyster bottom and describe substrate. The data may be outdated and not representative of the current habitat. Also, the habitats containing shell may be incorrect because in many areas the shell was actually covered by varying depths of sediment so it might not represent the habitat on the surface. Sand, sand and shell, mud, oyster rock, and mud and shell habitats for RR4 and RR5 were assigned high uncertainty because the data were based on a survey of public oyster grounds that could be outdated and not representative of the total bay habitat. The intertidal habitat for all risk regions were assigned high uncertainty because intertidal areas are the primary habitat for Virginia's eastern shore, but the exact location and areas were unknown. The hard-bottom habitat for RR4 and RR5 were assigned high uncertainty because that specific habitat type was not measured but could be present.

All values for exposure filter component A for larval current dispersal had high uncertainty because the reversion of triploid state to diploid state is possible, but only <1% of the Asian oyster would probably be able to produce normal gametes after being in the field for 3–4 years. All exposure filter component B values had low uncertainty for the aquaculture source, and some had medium or high uncertainty for larval current dispersal source. The exposure filter component B values for mud and hard bottom in risk region 1 to 5 had medium uncertainty because we only know that

Asian oyster can grow on mud and hard-bottom substrates in Japan and Pakistan, but there is not any information indicating whether or not they can grow on those types of substrates in the Chesapeake Bay. The exposure filter component B values had low uncertainty for all habitat types, but the sand and oyster rock in risk region 1 and the intertidal zone in risk region 2 to 5 had high uncertainty, because the minimum salinity required for reproduction is approximately 15 g/L (ppt) while the salinity in risk region 1 is in the range of 10–15 g/L (ppt) and the substrate type within the intertidal zone is also unknown.

All exposure filter component A values had low uncertainty except for the Asian oyster in hard-bottom habitat, the striped bass in sand and shell, mud and shell, and oyster rock habitat, and eelgrass in sand, sand and shell, mud, mud and shell, and intertidal zone. The Asian oyster in hard-bottom habitat had high uncertainty because it is unknown whether it is able to attach to hard-bottom substrates. The striped bass in sand and shell, mud and shell, and oyster rock habitats had medium uncertainty because the striped bass are known to be present near oyster reefs and are assumed to use the habitats containing oyster shells. The eelgrass in sand, sand and shell, mud, mud and shell, and intertidal zone had medium uncertainty because the dataset used to estimate the distribution of eelgrass contained all types of submerged vegetation not just eelgrass. All exposure filter component B values had low uncertainty based on the data available. The exposure filter component C values for blue crab, eastern oyster, and striped bass from both types of sources had medium uncertainty because the effects were probable only. The exposure filter component C value for the effects of larval current dispersal source of the stressor on the piping plover has high uncertainty because the effects are only suspected.

Sensitivity Analysis. We conducted a sensitivity analysis using the Crystal Ball® 2000 software. The analysis examines the sources of uncertainty, influenced by either the model sensitivity or parameter uncertainty (Goulet 1995, Warren-Hicks and Moore 1988). Model sensitivity is the influence of a parameter within a model, and parameter uncertainty is the influence of the range of possible parameter values. During sensitivity analysis, correlation coefficients are generated to rank each model parameter's contribution to predict uncertainty. A high rank correlation indicates that the uncertainty within the model parameter has great importance in influencing the uncertainty within the model.

Risk Characterization Results

The total risks scores for scenario 1 and 2 were identical in risk region (RR) 1 and different in the other RR. In scenario 2 (Table 12.1), RR2 had the highest overall risk in both scenarios. In scenario 1, RR1, RR4, and RR5 had beneficial effects; and in scenario 2, RR3 also had a beneficial effect. The overall risk is lower in scenario 2 when compared to scenario 1. The endpoints of blue crab and striped bass had negative risk scores in both scenarios, indicating the possibility of beneficial effects with Asian oyster introduction. The endpoints of eastern oyster, eelgrass, and piping

Table 12.1. Risk Due to Asian Oyster from Scenario 2

Risk Calculation Type	Endpoint	Blue Crab	Eastern Oyster	Striped Bass	Eelgrass	Piping Plover	Sum	Sum—Only Beneficial Effects Exhibited per Region	Sum—Only Undesirable Effects Exhibited per Region
Risk Region	RR 1	-132	96	-72	66	0	-42	-204	162
	RR 2	-60	63	-36	48	0	15	-96	111
	RR 3	-72	57	-48	48	0	-15	-120	105
	RR 4	-198	72	-144	72	36	-162	-342	180
	RR 5	-54	48	-48	0	0	-54	-102	48
	Sum	**-516**	**336**	**-348**	**234**	**36**	**-258**		
Habitat	Sand	-102	0	0	66	0	-36	-102	66
	Sand and shell	-96	108	-120	48	0	-60	-216	156
	Mud	-90	0	0	42	0	-48	-90	42
	Mud and shell	-120	96	-120	42	0	-102	-240	138
	Hard bottom	0	60	0	0	0	60	0	60
	Oyster rock	-108	72	-108	0	0	-144	-216	72
	Intertidal zone	0	0	0	36	36	72	0	72
	Sum	**-516**	**336**	**-348**	**234**	**36**	**-258**		

plover had positive risk scores, indicating the possibility of undesirable effects caused by the introduction of the Asian oyster. The risk scores of risk to endpoints in various habitats were different between the two scenarios in RR2, RR3, RR4, and RR5. The source of aquaculture contributed relatively more to the total risk than did the source of larval current dispersal. The scores of risk in habitats are positive only in hard bottom and in intertidal zone.

Uncertainty Analysis: Scenario 1

Subregions. The risk scores for different subregions consisted of both negative and positive values. The uncertainty analysis indicated that for each risk score, there is a possibility that it is positive instead of negative and vice versa. RR1 had a risk score of −42, indicating that there would be possible beneficial effects, but there is still a 30% probability that the risk score is actually a positive value. RR2 had a risk score of 30, but the distribution for the risk score indicated that it could be between −48 and 138. RR3 had a distribution with a range from −72 to 156, with a 50% probability that the risk score would be higher than the calculated risk score of 12. RR4 had a risk score of −84, and the range was between −234 and 126. RR5 had risk score distribution between −108 and 120. There is a 20% probability that the risk scores for RR4 and RR5 are actually positive.

Sources. The total risk scores in various subregions from aquaculture source were −114, and the range of the distribution was from −492 to 540. There is approximately a 40% probability that the actual risk score is positive.

Habitats. The total risk in sand habitat had a risk score of −36 and a distribution range from −78 to 120. The total risk in sand and shell habitat had a calculated risk score of −12, and there is a 50% probability that the actual risk score is higher than the calculated risk score. The risk in mud habitat had a range between −72 and 78, and there is a 20% probability that the actual risk score is positive. The total risk in mud and shell habitat was −48. The range is between −192 and 180, and the distribution is bimodal. The risk in hard-bottom habitat had a risk score of 48, and the risk in the intertidal zone is 36. There is a 0% probability that the actual risk score for hard-bottom habitat and intertidal zone is negative. The risk to oyster rock habitat had a calculated risk score of −72, and there is a 10% probability that the actual risk score is positive.

Species. The calculated risk to blue crab was −360 with a distribution range of −450 to 0. The risk to Eastern oyster is 222, and there is a 0% probability that the risk would be a negative value. The risk score for striped bass was −210 with a range of −300 to 0. The calculated risk score for eelgrass was 234 and there is a 0% probability that the risk score is negative. The calculated risk for piping plover is 0, and there is no uncertainty distribution for the piping plover.

Uncertainty Analysis: Scenario 2

Subregions. The ranges on the uncertainty distribution for all RR were wide and span across negative and positive risk values. In RR1, there is a 60% probability that the risk score will actually be higher than the predicted risk score of −42 from the RRM model. In RR2, there is about a 55% probability that the risk score will be higher, and there is a 45% probability that the risk score will be lower than the predicted score of 15. In RR3, there is about a 70% probability that the risk score will be larger than the predicted score of −15 and about a 50% probability that the risk score will actually be positive. In RR4 and RR5, there is about a 15% probability that the risk score will actually be positive.

Sources. The aquaculture uncertainty distribution has multiple nodes and a wide range of 1014. The predicted risk score was −114, and there is approximately a 65% probability that the risk score will be higher than the predicted score. The uncertainty distribution for larval current dispersal indicated that there is about a 15% probability that the risk score will be above 0.

Habitats. The uncertainty distributions for sand and mud were noticeably made up of multiple distributions. There is a 20% probability for sand and approximately a 15% probability for mud that the risk scores would be 0 or above. There is a 70% probability that the risk score for sand and shell habitat would be higher than the risk score of −60 as predicted in the RRM. There is a 0% probability that the risk score for hard bottom and the intertidal zone would be lower than 0.

Species. The uncertainty distribution for blue crab indicated two slightly overlapping distributions. There is an 80% probability that the risk score would actually be higher than the risk score of −516 predicted from the RRM. The risk score, however, has a 0% probability of being above 0. The uncertainty distribution for the eastern oyster and eelgrass indicated that there is a 0% probability that the actual risk scores would be below 0, whereas there is a 0% probability that the risk score for striped bass and piping plover would be above 0.

Sensitivity Analysis: Scenario 1
Subregions. The effects filter C that represents the effects from aquaculture source to blue crab (Effects C.AQ.BC) is the model parameter that contributed the most to uncertainty in all RR. RR1 and RR4 were slightly more sensitive to Effects C.AQ.BC relative to all other model parameters.

Sources. Aquaculture source was especially sensitive to Effects C.AQ.BC, with a rank correlation of 0.45. It was also sensitive to the effects C filter for aquaculture to eastern oyster and to striped bass, with rank correlations of 0.29 and 0.28.

Habitats. Effects C.AQ.BC was the dominant parameter in sand and mud. There was not any parameter that contributed most to uncertainty in mud and shell, hard bottom, and oyster rock. The effects filter from aquaculture to eastern oyster contributed the most to the hard-bottom habitat.

Species. There was a parameter that the endpoint of blue crab, eastern oyster, and striped bass was especially sensitive to. For blue crab, it was the effects filter C for aquaculture to blue crab, for eastern oyster it was the effects C filter for aquaculture to eastern oyster, and for the striped bass it was the effects filter C for aquaculture to striped bass. For the endpoints of eelgrass and piping plover, there was not a parameter that the endpoints were especially sensitive to.

Sensitivity Analysis: Scenario 2

Subregions. The Effects C.AQ.BC was the model parameter that contributed the most to uncertainty in all RR. Areas RR1 and RR4 were more sensitive to Effects C.AQ.BC than the other regions. RR1 and RR4 had Effects C.AQ.BC rank correlation values of 0.40 and 0.43, while the other regions had a rank correlation value 0.29.

Sources. Aquaculture and larval current dispersal were most sensitive to the parameters of Effects C.AQ.BC, effects filter C for aquaculture to eastern oyster, and effects filter C for aquaculture to striped bass. Those parameters were especially dominant in aquaculture, where the other parameters all had rank correlation values lower than |0.09|.

Habitats. Effects C.AQ.BC was the dominant parameter in sand and mud. There was not any parameter that contributed most to uncertainty in mud and shell, hard bottom, and oyster rock. The habitat ranking for intertidal zone in RR4, with a rank correlation value of 0.53, was the parameter that contributed most to the uncertainty in intertidal zone risks.

Species. Blue crab, eastern oyster, and striped bass had the effects C filter from aquaculture to endpoint as the parameter contributing most to uncertainty. Eelgrass was not particularly sensitive to any of the parameters. Piping plover was most sensitive to the effects C filter from larval current dispersal to piping plover.

Discussion of Asian Oyster Risk Assessment

From the risk characterization, RR2 had the highest overall risk whereas RR1, RR4, and RR5 had beneficial effects in both scenarios (Fig. 12.3). The uncertainty analysis must be considered along with the calculated risk scores. With the range of distributions for the various RR risk scores being very wide and with many of the distributions overlapping in large amounts, it would be inappropriate to conclude that any of the RR will especially have beneficial or undesirable effects. One way to reduce the uncertainty would be to gather more data regarding the effects of aquaculture to blue crab as indicated in the sensitivity analysis. That would reduce the uncertainty in RR1 and RR4, because those are the risk regions that are more sensitive to the effects filter from aquaculture to blue crab.

The risk from sources in scenario 2 indicated that both sources overall contributed beneficial effects. The uncertainty analysis indicated that both sources had high uncertainty and there is actually a high probability that the actual risk scores are much higher than the calculated risk scores. In aquaculture source, there is approximately a 30% probability that the actual risk score would be higher than 0, which would cause undesirable effects. The sensitivity analysis indicated that more data should be acquired for the effects of aquaculture to blue crab in order to reduce the uncertainty for the aquaculture source.

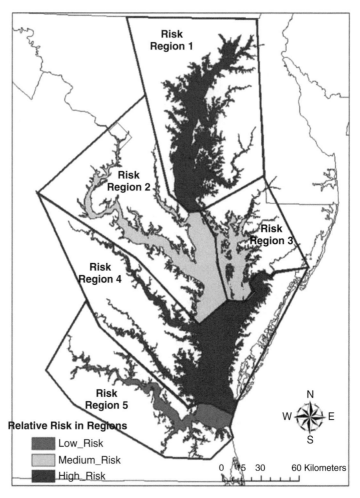

Figure 12.3. Distribution of relative risk for the Asian oyster in Chesapeake Bay.

In both scenarios, the endpoints of blue crab and striped bass had negative risk scores, indicating possible beneficial effects. The uncertainty analysis indicated that the distributions for the risk scores in both scenarios and endpoints had wide ranges. The distribution does not reach a positive risk score, however, indicating that the potential beneficial effects of Asian oyster might not be as large as predicted. There is a possibility that there would be no effect from the Asian oyster to the blue crab or striped bass, but there likely would not be a probability that there would be undesirable effects to blue crab or striped bass. The sensitivity analysis indicated that focus should be placed on predicting the effects of Asian oyster to the endpoints in order to reduce uncertainty.

There were undesirable effects of Asian oyster to the eastern oyster and eelgrass predicted under both scenarios. The uncertainty analysis indicated a large range for the risk score distributions. There is a possibility that the actual risks to those endpoints are not as large as expected. However, there is no plausibility that there would be potential beneficial effects to those endpoints. The sensitivity analysis indicated that focus should be placed on determining the effects Asian oyster to eastern oyster and eelgrass in order to reduce the uncertainty.

The exact risk scores for the different habitats between scenarios were not the same, but still had similar qualities in the risk distributions. The risk scores were negative for all habitats except for hard bottom and intertidal zone. The risk in habitat was highest in hard bottom and intertidal zone in both scenarios. The negative risk scores for habitats indicate that there are potential beneficial effects to the endpoints in those habitats. However, the uncertainty analysis must also be considered. The ranges shown in the uncertainty analysis for all the negative risk scores habitats were wide, extending from negative to positive risk scores. There is a probability that there are actually undesirable effects to the species in those habitats, which was consistent with the species risk results. The hard bottom and intertidal zone had positive risk scores, indicating undesirable effects. The uncertainty distributions showed that there is a possibility that the risks to species in those habitats might not be the exact calculated risk scores, but there is no expectation that there would be beneficial effects. More information should be acquired for the effects from aquaculture to blue crab in order to reduce the uncertainty in sand and mud habitats. An overlay plot of the risks in habitats showed that the habitats of hard bottom and intertidal zones were noticeably separated from the other habitats. This is very important to consider because it shows how different components interact differently in various habitats and it is essential to consider the differences in habitats when conducting a region risk assessment.

The sensitivity analysis indicated that the effects filter component C was the most sensitive parameter in the model. The effects filter component C considers whether or not the Asian oyster has potential undesirable or beneficial effects to the endpoints. Focus should be placed on gathering more information about how the Asian oyster affects the various endpoints in order to reduce the uncertainty with various risk scores and to be able to indicate the differences between various risk distributions.

CASE STUDY 2. NUN MOTH IN THE MID-ATLANTIC STATES

The second case study was for an organism that has not yet been introduced to North America, but may be introduced in cargo transported by ship or aircraft. The native range of the nun moth extends from Portugal to Japan south of $60°$ latitude (Novak 1976, USDA 1991). In the southern part of its range, the nun moth lives at higher elevations on conifers and broadleaf trees, whereas in the north it is found in lowlands, mainly on Norway spruce and Scotch pine (Novak 1976). The nun moth, *Lymantria monacha*, is one of the most damaging pests in Eurasia. The USDA Forest Service (1991) predicts that introduction of nun moth would lead to high mortality in North American forests. During the largest outbreak in history, which occurred between 1978 and 1984, 3.7 million hectares of Scots pine and Norway spruce forests in Poland were defoliated (Glowacka 1998).

Nun moth populations can grow on a wide range of host plants; eggs are laid in several clusters that can be spread over a wide area (Keena 2003). A USDA Animal and Plant Inspection Service draft risk assessment (USDA 2000) predicts that the moth could spread up to 15 km/year in their worst-case scenario.

Nun moth larvae feed preferentially on the young needles and male cones of conifers, but they are highly polyphagous and can also use the leaves of deciduous trees and shrubs (Keena 2003). Trees in the United States that could support an invading nun moth population include several species of spruce and oak, Scots pine with male cones, fir, and apple. Several other species found in North America, including western larch, were shown to provide forage for the nun moth, but with high mortality and slow development in larvae and lower fecundity rates in adults. These results are based on laboratory findings where 34 North American tree species were tested individually for ability to support nun moth larvae. It is possible that a mixture of the 21 species that were determined to be moderately likely or unlikely to support a population would be more effective in combination (Keena 2003).

Phenological synchrony between nun moth larvae and host plants is important. First and second instars cannot feed on growth from previous years because the needles are too tough and contain secondary compounds that can be harmful (Keena 2003). If the larvae hatch before budburst, they may feed on male cones or deciduous foliage, or they may disperse via wind (Keena 2003, USDA 2000).

The preference for conifer trees makes the nun moth a serious hazard; conifers generate new growth slowly, and they are more likely than hardwoods to die after defoliation (USDA 1991). Conifers are also more devastated than broadleaf trees because the nun moth destroys more needles than it consumes. Keena (2003) observed poor survival in laboratory when grown on a limited supply of larch because the larvae begin feeding at the base of the needle, resulting in most of the needle falling to the ground uneaten. In spruce, as little as 50% defoliation will cause mortality within a year (USDA 2000).

Problem Formulation

This portion of the risk assessment is presented in five subsections. After identification of potential sources of nun moth as eggs or larvae, since adults are unlikely to survive transport, we delineate the risk regions of interest, the habitat types, and assessment species. We organize this information into our working conceptual model.

Identification of Potential Sources. Nun moth eggs can be laid in cracks and crevasses in wood packing material, logs, and transport vessels from Europe and Asia. In the current risk assessment, we considered three potential sources of nun moth larvae to the study area: international airports, maritime ports, and natural dispersal. The relative risk model calculates risk from all sources at once, so in order to include the risk due to spread of the nun moth, we need to either use two scenarios or calculate the risk assuming that populations have become established. Solid wood-packing materials could become sources for nun moth populations near airports and maritime ports (USDA 2000). Once populations became established, natural dispersal would become an additional source of nun moth larvae.

Risk Regions. The study area includes five states—Pennsylvania, Virginia, West Virginia, Maryland, and Delaware—and encompasses several distinct landscape types. We defined seven risk regions using US EPA level III ecoregion data. Ecoregions have been defined for the United State based upon geological, physiographical, vegetative, climactic, soil, land use, wildlife, and hydrological information. Level III ecoregions is the most detailed level available for the entire United States, including 84 different regions (US EPA 1997).

Risk Region 1 is a combination of the glaciated Erie Drift Plains, characterized by low rounded hills, moraines, kettles, and areas of wetlands, plus the Western Allegheny Plateau, characterized by more rugged, forested hills. This region has mixed oak and mesophytic forests. In the northern portion of this region the weather is influenced by Lake Erie, which increases both the growing season and the winter snowfall.

Risk Region 2 consists of the North Central Appalachian ecoregion, with plateaus, high hills, and low mountains, with more forested areas than adjacent ecoregions. Land use in this region is primarily forestry and recreation.

Risk Region 3 is the southern section of the Northern Appalachian Plateau and Uplands ecoregion, a transitional area between (a) the urban and agricultural lowlands to the north and west and (b) the more mountainous, forested regions to the south and east. Much of the land is farmed and in pasture, but large areas remain forested in oak and northern hardwoods.

Risk Region 4, the Central Appalachian ecoregion, has rugged high hills and mountains. Appalachian oak and northern hardwood forests are the main land cover. This region has a cool climate and infertile soils. Streams have been polluted by coal mining.

Risk Region 5 consists of the Blue Ridge and the northern tip of the Ridge and Valley ecoregions. This region is the most diverse, with oak, northern hardwood, southeastern spruce–fir, hemlock, and oak–pine forests. The terrain is mountainous and rugged, with many forested slopes.

Risk Region 6 is a combination of the Northern Piedmont and the northern third of the Piedmont ecoregions. This area has been largely cultivated, although in the south it has successional pine and hardwood forests.

Risk Region 7 is a combination of the northern sections of the mid-Atlantic and Southeastern coastal plain ecoregions. Land cover consists of cropland, pasture, woodland, and forest. In the east, native vegetation is longleaf pine, oak and hickory, and southern mixed forest, while the west has loblolly and shortleaf pine, oak, and gum forest.

Habitats. The diverse land cover found in the Mid-Atlantic States region was divided into five habitat types. These were conifer forest, mixed forest, deciduous forest, woody wetlands, and rivers and streams. Geographical information system data was used to define habitat boundaries. In the conceptual model development, the impact on conifer, mixed, and deciduous forests are considered only as direct effects.

The nun moth may cause both direct effects (through removal of canopy cover and changes in litter composition) and indirect effects (though changes in water quality) on the woody wetlands. In the rivers and streams habitat, only indirect effects were considered.

Assessment Species. The assessment species used in this study were threatened or endangered species selected by Daniel Kluza at US EPA. These were (1) Swamp pink (*Helonias bullata*), (2) Dwarf wedgemussel (*Alasmidonta heterodon*), (3) Southern water shrew (*Sorex palustris punctulatus*), (4) Shenandoah salamander (*Plethodon shenandoah*), (5) Cheat Mountain salamander (*Plethodon nettingi*), (6) Northern flying squirrel subspecies (*Glaucomys sabrinus coloratus G. s. fuscus*), (7) Red-cockaded woodpecker (*Picoides borealis*), and (8) Duskytail darter (*Etheostoma percnurum*).

The swamp pink is listed as a federally threatened perennial plant species. It is found along streams in Virginia, Maryland, and Delaware, in meadow, cedar swamp, and forested wetlands. There is a strong correlation between the presence of swamp pink and several conifer species. This plant has limited seed dispersal and viability, and it spreads mainly through clonal rhizominal growth. These shade tolerant plants are inferior competitors in direct sunlight. The swamp pink requires a near-constant water level. The main threats to this plant include habitat loss due to wetland drainage and water-quality degradation through sedimentation (USFWS 1991b).

The federally endangered dwarf wedgemussel lives in areas of streams and rivers that have a muddy sand, sand, or sand and gravel substrate with low to moderate current and low turbidity. The historical range of the dwarf wedgemussel is Virginia, Maryland, and Pennsylvania; currently no populations exist in Pennsylvania. This mussel is sensitive to light penetration and dissolved oxygen. Threats to dwarf wedgemussel persistence include channelization, removal of shoreline vegetation, and polluted runoff from agriculture, industry, and homes (USFWS 1993b).

The southern water shrew is considered a vulnerable species in its range in Maryland, Pennsylvania, Virginia, and West Virginia. This shrew is found near streams in areas with low vegetation, rocks, and logs, which provide shelter, protection, and a high-humidity microclimate. The Southern water shrew is threatened by warming, siltation of streams, habitat loss, and toxicity due to pesticide control of forest insect pests (NatureServe 2005).

The main threat to the federally endangered Shenandoah salamander is interspecific competition with the red-backed salamander. The Shenandoah salamander is found in only three metapopulations within the Shenandoah National Park in Virginia (Griffis and Jaeger 1998). More draught-tolerant than other lungless salamanders, the Shenandoah salamander inhabits dry, rocky talus slopes above 800 m. Forest cover is required to maintain adequate moisture levels on the ground, and previous defoliation by the gypsy moth (*Lymantria dispar*) and the hemlock wooly adelgids (*Adelges tsugae*) have reduced the suitable habitat area. Defoliation may result in drying of the forest floor, as well as soil chemistry changes due to high composition of needles in the floor litter. Acid precipitation is an additional threat to the Shenandoah salamander. However, lowering soil pH, which may result as a combination of acid rain and defoliation, dramatically reduced red-backed salamander survival while effects on the

Shenandoah are not known (USFWS 1994). If the Shenandoah is less sensitive, its competitive ability may be enhanced.

The Cheat Mountain salamander, endemic to West Virginia, is federally listed as threatened. These salamanders live in spruce and mixed forests above 900 m (reported as 2980 ft) in cool, humid microclimates with moist soil and litter cover. Main food items include mites, springtails, beetles, flies, and ants. The home range of this salamander is $13-25$ m^2. In competition with other salamanders, the cheat mountain salamander will aggressively defend its territory, but it will usually not be successful. This salamander is probably limited to higher elevations due to a competitive disadvantage lower in its potential range. The main threats include habitat reduction through removal of forest canopy, fires, and alteration in the forests floor by road and trail development. Removal of canopy is the biggest factor affecting survival because increased sunlight alters the microhabitat. Roads and trails isolate diminishing populations (USFWS 1991a).

The northern flying squirrel subspecies, *G. s. coloratus* and *G. s. fuscus*, both federally endangered, are found in western Virginia and eastern West Virginia in boreal habitat, especially spruce–fir and northern hardwoods (USFWS 1990). Sites occupied by the squirrels have relatively more conifers, with little or no northern red oak. Understory components of forest habitat are not significant in determining habitat usage (Ford et al. 2004). Diet consists of tree buds, lichens, epigenous and hypogenous fungi, and beechnuts; at times, the population may be entirely supported by fungi (USFWS 1990). Mycorrhizal fungi spore dispersal, facilitated by squirrels, may contribute to tree health in high-altitude forests (Mitchell 2001). The individual home range requirement is $5-7$ hectares, and nesting habitat appears to be a limiting factor, with nests often containing several adults. These squirrels exist in fragmented relict populations, and they are at risk due to further habitat degradation that may result from insect pests. Risk also is possible due to the chemicals used to control insect pests, such as lindane which is used to control the balsam woody adelgid (*Adelges piceae*) (USFWS 1990).

The range of the federally endangered red-cockaded woodpecker is determined by the distribution of southern pines, with current populations in the study area being fragmented and isolated. These birds breed in family units, called groups, consisting of a monogamous pair and the male offspring of the previous year. The required breeding clusters are open stands of pines or savannahs with large pines at least 80 years old, with little to no hardwood understory. Birds forage on insects, including ants, beetles, wood-boring insects, and caterpillars, in pine or pine–hardwood forests at least 30 years old. Their diet also includes seasonal wild fruit. Home-range requirements vary greatly, between 40 and 160 hectares per group. The main threat is due to loss of older pine forests, along with growth of the hardwood mid-story due to fire suppression (USFWS 2003).

The federally endangered duskytail darter is very selective in microhabitat choice, using only pools with moderate to fast current where the substrate contains a mixture of pea gravel, cobble, and boulders (USFWS 1993a). These fish are only found in large creeks and rivers in areas with little to no siltation (Powers and Mayden 2003). Because of specific habitat requirements, duskytails don't disperse, and existing populations are fragmented and isolated (USFWS 1993a). Water-quality impairment, including siltation, is the major cause of decline (Powers and Mayden 2003, USFWS 1993a).

Nun Moth Conceptual Model Development. We built a conceptual model to illustrate the pathways from the stressor to the endpoints (Fig. 12.4). The model is in two parts reflecting direct and indirect effects. In this model, the nun moth can directly impact the woody wetland, conifer, deciduous, and mixed forest habitats, whereas it indirectly impacts the woody wetland and rivers and streams habitat.

In the calculation of risk, the exposure assessment is incorporated in the connection between the stressor and the habitats: If there is a high probability of introduction and reproduction, there will be a connection between the moth and the habitat. Each source is considered to have an equal chance of exposing any habitat within a particular risk region, so the source is not included in the model. In the case of the rivers and streams and woody wetland habitats, the stressors are indirect effects due to nun moth exposure in any other habitats. However, because the model only looks at one period, the potential for exposure to indirect effects is calculated at the same time as the potential for exposure to nun moth.

The effects assessment is diagramed as the connection between the habitats and the species, with possible effects listed after the species. General effects following defoliation include tree mortality, increased light penetration, soil temperature and water drainage, and decreased transpiration and moisture (Lovett et al. 2002; Russell et al. 2004). Then hypothesizing the effects, we draw from observations about the effects of gypsy moth, *Lymantria dispar*, that has altered canopy structure, increased understory growth, and affected avian nest site availability (Crooks 2002). In individual trees, changes in nutrient allocation under heavy defoliation may lead to limited mast production for many years in surviving trees (Lovett et al. 2002). In addition, the addition of green leaf litter and moth frass to the forest floor can alter the nitrogen cycle, possibly leading to (a) nitrogen loss from the system and (b) acidification or eutrophication in streams (Lovett et al. 2002). In a study of a poplar plantation, Russell et al. (2004) found that nitrogen redistributed by defoliation remained in the system, but noted that insect outbreaks in Appalachian Mountain forests were linked to elevated nutrient loads in rivers and nitrogen loss.

Analysis

A risk score combines the probability of establishment and effects to assessment species with the consequences of establishment; in the relative risk model, the probabilities of establishment and effects are represented by the filters, whereas the extent of the stressor at establishment and the consequences of effects are represented by the rankings. Risk scores were calculated by following the flow through the conceptual model for each risk region.

For the exposure assessment, the potential sources of nun moth larvae were ranked independently within each region. Source ranks were based on presence/absence criteria, whereas natural dispersal from established populations was considered a source

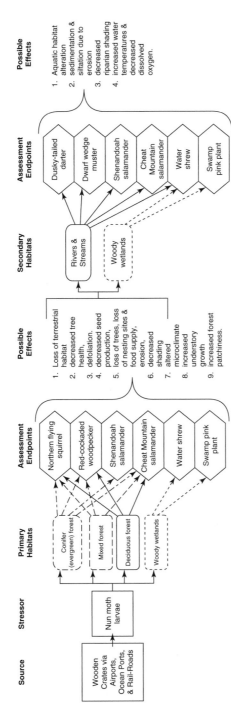

Figure 12.4. Nun moth conceptual model. The conceptual model is designed to incorporate direct and indirect effects of the nun moth infestation.

in all of the risk regions. Each of the five habitats was ranked according to the percentage of that habitat type in the risk region out of the total amount of that habitat type in the study area. The probability of habitat exposure was calculated by multiplying two filters: one representing the likelihood of introduction from the source and the other representing the potential for survival in the habitat. The values for the exposure filters were assigned using decision trees. The exposure assessment results in a risk score for introduction of nun moth larvae to each habitat from each source, within each risk region.

Effects risk scores were assigned using the results of the exposure assessment, with the addition of three filters. Each nonzero value for risk of exposure was multiplied by three exposure filters: one to indicate if the endpoint lives in and utilizes the habitat, one to indicate that there will be a temporal and spatial overlap between the endpoint and the stressor, and one to indicate whether it will be possible for the stressor to affect the endpoint. The first filter is based upon presence/absence of the endpoint within each habitat in each risk region. The second filter was assigned a value of one for all endpoints because it was assumed that defoliation could either directly or indirectly alter any habitat, and that these effects would be lasting. The third filter also received a value of one for each endpoint because all endpoints are threatened or endangered species, and further loss of habitat due to defoliation would likely effect population levels.

Uncertainty and Sensitivity Analysis. We used Monte Carlo analysis (Crystal Ball® 2000) to assess the variability in potential risk score values. Each rank and filter value was assigned an uncertainty level of high, medium, or low, along with a distribution of possible values. For example, the conifer forest rank value in risk region one was 4, but it was assigned with medium uncertainty. In the Monte Carlo risk score calculation, 80 out of 100 times this rank will take a value of 4; for the other 20 iterations, this rank will take a value of 2 for 10 iterations and a value of 6 for 10 iterations. To determine which input variables had the greatest impact on the risk score, we used a rank correlation sensitivity analysis (Crystal Ball® 2000).

Rank Value Uncertainty. All source ranks have low uncertainty except the rank for maritime ports in region six, which was assigned medium uncertainty. We were uncertain whether the ports at the top of region seven could be a direct source of nun moth larvae to region six. Habitat rank values were based on 10-year-old GIS data, so we assumed medium uncertainty associated with those values.

Filter Value Uncertainty. We did not use specific import data for each port, so all exposure filter A values for air and maritime ports have high uncertainty. The regions without a port source have a filter value of zero with low uncertainty. In addition, due to the large range of potential dispersal rates given by the USDA APHIS risk assessment (2000), natural dispersal exposure filter A values were assigned with high uncertainty in each region. All exposure filter B values were assigned with medium uncertainty. Effects filter A, indicating habitat usage by endpoints, was assigned a value of zero with low uncertainty if the endpoint does not utilize the habitat. However, if the endpoint does utilize the habitat type, it is assigned a filter value of one or zero with medium uncertainty. If the species is present in at least one county in a risk region, then it could be present throughout the risk region, with medium uncertainty. If the species is not found in any county in a risk region, then it will probably not be present in any habitat in that risk region and it is assigned a filter value of zero, with medium uncertainty. Effects filter B, representing that there will be temporal and spatial overlap between the stressor and species, was assigned a value of one with low uncertainty for all endpoints. Effects filter C was assigned a value of one, with low uncertainty for all species except the duskytail darter and the dwarf wedgemussel, which had medium uncertainty, indicating that all endpoints can potentially by impacted by defoliation.

Risk Characterization Results

The deterministic calculations (Table 12.2) and depiction of risk (Fig. 12.5) for developed for each of the seven risk regions. Natural breaks in the risk score, found using GIS, were used to define low (0–54), medium (55–252) and high (252–744) risk. The low risk regions were one, two, and three, with risk scores of 54, 6, and 0, respectively. The regions with medium risk were four and seven, with risk scores of 198 and 252, respectively. The high-risk regions were five and six with risk scores of 744 and 576, respectively.

Table 12.2. Deterministic Risk Scores for Each Risk Region, Habitat, Endpoint, and Source

Risk Region		Endpoints	
Risk region 7	252	SS	504
Risk region 6	576	RCW	444
Risk region 5	744	CMS	294
Risk region 4	198	NFS	240
Risk region 3	0	WS	120
Risk region 2	6	SP	102
Risk region 1	54	DW	90
		DTD	36
Habitats		Sources	
Woody wetlands	138	Natural dispersal	870
Rivers and streams	336	Maritime ports	294
Mixed forest	576	Airports	666
Deciduous forest	168		
Conifer forest	612		

Endpoint abbreviations: DTD, duskytail darter (*Etheostoma percnurum*); DW, dwarf wedgemussel (*Alasmidonta heterodon*); SP, swamp pink (*Helonias bullata*); WS, southern water shrew (*Sorex palustris punctulatus*); NFS, Northern flying squirrel subspecies (*Glaucomys sabrinus coloratus G.s. fuscus*); CMS, Cheat Mountain salamander (*Plethodon nettingi*); RCW, Red-cockaded woodpecker (*Picoides borealis*); SS, Shenandoah salamander (*Plethodon shenandoah*).

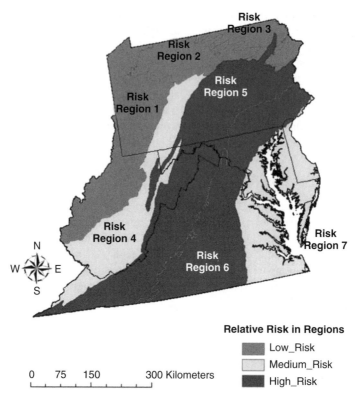

Figure 12.5. Relative risk due to nun moth. The areas of high risk correspond to the habitat range of the preferred species for the invasive.

The Shenandoah salamander and the red cockaded woodpecker had the highest risk scores; the Cheat Mountain salamander and northern flying squirrel had medium risk scores; and the water shrew, swamp pink, dwarf wedgemussel, and duskytail darter had low risk scores.

The habitats most at risk were the conifer and mixed forest types. The rivers and streams habitat had the next highest risk due to indirect effects, whereas the deciduous forest and the woody wetlands had the lowest risk scores.

The highest risk of nun moth larvae introduction came from natural dispersal. Airports and maritime ports came in second and third, respectively.

Uncertainty Analysis. Monte Carlo analysis produced probability distributions for each deterministic risk score calculated. Risk regions 1, 2, and 3 had deterministic risk scores that were less than or equal to the median risk score from their respective probability distributions. All other deterministic risk scores calculated were in the 90th percentile of the Monte Carlo-generated distribution, so there was at most a 10% chance for the risk score to occur.

All risk scores had a positive probability of being zero, and most had a positive probability of being close to zero, so distributions with high deterministic risk scores covered a very large range of values. The risk range of possible risk scores calculated for risk region 5, for example, was from a score of 0 to a score of 952.

Sensitivity Analysis. Most risk scores for region, source, habitat, and endpoint were sensitive to whether or not nun moth would be released. The scores for risk regions were most sensitive to the first exposure filter, which indicates whether or not one or more of the sources would release the stressor within the region. The highest rank correlation values were for 0.74, 0.77, and 0.54 for natural dispersal as a source in risk regions 2, 4, and 5, respectively. Risk region scores were also sensitive to whether or not the species exist in habitat in the risk region, and whether or not the nun moth would survive in the region.

Habitat risk scores were not sensitive to any specific input parameters, with no rank correlation value greater than 0.42. Habitat risk scores were most correlated with whether or not nun moth would be released by airports and natural dispersal, and whether or not the nun moth would survive in the habitat.

Endpoint risk scores were also insensitive to input parameters, with the highest rank correlation being 0.47 for uncertainty associated with the potential for the nun moth to affect the dwarf wedgemussel. The risk score for the swamp pink was most correlated with the likelihood that the nun moth will survive in the woody wetlands of risk region 7. All the other species were most sensitive to the likelihood that a source would release the stressor.

Source risk scores were sensitive to exposure filter A, which indicates the likelihood that the source will release the stressor. The risk score for airports was sensitive to whether or not airports would release the nun moth in risk region 5, with a rank correlation value of 0.68. The score for maritime ports was sensitive to the uncertainty in whether or not ports in region 7 could be direct sources to region 6. The score for natural dispersal was sensitive to whether or not the nun moth could disperse to region 5.

Discussion of Nun Moth Risk Assessment

The risk regions most at risk in this study were risk regions 5 and 6. These regions, which contain the Blue Ridge, ridge and valley, and piedmont ecoregions (US EPA 1997), have the largest amount of high-risk conifer and mixed forest habitat. More detail could be incorporated into the risk assessment to include the fact that risk region 5 contains a varied topography, which may affect the nun moth rate of spread. Conifer and mixed forests were the habitats most at risk in these regions. The assessment species most at risk were the Cheat Mountain and Shenandoah salamanders in region 5 and the Shenandoah salamander in risk region 6; these species use the largest number of habitats. Risk region 3, which had the lowest risk of exposure, has a zero risk because none of the assessment endpoints are found there. Risk region 1, which had the third highest risk of exposure, is also a low-risk region because it has only a moderate amount of habitat and just one assessment species is present. Risk scores to regions in this model are determined by the amount and type of habitat, and the number of assessment species present, so that highest risk region is the most-to-lose region.

Although the risk scores to regions follow the ranking patterns for habitats and assessment species and not the pattern of exposure risk, all risk regions are most sensitive to the likelihood that the nun moth will be released by sources in that region, indicating that to narrow the range of possible risk scores, more information about potential sources is necessary. Jensen (1991) states that forest type and quality, as well as soil type, could be used to estimate susceptibility to an outbreak; incorporating this level of detail into the current risk assessment would lower the uncertainty associated with the survival. More closely examining the materials coming through airports and maritime ports as well as wood treatment procedures in use in the study area would lower the uncertainty associated with source rankings and exposure filter A values. However, the exposure filter value for natural dispersal had the largest influence on most of the risk scores, so further study should attempt to portray the potential for nun moth dispersal in North American habitats more accurately.

INVASIVE SPECIES RISK ASSESSMENT SUMMARY AND CONCLUSIONS

The risk assessments for the Asian oyster and nun moth along with the European Green crab (Colnar and Landis 2007) are examples of detailed risk assessments for invasive species that are comparable in form to ecological risk assessments for contaminants. The case studies of this chapter incorporate large spatial and temporal scales, covering the largest estuary and one of the most populated areas of the United States. Finally, the risk assessments illuminate areas of further investigation in order to reduce uncertainty.

These risk assessments followed the form of other risk assessments performed because of contaminant or other concerns such as Cherry Point (Hart Hayes and Landis 2004) or Codorus Creek, PA (Obery and Landis 2002).

1. The risk assessments in this report each had detailed conceptual models in that described cause–effect relationships.
2. The conceptual models were developed using assessment species deemed to be culturally important attributes of each system.
3. The risk assessments were spatially explicit, exploring patterns within the landscape of interest.
4. The risk calculation was performed following the relative risk model for regional risk assessment that has been used at a variety of sites around the world.
5. The uncertainty and sensitivity analyses were performed using a variety of methods including a detailed Monte Carlo analysis.
6. The sensitivity analyses pointed to features of each conceptual model and risk calculation that required additional description or data.

7. The results were portrayed spatially and with distribution surrounding each of the risk scores. Unlike many chemical-oriented risk assessments, these risks were not based upon a quotient method,[1] but attempt to reflect multiple pathways and combinations of direct and indirect effects that may affect an endpoint.

That risk assessment can be applied to the estimation of the impacts of invasive species in a manner comparable to other environmental stressors has been demonstrated. However, there are clear and important gaps in our knowledge as demonstrated by the uncertainty and sensitivity analyses of the studies.

In each of the case studies, uncertainty was present due to a lack of detailed knowledge about fundamental features of the invasive and the receiving environment. In the case of the Asian oyster, it was not clear what the rate of transition from triploid to fertile diploid would be. Even at a very low frequency of occurrence, the reversion could be important given the number of gametes produced. Specific locations and extent of habitats in the Chesapeake Bay were not available. In the nun moth case study, information on rates of transport of invasive insects and the dispersal rates of the nun moth from an entry site were not known.

A clear pattern for each of the risk assessments in this study and in other cases that we have investigated is the lack of physiological, life history, and ecological information available for each of the invasive species and the assessment species. Although Chesapeake Bay and the Mid-Atlantic states are home to numerous research institutions and funded projects, fundamental questions about these systems remain regarding characteristics of the ecological structure and the biology of the components. No refinement of a risk assessment methodology can make up for the lack of basic information.

ACKNOWLEDGEMENT

This research was supported by US EPA grant no. 1-54068.

REFERENCES

Ahmed M, Barkati S, Sanaullah M. 1987. Spatfall of oysters in the Gharo-phitti salt water creek system near Karachi, Pakistan. *Pakistan J Zool* **19**:245–252.

Allard P, Fairbrother A, Hope BK, Hull RN, Johnson MS, Kapustka L, Mann G, McDonald, B, Sample BE. 2010. Recommendations for the development and application of wildlife toxicity reference values. *Integr Environ Assess Manage* **6**:28–37.

---

[1]Note that Hope (2009) and Allard et al. (2010) are drawing a distinction that the HQ is useful only in scoping and screening assessments and that to have true risk assessments requires a probability of a specified magnitude of effects.

Amemiya I. 1928. Ecological studies of Japanese oyster, with special reference to the salinity of their habitats. *Imperial University, Tokyo J College Agric* **9**:333–382.

Andersen MC, Adams H, Hope B, Powell M. 2004a. Risk assessment for invasive species. *Risk Anal* **24**:787–794.

Andersen MC, Adams H., Hope B, Powell M. 2004b. Risk analysis for invasive species: General framework and research needs. *Risk Anal* **24**:893–900.

ANSTF (Aquatic Nuisance Species Task Force). 1996. *Generic Nonindigenous Aquatic Organisms Risk Analysis Review Process: For Estimating Risk Associated with the Introduction of Nonindigenous Aquatic Organisms and How to Manage for that Risk*. US Government Printing Office, Washington, DC, 36 pp.

Bahr LM, Lanier WP. 1981. *The Ecology of Intertidal Oyster Reefs of the South Atlantic Coast: A Community Profile*. US Fish and Wildlife Service, Office of Biological Service, Washington, DC, 105 pp.

Bartell SM, Nair SK. 2004. Establishment of risks for invasive species. *Risk Anal* **24**:833–845.

Bent AC. 1929. Life histories of North American shorebirds. *US Natural Museum Bull* **146**:236–246.

Breese WP, Malouf RE. 1977. Hatchery rearing techniques for the oyster *Crassostrea rivularis*. *Aquaculture* **12**:123–126.

Cai Y, Deng C, Lui Z. 1992. Studies on the ecology of *Crassostrea rivularis* in Zhanjiang Bay. *Tropic Oceanology/Redai Haiyang, Guangzhou* **11**(3): 37–44.

Carriker MR, Gaffney PM. 1996. The Eastern oyster: *Crassostrea virginica*. In Kennedy VS (Ed.), *A Catalogue of Selected Species of Living Oysters of the World*. College Park, MD, Sea Grant College Program, Chapter 1.

Coen LD, Luckenbach MW. 2000. Developing success criteria and goals for evaluating oyster reef restoration: Ecological function or resource exploitation? *Ecol Eng* **15**:323–343.

Colnar AM, Landis WG. 2007. Conceptual model development for invasive species and a regional risk assessment case study: The European Green Crab, *Carcinus maenas*, at Cherry Point, Washington USA. *Hum Ecol Risk Assess* **13**:120–155.

Crooks JA. 2002. Characterizing ecosystem-level consequences of biological invasions: The role of ecosystem engineers. *Oikos* **97**:153–166.

Cuthbert FJ, Scholtens B, Wemmer LC, McLain R. 1999. Gizzard contents of piping plover chicks in northern Michigan. *Wilson Bull* **111**:121–123.

Deines AM, Chen V, Landis WG. 2005. Modeling the risks of non-indigenous species introductions using a patch-dynamics approach incorporating contaminant effects as a disturbance. *Risk Anal* **6**:1637–1651.

Doak DF, Bigger D, Harding EK, Marvier MA, O'Malley RE, Thompson D. 1998. The statistical inevitability of stability-diversity relationships in community ecology. *Am Naturalist* **151**:264–276.

Fay CW, Neves RJ, Pardue GB. 1983. *Species Profiles: Life Histories and Environmental Requirements of Coastal Fishes and Invertebrates (Mid-Atlantic)-Striped Bass*. US Fish and Wildlife Service, Division of Biological Services, FWS/OBS-82/11.8. U.S. Army Corps of Engineers, TR EL-82-4, 36 pp.

Ferland CL, Haig SM. 2002. *International Piping Plover Census*. US Geological Survey, Forest and Range Ecosystem Science Center, Corvallis, OR. 293 pp.

Ford WM, Stephenson SL, Menzel JM, Black DR, Edwards JW. 2004. Habitat characteristics of the endangered Virginia northern flying squirrel (*Glaucomys sabrinus fuscus*) in the central Appalachian Mountains. *Am Midl Nat* **152**:430–438.

Glowacka B. 1998. The control of the nun moth (*Lymantria monacha* L.) in Poland: A comparison of two strategies. In McManus ML, Liebhold AM (Eds.), *Population Dynamics, Impacts, and Integrated Management of Forest Defoliating Insects*. USDA Forest Service general technical report NE-247, pp. 108–115.

Goulet JM. 1995. Application of Monte Carlo uncertainty analysis to ecological risk assessment. Master's thesis, Western Washington University, Bellingham, WA.

Griffis MR, Jaeger RG. 1998. Competition leads to an extinction-prone species of salamander: Interspecific territoriality in a metapopulation. *Ecology* **79**:2494–2502.

Hart Hayes E, Landis WG. 2004. Regional ecological risk assessment of a near shore marine environment: Cherry Point, WA. *Hum Ecol Risk Assess* **10**:299–325.

Hewitt CL, Hayes KR. 2002. Risk assessment of marine biological invasions. In Lepplakoski E, Gollasch S, Olenin S (Eds.), *Invasive aquatic species of Europe: Distribution, Impacts and Management*. Kluwer Academic Publishers, Dordrecht, The Netherlands, pp. 456–466.

Hirase S. 1930. On the classification of Japanese oysters. *Japanese J Zool* **3**:1–65.

Hope 2009.

Jensen TS. 1991. Integrated pest management of the nun moth, *Lymantria monacha* (Lepidoptera: Lymantriida) in Denmark. *Forest Ecol Manage* **39**:29–34.

Keena MA. 2003. Survival and development of Lymantria monacha (Lepidoptera: Lymantriidae) on North American and introduced Eurasian tree species. *J Econ Entomol* **96**:43–52.

Kolar CS, Lodge DM. 2001. Progress in invasion biology: Predicting invaders. *Trends Ecol Evol* **16**:199–204.

Kolar CS, Lodge DM. 2002. Ecological predictions and risk assessment for alien species. *Science* **298**:1233–1236.

Kolar CS. 2004. Risk assessment and screening for potentially invasive species. *NZ J Marine Freshwater Res* **38**:391–397.

Landis WG. 2004. Ecological risk assessment conceptual model formulation for nonindigenous species. *Risk Anal* **24**:847–858.

Landis WG, Wiegers JK. 1997. Design considerations and a suggested approach for regional and comparative ecological risk assessment. *Hum Ecol Risk Assess* **3**:287–297.

Landis WG, Wiegers JK. 2005. Introduction to the regional risk assessment using the relative risk model. In Landis WG (Ed.), *Regional Scale Ecological Risk Assessment Using the Relative Risk Model*. CRC Press, Boca Raton, FL, pp. 11–36.

Landis WG, Duncan PB, Hart Hayes E, Markiewicz AJ, Thomas JF. 2004. A regional assessment of the potential stressors causing the decline of the Cherry Point Pacific herring run and alternative management endpoints for the Cherry Point Reserve (Washington, USA). *Hum Ecol Risk Assess* **10**:271–297.

Lovett GM, Christenson LM, Groffman PM, Jones CG, Hart JE, Mitchell MJ. 2002. Insect defoliation and nitrogen cycling in forests. *Bioscience* **52**:335–341.

Mack RN, Simberloff D, Lonsdale WM, Evans H, Clout M, Bazzaz FA. 2000. Biotic invasions: Causes, epidemiology, global consequences, and control. *Ecol Appl* **10**:689–710.

Magnin E, Beaulieu G. 1967. Striped bass of the St. Lawrence River. *Nature Canada* **94**:539–555.

Manooch CS III. 1973. Food habits of yearling and adult striped bass, *Marone saxatilis*, from Albemarle Sound, North Carolina. *Chesapeake Sci* **14**:73–86.

Marvier M, Kareiva P, Neubert MG. 2004. Habitat destruction, fragmentation, and disturbance promote invasion by habitat generalists. *Risk Anal* **24**:869–878.

McLane WM. 1955. The fishes of the St. John's River system. Doctoral dissertation, University of Florida, Gainesville (FL), 361 pp.

Mitchell D. 2001. Spring and fall diet of the endangered West Virginia northern flying squirrel (*Glaucomys sabrinus fuscus*). *Am Midl Nat* **146**:439–443.

Moraes R, Landis WG, Molander S. 2002. Regional risk assessment of a Brazilian rain forest reserve. *Human Ecol Risk Assess* **8**:1779–1803.

NatureServe. 2005. *NatureServe Explorer: An Online Encyclopedia of Life*. www.natureserve. org/explorer. Accessed August 20, 2005.

Newcombe CL. 1945. *The Biology and Conservation of the Blue Crab, Callinectes sapidus Rathbun*. Virginia Fish Lab Edited Series, Number 4: Gloucester Point, VA. 39 pp.

Novak V, Hroznika F, Stary, B. 1976. *Atlas of Insects Harmful to Forest Trees*, Vol. **1**. Elsevier Scientific, New York, 125 pp.

NRC (National Research Council). 2004. *Nonnative Oysters in the Chesapeake Bay*. The National Academies Press, Washington, DC.

NSTC (National Science and Technology Council). 1999. *Ecological Risk Assessment in the Federal Government*. Committee on the Environment and Natural Resources CENR/5-99/001.

Obery AM, Landis WG. 2002. A regional multiple stressor risk assessment of the Codorus Creek watershed applying the relative risk model. *Human Ecol Risk Assess* **8**:405–428.

Patel SK, Jetani KL. 1991. Survey of edible oysters from the Saurashtra coast. *J Curr Biosci* **8**:79–82.

Piers H. 1923. The blue crab: Extension of its range northward to near Halifax, Nova Scotia. *Proc Nova Scotian Inst Sci* **15**:83–90.

Powers GLJ, Mayden RL. 2003. Threatened fishes of the world: *Etheostoma percnurum* Jenkins 1993 (Percidae). *Environ Biol Fishes* **67**:358–358.

Rao KS. 1987. Taxonomy of Indian oysters. *Central Marine Fisheries Res Inst Bull* **38**:1–6.

Ricciardi A, 2006. Patterns of invasion in the Laurentian Great Lakes in relation to changes in vector activity. *Diversity and Distributions* **12**:425–433.

Riccardi A, Cohen J. 2007. The invasiveness of an introduced species does not predict its impact. *Biol Invasions* **9**:309–315.

Russell CA, Kosola KR, Paul EA, Robertson GP. 2004. Nitrogen cycling in poplar stands defoliated by insects. *Biogeochemistry* **36**:365–381.

Scattergood LW. 1960. Blue crabs in Maine. *Maine Field Natur* **16**(3): 59–63.

Sellens L, Markiewicz AJ, Landis WG. 2007. *Risk Evaluation of Invasive Species Transport Across the US–Canada Border in Washington State*. Report prepared for the Border Policy Research Institute, Western Washington University, Bellingham, WA. US Department of Transportation, Office of the Secretary, DTOS59-05-C-00016, May 31, 2007, 50 pp. http://www.ac.wwu.edu/~ietc/BPRIinvasiverisk.pdf.

Shaffer F, Laporte P. 1994. Diet of piping plovers on the Magdalen Islands, Quebec. *Wilson Bull* **106**:531–536.

Shea K, Chesson P. 2002. Community ecology theory as a framework for biological invasion. *Trends Ecol Evol* **17**:170–176.

Smith, GF, KN., Greenhawk, DG. Bruce, EB Roach, and SJ. Jordan. 2001. *J Shellfish Res* **20**(1):197–206.

Thayer, GW, Adams, SW, and LaCroix MW. 1975. Structural and functional aspects of a recently established *Zostera marina* community, in Cronin, LE (Ed.) *Estuarine Research*, vol. 1. Academic Press, New York, pp. 517–540.

Thomas J. 2005. Codorus Creek: Use of the relative risk model ecological risk assessment as a predictive model for decision making. In Landis WG (Ed.), *Regional Scale Ecological Risk Assessment Using the Relative Risk Model*. CRC Press, Boca Raton, FL, pp. 143–158.

Tschang S, Tse-kong L. 1956. A study on Chinese oysters. *Acta Zool Sin* **8**:65–93.

USDA (United States Department of Agriculture) Forest Service. 1991. *Pest Risk Assessment of the Importation of Larch from Siberia and the Soviet Far East*. Forest Service Miscellaneous Publication No. 1495, September 1991, p. 27.

USDA (United States Department of Agriculture). 2000. *Pest Risk Assessment for Importation of Solid Wood Packing Materials into the United States*. USDA Animal and Plant Health Inspection Service and Forest Service Raleigh, NC.

US EPA (United States Environmental Protection Agency), Chesapeake Bay Program, Department of Interior, United States Fish and Wildlife Service. 1996. *Chesapeake Bay: Introduction to an Ecosystem*. Chesapeake Bay Program, Annapolis, MD.

US EPA (United States Environmental Protection Agency). 1997. *An Ecological Assessment of the United States Mid-Atlantic Region: A Landscape Atlas*. EPA 600/R-97/130 Washington, DC.

USFWS (United States Fish and Wildlife Service). 1990. *Appalachian Northern Flying Squirrels (Glaucomys sabrinus fuscus and Glaucomys sabrinus coloratus) Recovery Plan*. Newton Corner, MA.

USFWS (United States Fish and Wildlife Service). 1991a. *Cheat Mountain Salamander (Plethodon nettingi) Recovery Plan*. Newton Corner, MA, 35 pp.

USFWS (United States Fish and Wildlife Service). 1991b. *Swamp Pink (Helonias bullata) Recovery Plan*. Newton Corner, MA, 56 pp.

USFWS (United States Fish and Wildlife Service). 1993a. *Duskytail Darter (Etheostoma percnurum) Recovery Plan*. Atlanta, GA, 25 pp.

USFWS (United States Fish and Wildlife Service). 1993b. *Dwarf Wedge Mussel (Alamidonta heterodon) Recovery Plan*. Hadley, MA, 52 pp.

USFWS (United States Fish and Wildlife Service). 1994. *Shenandoah Salamander (Plethodon Shenandoah) Recovery Plan*. Hadley MA, 36 pp.

USFWS (United States Fish and Wildlife Service). 2003. *Red-Cockaded Woodpecker (Picoides Borealis) Recovery Plan*, second revision. Atlanta, GA.

Vermeij GJ. 1996. An agenda for invasion biology. *Biol Conserv* **78**:3–9.

Wardle DA. 2001. Experimental demonstration that plant diversity reduces invisibility— Evidence of a biological mechanism or a consequence of sampling effect? *Oikos* **95**:161–170.

Warren-Hicks WJ, Moore DRJ. 1998. *Uncertainty Analysis in Ecological Risk Assessment*. SETAC Press, Pensacola, FL, 277 pp.

Wiegers JK, Feder HM, Mortensen LS, Shaw DG, Wilson VJ, Landis WG. 1998. A regional multiple-stressor rank-based ecological risk assessment for the fjord of Port Valdez, Alaska. *Hum Ecol Risk Assess* **4**:1125–1173.

Wu J, David JL. 2002. A spatially explicit hierarchical approach to modeling complex ecological systems: theory and applications. *Ecol Modelling* **153**:7–26.

Wu J, Loucks OL. 1995. From balance of nature to hierarchical patch dynamics: A paradigm shift in ecology. *Q Rev Biol* **70**:439–466.

Wu, XZ, Pan JP. 2000. An intracellular prokaryotic microorganism associated with lesions in the oyster, *Crassostrea ariakensis*. *J Fish Dis* **23**:409–414.

Zhou M, Allen SK Jr. 2003. A review of published work on *Crassostrea ariakensis*. *J Shellfish Res* **22**:1–20.

13

ECOLOGICAL RISK ASSESSMENT OF THE INVASIVE *SARGASSUM MUTICUM* FOR THE CHERRY POINT REACH, WASHINGTON

Ananda Seebach, Audrey M. Colnar, and Wayne G. Landis

INVASIVE SPECIES RISK ASSESSMENT

Colnar and Landis (2007) developed a risk assessment model for examining the risk due to nonindigenous species at large scales. The initial case study was for the European green crab (Colnar and Landis 2007). Asian oyster and nun moth (see Chapter 12) presents two additional case studies with invasive species of very different characteristics. The Asian oyster was a planned and intentional introduction to the Chesapeake Bay. The nun moth, a European species with characteristics similar to those of the gypsy moth, was studied in the Mid-Atlantic states.

This chapter presents a detailed case study performed during the development of the risk assessment process for invasive species. The assessment process is based upon the Relative Risk Model (RRM) (Landis and Wiegers 1997, 2005; Wiegers et al. 1998), one approach currently used to predict the risks of impacts at a regional scale. The RRM quantitatively ranks sources of stressors and habitats by using Geographic Information Systems (GIS) to analyze spatial datasets to determine risk at the regional level. This RRM has proven very useful in many circumstances including the multi-stressor risk assessments for the Fjord of Port Valdez, Alaska (Wiegers et al. 1998), Cherry Point, Washington (Hart Hayes and Landis 2004), and many others. These risk

Environmental Risk and Management from a Landscape Perspective, edited by Kapustka and Landis
Copyright © 2010 John Wiley & Sons, Inc.

assessments have only been performed for chemicals and other abiotic stressors and have not considered organisms, specifically invasive species, as stressors.

The marine brown alga, *Sargassum muticum* (Yendo) Fensholt (sargassum), provides a different situation than the other case studies. It is widely spread within the region and is extant within the Cherry Point reach. There is much concern regarding the potential impacts of sargassum upon native nearshore intertidal and subtidal organisms. There are two critical characteristics that make this case study distinctive. First, the life history of sargassum, especially its ability to reproduce asexually, changes the dynamics of invasion compared to our first three test cases. Second, the small- and large-scale dynamics of the Cherry Point study area influence the habitat suitability and patch dynamics of sargassum, which in turn affect the success of its invasion.

The description of the risk assessment process is detailed in this chapter and is a template for the methods applied to the nun moth and the Asian oyster assessments. Background on the hierarchical nature of this approach is presented along with details of the general process. The next sections introduce details of the biology and life history of sargassum and present the methods used for this risk assessment. The risk calculations and the associated uncertainties are then presented. Unique in this study is the result that sargassum provides a negative risk or a potential benefit to several of the assessment species at the Cherry Point region. Finally, a brief comparison is made to the risks to the region posed by European green crab.

Introduction to *Sargassum muticum*

Sargassum has been introduced, either directly or indirectly from its indigenous Japanese habitat to North America, Europe, and the Mediterranean Sea. Since its introduction during the early 1900s, sargassum has successfully established populations in the subtidal and intertidal zones along the Pacific Coast of North America and along both sides of the English Channel. For this ecological risk assessment, we evaluated the risk of impacts from sargassum to selected assessment species under current conditions and considered the possibility of its future spread within the Cherry Point region.

Sargassum was originally recognized in the 1920s and was first documented as a distinct species by D. Fensholt in 1951 (Druel 2000). It may have been accidentally introduced to North America as early as 1902 with shipments of Japanese oysters that were imported for aquaculture (Britton-Simmons 2004). Since its arrival, sargassum has successfully spread along the North American coastline from Alaska to Mexico at an average rate of approximately 60 km per year (Farnham et al. 1981) and with a possible maximum rate of 1100 km per year (Deysher and Norton 1982).

Vectors for the successful spread and establishment of sargassum populations include dispersal with ships and pleasure craft through entanglement in anchor chains or propellers (Jansson 2000). It is also frequently introduced with shellfish importation. Sargassum plants can survive up to three months while drifting long distances by wind-generated surface currents, and they are often fully fertile and able to recolonize new areas on arrival (Farnham et al. 1981). Because it is monoecious with the ability for self-fertilization, its floating fertile material is a very effective method of dispersing zygotes (Critchley et al. 1983).

Sargassum grows by attaching its conical holdfast to a solid surface, such as rock and shell substrata, and growth occurs between May and late September. Plants can reach 6 m in length and are able to form large canopies due to their extensive branching and buoyancy (Critchley 1983a, Giver 1999). Receptacles (reproductive structures) and air vesicles arise from the laterals of each frond, with both male and female gametangia being present in conceptacles within the same receptacle (Critchley 1983a). Up to 24 conceptacles can develop in a single receptacle. Additional fronds develop with age; and each frond is equally fertile during the reproductive season (June to September), thereby enhancing reproductive output (Critchley 1983b). The fronds and reproductive structures are cast-off from the perennial holdfast towards the end of the reproductive cycle and can float considerable distances due to the numerous air vesicles (Critchley et al. 1983). During the autumn months the plants die back and over-winter.

The preferred habitat of sargassum is a hard substrate, such as rock-cobble, mollusk shells, and man-made structures located in the lower intertidal and upper subtidal zones (den Hartog 1997). It can also grow in mudflats and sandy habitats when the substrate consists of flintstones, rocks, or broken shells, but only when there is not exposure to desiccation (Farnham et al. 1981). Sargassum prefers protected to moderately exposed rocky areas and tide pools where direct wave exposure is limited.

This brown alga is very tolerant to variations in salinity. A salinity of 34 ppt is optimal for growth, but plants have survived in salinities as low as 6 ppt (Jansson 2000). Sargassum is also extremely tolerant to variations in temperature. It can grow within a range of 5°C to 25°C. Plants have even survived long periods in water temperatures of -1.4°C (Jansson 2000).

The most extensive stands of sargassum have become established in bays (Deysher and Norton 1982, Karlsson and Loo 1999), where the plants have been able to remain attached to the substrate for a period of several years. Its morphological plasticity in the production of main branches and branch elongation reduces interference from neighboring plants, allowing it to grow in locally dense populations with low mortality (Arenas et al. 2002). Given optimal growing conditions, sargassum is able to form populations covering many hundreds of hectares (Critchley et al. 1983).

Damage to plants and declines in sargassum populations are most often caused by exposure to summertime desiccation, winter storms, log abrasion, and freezing temperatures. Grazers can also limit its growth. In turbid waters, overlapping leaves can collect sediment, which promotes fungal attack and can lead to the loss of apical dominance (Critchley 1983a).

One benefit of this brown alga's growth is its ability to support an extensive community of micro- and macro-fauna (Norton and Benson 1983). It has been shown to support more epiphytic biomass per gram of algal tissue than the native macroalga it most often displaces, *Laminaria saccharina* (Giver 1999). Overall, epibiont diversity and abundance appears to increase in areas invaded by sargassum (Giver 1999). Browsers benefit from the detritus and diatoms associated with it, and grazers use it as a primary food source. Moreover, sargassum provides refuge for several fish species as well as juveniles of both the red rock and Dungeness crab (Giver 1999).

Conversely, sargassum competes with seagrass (Zostera) and other macroalgae for space and light, including Alaria, Bifurcaria, Cystoseira, Fucus, Gracilaria, Halidrys, Laminaria, Macrocystis, Neorhodomela, Nereocystis, Rhodomela, Scytosiphon, and Ulva. Its ability to alter sedimentation processes, reduce light penetration, decrease water movement, and reduce oxygen levels makes it a very effective competitor. In cases where both sargassum and *Rhodomela larix* were present and reproductive, sargassum was able to dominate quite effectively (DeWreede 1983). In areas where the intertidal and subtidal substrate was a mix of sand, gravel, and stones, sargassum restricted the growth of permanent, native communities of eelgrass (*Zostera marina*) (den Hartog 1997). On the coast of Spain a 50% increase in sargassum cover caused a significant decline in the native leathery alga *Bifurcaria bifurcata* (Viejo 1997). Moreover, sargassum has replaced native *Halidrys siliquosa* in deep pools and channels in England, as well as native *Z. marina* and *L. saccharina* populations along the French Atlantic coast (Viejo 1997).

HIERARCHICAL INVASIVE RISK MODEL (HIRM)

The HIRM derivative of the RRM was developed as part of a previous program to provide frameworks to incorporate the differences between conventional stressors and invasive species (Landis 2004). The hierarchical patch dynamics paradigm (HPDP) (Wu and Loucks 1995, Wu and David 2002) was incorporated into the RRM process. This process is extensively described in Colnar and Landis (2007) and Landis et al. (Chapter 12, this volume) and is summarized below.

As with the original RRM, uncertainty analysis is performed using a number of methods including a Monte Carlo analysis and the sensitive parameters are identified. Maps are used to compare risk rankings with charts delineating habitats or endpoints at the highest risk.

Problem Formulation

In this study we used the six risk region boundaries that were defined defined by Hart Hayes and Landis (2004, 2005) and Colnar and Landis (2007) (Fig. 13.1). Similarly, we used the nine assessment species/taxa from those studies: (1) Coho salmon (*Onchorhynchus kisutch* (Walbaum)), (2) juvenile Dungeness crab (*Cancer magister*), (3) English sole (*Parophrys vetulus*), (4) great blue heron (*Ardea herodias*), (5) common littleneck clam (*Protothaca staminea*), (6) surf smelt embryos (*Hypomesus pretiosus*), (7) Pacific herring (*Clupea harengus pallasi*), (8) eelgrass (*Zostera marina*) and (9) macro-algae. And we used seven of the ten habitat types: (1) gravel-cobble intertidal, (2) sandy intertidal, (3) mudflats, (4) eelgrass, (5) macroalgae, (6) soft bottom subtidal, (7) water column; stream, wetland, and forest habitats were excluded from this study because they represented the terrestrial component of the Cherry Point study area, which does not apply in this marine-based risk assessment.

Several sources have been responsible for the accidental introduction of aquatic nonindigenous species such as sargassum. Based on the scientific literature concerning

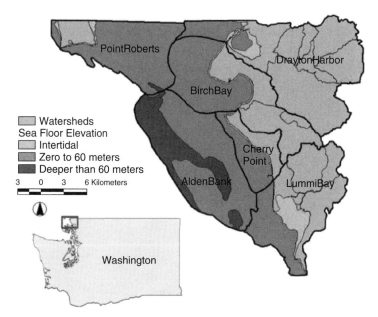

Figure 13.1. Map of the study area at Cherry Point, WA with the risk regions designated.

current and historical sargassum sources, we identified five possible sources of its introduction to the area that will be evaluated in this study. These sources include (1) aquaculture shipments, (2) ballast water, (3) recreational boats, (4) live seafood shipments, and (5) passive current dispersal.

From these layers of information, we developed a hierarchical conceptual model that had been designed previously (Colnar 2004). This model not only allows for the exploration of each potential pathway leading to impacts, but also illustrates the invasion process. It also addresses the concept of scale and the factors that influence invasion at each scale. This concept originated from the hierarchical patch dynamic paradigm (HPDP) (Wu and Loucks 1995, Wu and David 2002) in which the system is evaluated using three levels or scales: local, focal and regional (Fig. 13.2). In this model each scale, beginning with the local scale, has an increasing spatial extent, which in turn affects the types of factors influencing the invasion process. The local scale of the conceptual model is nested within the focal scale, which is in turn integrated into the regional scale (Fig. 13.2). It is important to note that the relationship between the scales is relatively symmetric and does not imply top-down or bottom-up control.

The local scale has the smallest spatial scale and provides the mechanistic processes for the overall model (Fig. 13.3). At this scale, introduction of sargassum occurs in discrete habitat patches within the Cherry Point area. The effects (Fig. 13.4) are due to interactions of sargassum with the endpoints within each patch. The effects may be either potentially undesirable (e.g., resource competition) or potentially beneficial (e.g., increased shelter). The factors influencing exposure are more localized with respect to habitat patches so that small factors, such as a freshwater discharge

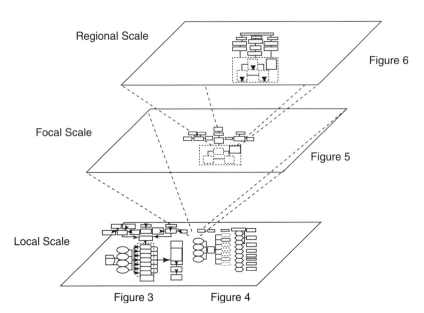

Figure 13.2. Schematic diagram of the hierarchical conceptual model for Cherry Point and the integration of the local, focal, and regional scales.

pipe, may influence the localized temperature and salinity, as well as possible contaminant load within a particular habitat patch. The effects, which include both potentially beneficial and potentially undesirable effects, were determined based on information from the scientific literature.

All of the interactions occurring within each patch at the local scale were then integrated together to represent the focal scale (Fig. 13.5). In this study, the focal scale is the spatial extent of the Cherry Point study region and includes the overall habitats, associated endpoint populations, and all sargassum patches, as well as interactions between the patches due to transport vectors and current dispersal. Additionally, all relevant habitat suitability parameters at the focal scale are generalized for the region to include average ranges and interactions at the metapopulation level. The factors influencing exposure at this scale (average temperature, salinity, interactions with endpoints at the population and metapopulation level) are generalized for the entire region, whereas the effects are considered at the population level within Cherry Point.

The regional scale (Fig. 13.6) has the largest spatial scale and provides a general context for the overall model. The spatial extent is that of the entire Pacific coast of the United States. Consequently, this scale considers all populations of sargassum on the eastern Pacific coast and their interactions through transport vectors (such as ballast water) and passive current dispersal. The factors influencing exposure are also of a much larger scale and include oceanic processes and climate regimes, which in turn influence events such as the Pacific Decadal Oscillation (PDO) and *El Nino/La Nina* events.

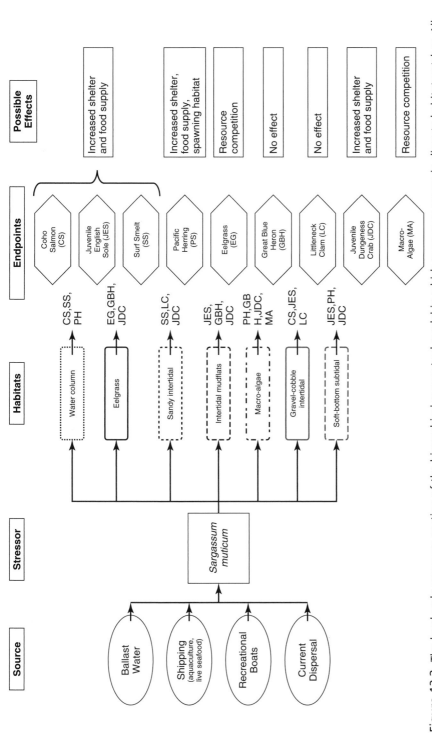

Figure 13.3. The local scale exposure portion of the hierarchical conceptual model in which sargassum occurs in discrete habitat patches. All sources and habitats as well as the related local habitat suitability parameters that influence survival and growth are included. Each arrow represents an individual pathway from the source and habitat to the survival of the sargassum populations.

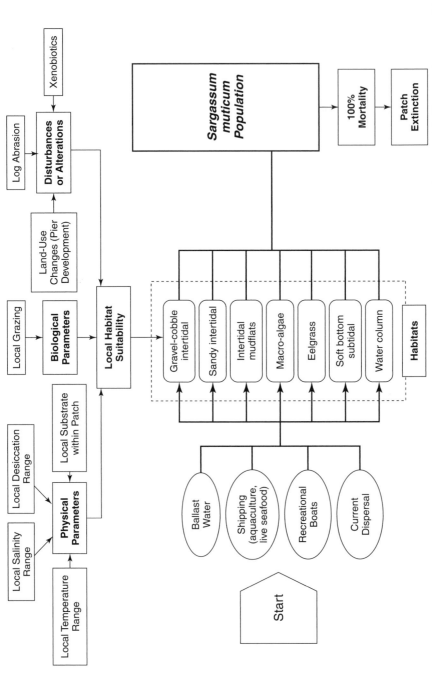

Figure 13.4. The local scale effects portion of the hierarchical conceptual model. This model illustrates the potential effects pathways resulting from local patch overlap and subsequent interaction with sargassum. The effects may be either potentially unesirable (e.g., resource competition) or potentially beneficial (e.g., increased shelter).

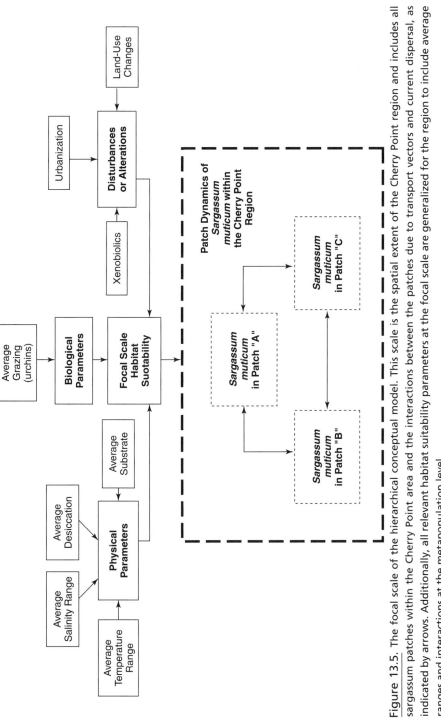

Figure 13.5. The focal scale of the hierarchical conceptual model. This scale is the spatial extent of the Cherry Point region and includes all sargassum patches within the Cherry Point area and the interactions between the patches due to transport vectors and current dispersal, as indicated by arrows. Additionally, all relevant habitat suitability parameters at the focal scale are generalized for the region to include average ranges and interactions at the metapopulation level.

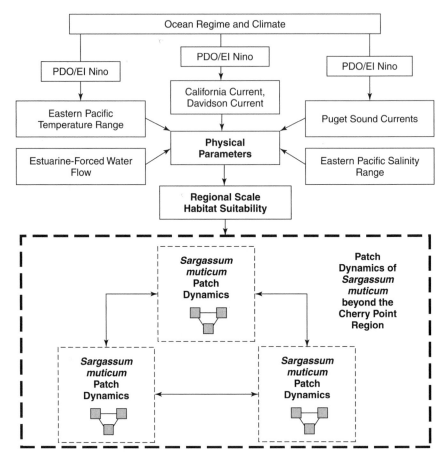

Figure 13.6. The regional scale of the hierarchical conceptual model. This scale includes all sargassum patches within the entire Pacific coast region of the United States and the interactions between the patches through transport vectors and current dispersal, as indicated by arrows. The large-scale parameters that influence regional scale habitat suitability are also included.

ANALYSIS

To analyze the risk of exposure and effects, we used the RRM methods to evaluate each possible source–habitat–endpoint pathway identified in the conceptual model. The sources and habitats were ranked and the combinations filtered using specific exposure and effects criteria.

Development of Habitat and Source Ranks

We incorporated the habitat ranking scheme developed by Hart Hayes and Landis (2004, 2005). The habitat ranks ranged from zero to six in increments of two, where

Table 13.1. Sargassum EcoRA Source Rank Criteria

Source	Ranking Criteria	Range	Ranks
Aquaculture shipments	Presence/absence of aquaculture operation	Absent	0
		Present	6
Ballast water	Presence/absence of vessels discharging ballast water	Absent	0
		Present	6
Live seafood shipments	Presence/absence of live seafood shipments	Absent	0
		Present	6
Passive current dispersal	Presence/absence of currents suitable for transport of sargassum	Absent	0
		Present	6
Recreational boats	Presence/absence of recreational boats	Absent	0
		Present	6

zero represents the lowest potential for exposure (no habitat present) and six represents the highest potential (relatively largest amount of habitat present).

Alternatively, the source ranks were based solely on presence/absence of each source with zero representing absence of source (low exposure potential) and six indicating presence of source (high exposure potential) (Table 13.1). Only zero and six were used as the minimum and maximum source ranks, respectively.

Development of Exposure and Effects Filters

The exposure filters, which evaluate the source–habitat pathways, were developed based on the following components, adapted from Wiegers et al. (1998):

- Will the source release the stressor?
- Will the stressor then occur and persist in the habitat? (For example, is the habitat suitable to allow for survival and growth of the stressor?)

We created a decision tree (Fig. 13.7) to aid in the assignment of the filter value related to exposure filter Component A, which considers whether or not the source will release the stressor. A value of zero indicated the source would most likely not release the stressor, whereas a value of one indicated that release would occur.

A decision tree was also designed to assign the filter values related to exposure filter Component B, which evaluates habitat suitability (Fig. 13.8). This decision tree was based on the habitat suitability parameters identified in the conceptual model. The values assigned were 0, 0.5, or 1, with a 0 indicating unsuitable habitat, 0.5 representing moderately suitable habitat, and 1 representing highly suitable habitat.

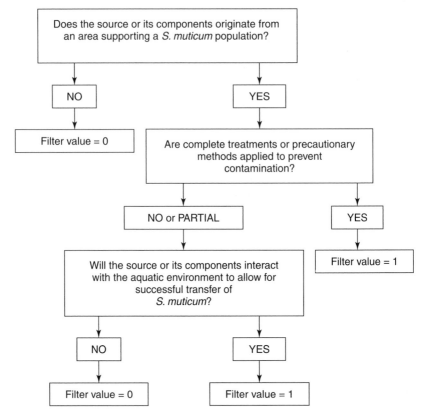

Figure 13.7. Decision tree used to determine the value for exposure filter Component A (e.g., Will the source release sargassum?).

Disturbances were not considered as parameters in this analysis because there is not enough evidence available to determine the role of disturbances in invasion (Ruiz et al., 1999).

Similar to the exposure filters, the effects filters were developed based on three components, partially adapted from previous risk assessments (Wiegers et al. 1998, Hart Hayes and Landis 2004). The three effects filter components included the following:

- Does the endpoint occur in and utilize the habitat?
- Is there seasonal overlap in habitat usage between the stressor and the endpoint?
- Are effects (either beneficial or undesirable) to the endpoint possible from interaction with the stressor?

To assign values for whether the endpoint utilized the habitat (effects filter Component A), we incorporated a previously developed ranking scheme (Hart Hayes and

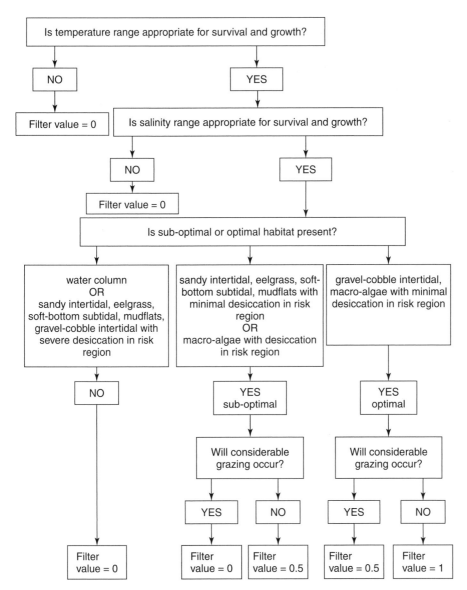

Figure 13.8. Decision tree used to determine the value for exposure filter Component B (e.g., Is the habitat suitable enough to allow for survival and growth of sargassum?).

Landis 2004, 2005) in which a 0 indicated that the endpoint did not use the habitat, a 0.5 indicated that the endpoint used the habitat only marginally, and a 1 indicated that the endpoint completely used the habitat.

Seasonal overlap between the stressor and each endpoint (effects filter Component B) was addressed by assigning a 0 if overlap was not possible, whereas a value of 1 was assigned if overlap was possible. All endpoints were assigned a value of 1 because seasonal habitat overlap was expected with sargassum, indicated by the life history of each organism.

Finally, the possible effects (effects filter Component C) were addressed by assigning a 0 for no effects possible, a positive value (+1) for potentially undesirable effects, and a negative value (−1) for potentially beneficial effects. In previous risk assessments using the relative risk model, positive values indicated risk (Hart Hayes and Landis 2004, 2005). Therefore in this risk assessment a positive value (+1) was chosen to represent potentially undesirable effects, which lead to risk. Conversely, a negative value (−1) was used to represent reduced risk associated with possible beneficial effects.

Risk Characterization

Using the relative risk model, all ranks were converted into a point system. To generate the exposure and effects filters for each pathway, the components of the exposure and effects filters were multiplied. Finally, the source and habitat ranks were integrated with the exposure and effects filter products to generate the following predictions: (1) risk to assessment endpoints, (2) risk in risk regions, and (3) contribution of each source to risk.

Uncertainty Analysis

The sources and amounts of uncertainty within each component of the RRM were identified and addressed using Monte Carlo analysis. Using methods similar to that of Hart Hayes and Landis (2004), we classified the uncertainty for each rank and filter component as low, medium, or high based on the amount of confidence within each assigned value. We then assigned discrete statistical distributions to represent the uncertainty within the ranks and filter components with medium and high classifications according to specific criteria (Table 13.2). The ranks and filter components with low uncertainty classifications were not assigned a distribution, but instead retained their original value. The habitat rank and effects filter Component A uncertainty classifications determined by Hart Hayes and Landis (2004, 2005) was used in this study.

We ran the Monte Carlo simulations for 1000 iterations using Crystal Ball® 2000 software as a macro in Microsoft® 2002 Excel. The output was generated in the form of statistical distributions representing the range of possible final risk scores for each assessment endpoint, risk region, and source. The final Monte Carlo risk score outputs included both positive and negative values. Positive values indicate relative risk, whereas negative values indicate reduced relative risk.

We then conducted a sensitivity analysis using the Crystal Ball® 2000 software as well. During sensitivity analysis, correlation coefficients were generated to rank the relative risk model parameters according to their contribution to prediction uncertainty. Consequently, a high rank correlation indicates that the uncertainty within the model parameter has great importance in influencing the uncertainty within the model.

Table 13.2. Example Distributions for Uncertainty Analysis for Two Elements of the Risk Assessment

Uncertainty Analysis Input Distributions for the Effects Filter Component A Values (Does the endpoint occur in and utilize the habitat?), Originally Designed by Hart Hayes and Landis (2004)

Assigned Value	Uncertainty	Probability (%) for Each Possible Value		
		0	0.5	1
0	Medium	80	10	10
	High	60	20	20
0.5	Medium	0	80	20
	High	0	60	40
1	Medium	0	20	80
	High	0	40	60

Uncertainty Analysis Input Distributions for the Effects Filter Component C Values (Are Effects, Either Beneficial or Undesirable, to the Endpoint Possible from Interaction with *Sargassum muticum*?)

Assigned Value	Uncertainty	Probability (%) for Each Possible Value		
		0	1	−1
0^a	Medium	80	20	0
	High	60	40	0
0^b	Medium	80	0	20
	High	60	0	40
+1	Medium	20	80	0
	High	40	60	0
−1	Medium	20	0	80
	High	40	0	60

[a] If undesirable effects are possible.
[b] If beneficial effects are possible.

RESULTS

The results from this study are presented in three subsections: risk characterization, uncertainty, and sensitivity analysis.

Risk Characterization

The risk characterization phase yielded overall final risk scores for each assessment endpoint, risk region and source.

Assessment Endpoints. The greatest risk was to the macro-algae and eelgrass assessment species, as indicated by the positive values (Table 13.3). Conversely, the

Table 13.3. Relative Risk to Each Assessment Endpoint Within Each Risk Region

Risk Region	Assessment Species[a]								
	CS	EG	GBH	DC	ES	LC	MA	PH	SS
Birch Bay	0	24	0	−120	−60	0	72	−96	−48
Cherry Point	0	18	0	−54	−27	0	72	−72	−90
Drayton Harbor	−6	0	0	−12	−6	0	12	−12	0
Alden Bank	−6	0	0	−6	0	0	12	0	0
Lummi Bay	−9	0	0	−54	−36	0	18	−54	0
Point Roberts	−12	0	0	−12	0	0	24	−24	0
Final assessment endpoint risk score	−33	42	0	−258	−129	0	210	−258	−138

[a]CS, coho salmon; EG, eelgrass; GBH, great blue heron; DC, juvenile Dungeness crab; ES, juvenile English sole; LC, adult Littleneck clam; MA, macro-algae; PH, Pacific herring; SS, surf smelt embryos.

Coho salmon, surf smelt, juvenile English sole, Pacific herring, and juvenile Dungeness crab actually received negative risk scores, due to the possibility of beneficial effects occurring to each of these endpoints. The great blue heron and littleneck clam assessment species did not exhibit any apparent risk.

Risk Regions. All of the risk regions, except for Alden Bank, received negative risk scores (Table 13.4). These negative values are the result of the summation of both the positive and negative risk scores for each assessment endpoint within each region. Independent of the positive and negative values, a larger risk score indicated the possibility of more potential effects occurring to the assessment endpoints, which can be either beneficial or undesirable. The Birch Bay risk region received the highest negative risk score, suggesting that the majority of effects are going to occur within this region.

Table 13.4. Relative Risk in Each Risk Region as Sorted by Assessment Endpoint

Assessment Endpoint	Risk Region					
	Alden Bank	Birch Bay	Cherry Point	Drayton Harbor	Lummi Bay	Point Roberts
Coho salmon	−6	0	0	−6	−9	−12
Eelgrass	0	24	18	0	0	0
Great blue heron	0	0	0	0	0	0
Juvenile Dungeness crab	−6	−120	−54	−12	−54	−12
Juvenile English sole	0	−60	−27	−6	−36	0
Littleneck clam	0	0	0	0	0	0
Macro-algae	12	72	72	12	18	24
Pacific herring	0	−96	−72	−12	−54	−24
Surf Smelt	0	−48	−90	0	0	0
Final risk region risk score:	0	−228	−153	−24	−135	−24

Table 13.5. Relative contribution of Each Source to Risk

Risk Region	Source[a]				
	AS	BW	RB	LSS	PCD
Birch Bay	0	0	−144	0	−156
Cherry Point	0	−75	−75	0	−75
Drayton Harbor	0	0	−18	0	−18
Alden Bank	0	0	0	0	−12
Lummi Bay	−51	0	−51	0	−51
Point Roberts	0	0	−24	0	−24
Final source risk score:	−51	−75	−312	0	−336

[a] AS, aquaculture shipments; BW, ballast water; RB, recreational boats; LSS, live seafood shipments; PCD, passive current dispersal.

Sources. The sources that contribute the most to the introduction of sargassum are passive current dispersal and recreational boats (Table 13.5). The source risk scores are negative values due to the original summation of both the positive and negative risk scores for each assessment species. The sources contribute to risk, but do not represent the risk of effects to the species.

Uncertainty Analysis

The Monte Carlo analysis produced probability distributions for each assessment species, risk region, and source risk score. The distributions show all possible risk scores and the probability of those scores occurring as a result of the uncertainty within the model inputs.

Assessment Species. Of the assessment species potentially exhibiting beneficial effects, the juvenile Dungeness crab and Pacific herring uncertainty distributions each had a mode of zero and the widest range of all endpoints with no more than a 20% chance of each risk score occurring (Fig. 13.9a). This indicated a large amount of variability and thus less confidence within the RRM score for these species. The distribution for the eelgrass, which exhibits potentially undesirable effects, has a mode of 42, which is equal to the risk score predicted by the RRM. This distribution was the narrowest for all assessment species, indicating slightly more confidence in the RRM score. However, there is no more than a 20% chance of each risk score occurring, and thus much uncertainty remains regarding this species.

Risk Regions. The uncertainty distribution for the Birch Bay risk region was considerably wider than any of the other risk regions, having a width ranging from −397 to 78 and a mode of −228 (Fig. 13.9b). This distribution had no more than a 5% probability of each risk score occurring, suggesting the least confidence in the RRM score for this region. Conversely, the Alden Bank uncertainty distribution was the narrowest with a width only ranging from −10 to 13 and a 72% probability of the risk score equaling zero, which was also the mode (Fig. 13.9b). This indicated low uncertainty in the RRM score.

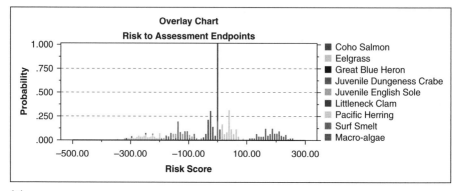

(a)

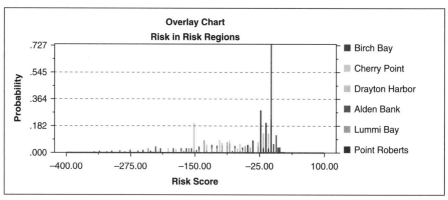

(b)

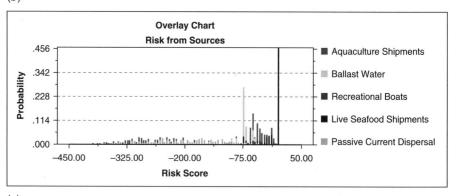

(c)

Figure 13.9. Monte Carlo uncertainty distributions for relative risk to (a) each assessment species, (b) each risk region, and (c) each source.

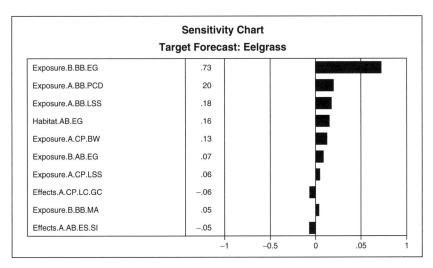

Sensitivity Chart
Target Forecast: Eelgrass

Exposure.B.BB.EG	.73	
Exposure.A.BB.PCD	20	
Exposure.A.BB.LSS	.18	
Habitat.AB.EG	.16	
Exposure.A.CP.BW	.13	
Exposure.B.AB.EG	.07	
Exposure.A.CP.LSS	.06	
Effects.A.CP.LC.GC	−.06	
Exposure.B.BB.MA	.05	
Effects.A.AB.ES.SI	−.05	

Figure 13.10. The sensitivity analysis output example for eelgrass. The highest correlated parameter is for the exposure component for the habitat to be compatible with sargassum for the Birch Bay risk region, with eelgrass habitat. The correlation coefficient is 0.73.

Sources. The uncertainty distributions for passive current dispersal and recreational boats were the widest of all source distributions with no more than a 4% chance of each risk score occurring (Fig. 13.9c). The modes for passive current dispersal and recreational boats were −168 and −252, respectively. This indicates a large amount of variability and thus low confidence in the RRM scores for these sources. The live seafood shipment uncertainty distribution had approximately a 46% chance of risk score equaling a zero, which was also the mode (Fig. 13.9c), indicating slightly more confidence in the RRM score compared with the other sources.

Sensitivity Analysis

The sensitivity analysis produced rank correlations for each assessment species, risk region, and source risk score. The rank correlations indicate whether the uncertainty within any model parameters influences the uncertainty within the final risk scores (e.g., Fig. 13.10, eelgrass).

Assessment Species. The uncertainty within the effects filter Component C, which considers whether effects to the assessment species are possible, was highly correlated with the uncertainty in the risk scores for the following assessment species: coho salmon, juvenile Dungeness crab, juvenile English sole, Pacific herring, and surf smelt. The rank correlations ranged from 0.48 to 0.51. The uncertainty within the exposure filter Component B, which considers habitat suitability for sargassum, was highly correlated with the uncertainty in the eelgrass risk score (rank correlation = 0.73). The macro-algae, littleneck clam, and great blue heron did not have one model parameter that exhibited a dominant effect upon the uncertainty within the risk scores, but instead, several model parameters contributed to the uncertainty.

Risk Regions. The model parameter that exhibited a dominant effect upon the uncertainty within the risk score for Alden Bank was the habitat rank for eelgrass. The Lummi Bay uncertainty was most influenced by the uncertainty within the exposure filter Component B (rank correlation = 0.64), which considers habitat suitability for sargassum. The Birch Bay, Cherry Point, Drayton Harbor, and Point Roberts risk regions did not have one model parameter that exhibited a dominant effect upon the uncertainty within the risk scores, but instead, several model parameters contributed to the uncertainty.

Sources. The uncertainty within the exposure filter Component A, which considers whether the source releases sargassum, was highly correlated with the uncertainty in the risk scores for aquaculture shipments and ballast water, having rank correlations of 0.47 and 0.50, respectively. Recreational boats, live seafood shipments, and passive current dispersal did not have one model parameter that exhibited a dominant effect upon the uncertainty within the risk scores, but instead, several model parameters contributed to the uncertainty.

DISCUSSION AND CONCLUSIONS

This study describes the results of an ecological risk assessment of the Japanese brown alga, *Sargassum muticum*, in the Cherry Point, Washington region. As a part of this risk assessment, we used the relative risk model to evaluate the risk of impacts to selected endpoints for current conditions, as well as the possibility of future spread within the Cherry Point region.

The macro-algae and eelgrass endpoints are the most at risk, which was expected because sargassum has been shown to compete for available habitat and light with these organisms (Viejo 1997). In contrast to the macro-algae and eelgrass, the coho salmon, surf smelt, juvenile English sole, Pacific herring, and juvenile Dungeness crab actually exhibit reduced risk, which is due to the potential beneficial effects of sargassum including possibly providing shelter, food supply, and spawning habitat (Giver 1999) for these species. The results concerning potentially beneficial effects, however, should be interpreted with caution because this study only identifies short-term effects. The long-term effects from interactions with sargassum are unknown.

Of all the risk regions, the Birch Bay risk region had the most negative risk score, suggesting that within this region the majority of effects are going to occur. This may be due to the Birch Bay risk region possessing a variety of habitats that possibly support a greater abundance of the endpoints, specifically the juvenile Dungeness crab, which may exhibit potential beneficial effects from sargassum.

Passive current dispersal and recreational boats were the sources that contributed the most to risk. This was expected as sargassum has been frequently shown to drift long distances by surface currents and is a likely mode of distribution (Farnham et al. 1981). Recreational boats have also been documented as a vector for sargassum dispersal by way of entanglement (Jansson 2000).

These risk scores are initially useful in determining which risk regions, sources, and endpoints should receive more immediate attention by environmental managers;

however, uncertainty analysis was necessary to determine the amount of confidence in the risk scores. Monte Carlo analysis offers a way to characterize the uncertainty within the risk scores and also identify the model parameters, if any, that influence the uncertainty the most. More importantly, these analyses identified areas toward which environmental managers need to focus their research efforts to reduce the uncertainty and produce more accurate predictions of risk.

The variability in the risk scores for juvenile Dungeness crab and Pacific herring is mostly due to uncertainty concerning the beneficial effects of the sargassum to these endpoints. Further research concerning these effects may lead to a reduction in the risk score uncertainty and clarify whether or not beneficial effects are possible or if the endpoints remain unaffected.

Compared to the Dungeness crab and herring, there is less uncertainty in the eelgrass risk scores; however, much uncertainty still remains. This is due to the uncertainty regarding whether or not sargassum can live in the eelgrass habitat and cause effects. Survival of sargassum in the eelgrass habitat is questionable because it is unknown whether or not patches of solid substrate, which are essential for attachment, are present within this habitat. Additionally, it is unknown whether grazers will have a considerable impact upon the sargassum population. To reduce this uncertainty, detailed eelgrass habitat analysis and research concerning possible grazers of sargassum specific to the study area is necessary.

The Birch Bay risk region score had the greatest uncertainty, however, the lack of confidence in this risk score was due to uncertainty concerning multiple model parameters, including exposure and effects components. Much of this uncertainty can be attributed to lack of data concerning effects to endpoints, as well as habitat suitability within this risk region. More research needs to be conducted regarding potential effects to endpoints present in the Birch Bay region. Moreover, habitats and substrates need to be better quantified to reduce the uncertainty concerning habitat suitability.

Similar to the Birch Bay risk region, the risk scores for the sources, passive current dispersal, and recreational boats had the greatest uncertainty; however, no individual model parameter contributed the most to the uncertainty. It is difficult to estimate the direction of current flow during optimal time of sargassum drift; therefore, estimating the spread and locations of establishment of this species is uncertain. Further research regarding local currents and optimal time of drift in the Cherry Point study area is warranted.

The spread of sargassum associated with recreational boats is highly uncertain due to the lack of data regarding the possibility of entanglement and hull-fouling. Additionally, there is a lack of data regarding whether or not the recreational boats using the marinas located within the study area originate from an area that supports a population of this brown alga. A database of the origins of the recreational boats would be especially useful in determining whether sargassum will be introduced via this source.

This study provides a better understanding of the potential effects to the specified assessment endpoints caused by the possible spread and establishment of sargassum throughout the Cherry Point study area. Even though the risk from sargassum is now better quantified, much uncertainty still remains. As revealed in a previous risk assessment of the European green crab, *Carcinus maenas*, in Cherry Point, Washington (Colnar 2004), uncertainty in risk assessments of invasive species is not uncommon. This risk assessment not only provides environmental managers with risk predictions concerning sargassum spread and effects, but also uses uncertainty analysis to suggest specific research necessary to reduce the uncertainty within the risk scores, which can assist in the management of this nonindigenous species.

ACKNOWLEDGEMENTS

This research was supported by US EPA grant no. 1-54068.

REFERENCES

Arenas F, Viejo RM, Fernández C. 2002. Density-dependent regulation in an invasive seaweed: responses at plant and modular levels. *J Ecol* **90**:820–829.

Britton-Simmons, K. H. 2004. Direct and indirect effects of the introduced alga *Sargassum muticum* on benthic, subtidal communities of Washington State, USA. *Marine Ecol Progress Ser* **277**:61–78.

Colnar AM. 2004. Regional risk assessment of the European green crab *Carcinus maenas* in Cherry Point, Washington. Master of Science thesis. Western Washington University, Bellingham, Washington.

Colnar AM, Landis WG. 2007. Conceptual model development for invasive species and a regional risk assessment case study: the European Green Crab (*Carcinus maenas*) at Cherry Point, Washington USA. *Hum Ecol Risk Assess* **13**:120–155.

Critchley AT. 1983a. *Sargassum muticum*: A morphological description of European material. *J Marine Biol Assoc United Kingdom* **63**:813–824.

Critchley AT. 1983b. The establishment and increase of *Sargassum muticum* (Yendo) Fensholt populations within the Solent area of Southern Britain. II. An investigation of the increase in canopy cover of the alga at low water. *Bot Marina* **26**:547–552.

Critchley AT, Farnham WF, Morrell SL. 1983. A chronology of the new European sites of attachment for the invasive brown alga, *Sargassum muticum*, 1973–1981. *J Marine Biol Assoc United Kingdom* **63**:799–811.

den Hartog C. 1997. Is *Sargassum muticum* a threat to eelgrass beds? *Aquat Bot* **58**:37–41.

DeWreede RE. 1983. *Sargassum muticum* (Fucales, Phaeophyta): Regrowth and interaction with *Rhodomela larix* (Ceramiales, Rhodophyta). *Phycologia* **22**:153–160.

Deysher L, Norton TA. 1982. Dispersal and colonization in *Sargassum muticum* (Yendo) Fensholt. *J Exp Marine Biol Ecol* **56**:179–195.

Druel LD. 2000. *Pacific Seaweeds*. Harbour Publishing, Madeira Park, British Columbia.

Farnham W, Murfin C, Critchley A, Morrell S. 1981. Distribution and control of the brown alga *Sargassum muticum*. *In Xth International Seaweed Symposium*. Walter de Gruyter & Co., Berlin.

Giver KJ. 1999. Effects of the invasive seaweed *Sargassum muticum* on native marine communities in northern Puget Sound, Washington. Master of Science thesis, Western Washington University, Bellingham, Washington.

Hart Hayes E, Landis WG. 2004. Regional ecological risk assessment of a near shore marine environment: Cherry Point, WA. *Hum Ecol Risk Assess* **10**:299–325.

Hart Hayes E, Landis WG. 2005. The ecological risk assessment using the relative risk model and incorporating a Monte Carlo uncertainty analysis, In Landis WG (Ed.), *Regional Scale Ecological Risk Assessment Using the Relative Risk Mode*. CRC Press, Boca Raton, FL, Chapter 13., pp. 257–290.

Jansson K. 2000. The marine environment. In Weidema IR (Ed.), *Introduced Species in the Nordic Countries*. Nord **2000**:13. Nordic Council of Ministers, Copenhagen, pp. 43–86.

Karlsson J, Loo LO. 1999. On the distribution and the continuous expansion of the Japanese seaweed—*Sargassum muticum*—in Sweden. *Bot Marina* **42**:285–294.

Landis WG. 2004. Ecological risk assessment conceptual model formulation for nonindigenous species. *Risk Anal* **24**:847–858.

Landis WG, Wiegers JK. 1997. Design considerations and a suggested approach for regional and comparative ecological risk assessment. *Hum Ecol Risk Assess* **3**:287–297.

Landis WG, Wiegers JK. 2005. Introduction to the regional risk assessment using the relative risk model. In Landis WG (Ed.), *Regional Scale Ecological Risk Assessment Using the Relative Risk Model*. CRC Press, Boca Raton, FL, Chapter 2., pp 11–36.

Norton TA, Benson TR. 1983. Ecological interactions between the brown seaweed *Sargassum muticum* and its associated fauna. *Marine Biol* **75**:169–177.

Ruiz GM, Fofonoff P, Hines AH, Grosholz ED. 1999. Non-indigenous species as stressors in estuarine and marine communities: Assessing invasion impacts and interactions. *Limnol Ocean* **44**:950–972.

Viejo RM. 1997. The effects of colonization by *Sargassum muticum* on tidepool macroalgal assemblages. *J Marine Biol Assoc United Kingdom* **77**:325–340.

Wiegers JK, Feder HM, Mortensen LS, Shaw DG, Wilson VJ, Landis WG. 1998. A regional multiple-stressor rank-based ecological risk assessment for the fjord of Port Valdez, Alaska. *Hum Ecol Risk Assess* **4**:1125–1173.

Wu J, Loucks OL. 1995. From balance of nature to hierarchical patch dynamics: A paradigm shift in ecology. *Q Rev Biol* **70**:439–466.

Wu J, David JL. 2002. A spatially explicit hierarchical approach to modeling complex ecological systems: Theory and applications. *Ecol Modelling* **153**:7–26.

14

INTEGRATED LABORATORY AND FIELD INVESTIGATIONS: ASSESSING CONTAMINANT RISK TO AMERICAN BADGERS

Dale J. Hoff, Deborah A. Goeldner, and Michael J. Hooper

Risk posed by chemicals to wildlife species inhabiting hazardous waste sites is a function of exposure and toxicity. To estimate exposure, many EcoRAs compile ecological data from the scientific literature on parameters such as the expanse of the home range, composition of the diet, and rates of ingestion. These exposure parameters often come from natural history studies of wildlife inhabiting the landscapes of National Wildlife Refuges, wilderness, and other natural areas not likely to be similar to hazardous waste sites. Relative to pristine settings, hazardous waste sites have both chemical and nonchemical stressors in a landscape disturbed by anthropogenic activities. These perturbations can include chemical contamination, large industrial building structures, roadways and the traffic that goes with them, artificial water containment structures, and altered habitat structure and function. Collectively, these factors may influence the movement of animals and their exposure to chemicals in the contaminated landscape (see Kapustka et al. 2004). Because anthropogenic influences on each site are different, each hazardous waste site may have unique characteristics profoundly impacting behavior and ecology of wild species inhabiting the area. Site-specific behavioral and ecological data are, therefore, critical for accurate exposure estimates in EcoRAs, as well as a broader understanding of how wildlife may be impacted by nonchemical stressors.

The Rocky Mountain Arsenal Landscape

The Rocky Mountain Arsenal National Wildlife Refuge (Arsenal, Refuge) is located in southwest Adams County near Denver, Colorado (see ftp site for Fig. S14.1). The Arsenal was a military installation where, for over four decades, the United States Army (Army) produced, stored, and eventually demilitarized chemical and incendiary weapons. Munitions filling operations ceased in 1969. Since 1970, the Army's primary activity at the Arsenal has been the demilitarization of chemical warfare agents and the identification of areas having unacceptable risk from the contaminants (Biota Remedial Investigation 1989).

The 6900-ha site, approximately 16 km northeast of downtown Denver, is surrounded by suburbs of Denver. Movement of badgers on and off of the site during the time of this study was limited to the north Arsenal boundary ($\sim$6.4 km). Urban development associated with the Denver International Airport continues to encroach upon the free movement of wild badgers on and off the site today.

Since World War II, private corporations leased facilities on the Arsenal to produce chemical pesticides. Several different companies produced chemicals in the South Plants area of the Arsenal from 1947 to 1982. Principal among them was Shell Chemical Company (SCC), which produced organochlorine insecticides, herbicides, nematocides, adhesives, anti-icers, curing agents, cutting oil additives, gear oil additives, and lubrication greases. Ineffective waste storage and disposal practices contaminated areas of the Arsenal with a multitude of chemicals. The Arsenal was placed on the United States Environmental Protection Agency's (US EPA) Superfund National Priorities List in 1987. All production on the site was stopped in 1982, but SCC held the lease on the property until 1987 (Biota Remedial Investigation 1989).

Despite the extensive military and industrial use of the Arsenal in the past, a high diversity of wildlife species exists on the site. The variety of site activities has led to a highly heterogeneous ecosystem (see ftp site for Fig. S14.2). The mix of nondisturbed and disturbed areas undergoing various stages of succession comprise a continuum of habitats that enhances ecological diversity (Warren et al. 2007). Further, since 1942, the entire area has been fenced with limited public access. Habitat heterogeneity and lack of public use has led to wildlife diversity rivaling any location on the Rocky Mountain's Front Range corridor, including existing government parks and preserves (Biota Remedial Investigation 1989). In 1992, the United States Congress signed a bill designating the Arsenal as a National Wildlife Refuge pending clean-up of its Superfund components. Currently, the site has undergone extensive remediation of contaminated soils and destruction of chemical manufacturing facilities through the joint efforts of the Army, SCC, and the United States Fish and Wildlife Service (US FWS). As of 2006, nearly 5000 ha have been remediated and transferred from the Army to the USFWS as one of the nation's largest urban wildlife refuges. Expanded discussions, including site historical and remedial photos, can be found at http://www.rma.army.mil/(accessed July 23, 2009). Our study occurred prior to significant remediation of contaminated soils from 1993 to 1996.

The Arsenal, part of the High Plains Grassland of Colorado, is dominated by shortgrass prairie. However, vegetation on the Arsenal also includes species of mid-to-tall grass prairies found on mesas to the east and west. Of the Arsenal's 6900 hectares, the central "core area" of the site had the majority of industrial plants and evaporative basins (see ftp site for Fig. S14.1, Sections 1, 2, 25, 26, 35, and 36). Areas in the north and east were dominated by gently rolling hills with grassy slopes intermixed with disturbed vegetation type and large areas revegetated with crested wheatgrass (*Agropyron cristatum*). Grasslands were also predominant in the southern sections, but included areas of wetlands and tree cover (see ftp site for Fig. S14.2).

Though at least 666 chemicals were attributed to activities of the Army and lessees on the Arsenal, a vast database of on-site chemical concentrations in abiotic media and biota prior to the start of remedial activities facilitated additional focused wildlife studies on seven chemicals of concern (CoC): dieldrin, aldrin, endrin, DDT, DDE, mercury (Hg), and arsenic (As) (Stollar and Associates, Inc. 1992). The Detection frequencies of CoC in the Arsenal terrestrial biota (plants, invertebrates, vertebrates) were: dieldrin, 60%; aldrin, 2.6%; endrin, 5.6%; DDT, 1.3%; DDE, 7.3%; Hg, 8.3%; and As, 6.9% (Stollar and Associates, Inc. 1992). The high frequency of detection of dieldrin in Arsenal biota identified it as the focal CoC for assessing risks to resident biota.

Significant concentrations of dieldrin were found in Arsenal surface soil near industrial plants and evaporation basins (Fig. S14.3). The highest concentrations of dieldrin (up to 15,000 mg/kg dry weight) occurred in soils from the core areas and evaporation basins. Soils in the peripheral areas of the Arsenal, however, generally contained concentrations of dieldrin in soil that were less than 0.025 mg/kg dry weight.

Receptor of Potential Concern

American badger (*Taxidea taxus*) was chosen as a receptor of concern for several reasons. First, badger natural history characteristics result in frequent, intimate contact with contaminated soils due to their intensive burrowing (fossorial) behavior. Furthermore, badgers are mammalian tertiary predators feeding at the highest trophic level within a fossorial food web that was known to be contaminated with dieldrin. Finally, there had been a documented mortality of a badger from dieldrin poisoning on the Arsenal.

MATERIALS AND METHODS

American badgers were fitted with radio transmitters to track their movement throughout the Arsenal landscape. Home ranges of radio-tagged individuals were used as geographical references for estimates of exposure to dieldrin. Controlled laboratory dosing studies were conducted to generate dieldrin toxicological profiles of badgers. These chemical and species-specific toxicity profiles were then used to assess the risk of badgers that were exposed to dieldrin in the field and then captured on the Arsenal. Finally, the use of radio telemetry allowed for collection of movement and demographic data, which were compared to similar information from other western locations. The relative impact of contaminant exposure and landscape characteristics on the badger population was determined. Detailed descriptions of the laboratory dosing studies, field techniques, and statistical procedures are presented in appendix format on the ftp site accompanying this book at ftp.wiley.com/public/sci_tech_med/environmental_risk.

Resident Population Characteristics

Badgers captured on the site were fitted with peritoneal implant radio transmitters (Advanced Telemetry Systems Inc., Isanti, MN). Each individual was assigned a unique identification code that was tattooed inside the upper right lip before releasing the animal back at the site of capture. Subsequently, searches for radio-tagged individuals were conducted five nights per week (primarily February–November) from dusk to dawn. Signal triangulation was used to assign Universal Transverse Mercator (UTM) grid coordinates to the nearest 10 m.

Home ranges of Arsenal badgers defined as "the area traversed by the individual in its normal activities of food gathering, mating and caring for young" (Burt 1943) were expressed as layers in format compatible with ARC/INFO Geographic Information Systems (GIS) software. Mating, gestation, lactation caring for young, and dispersal occur at temporally distinct and separate periods of the year for badgers. Four time periods of the year were distinguished to reflect changes in behavior of adult badgers:

March, April, May: Females gave birth and cared for nonmobile altricial young, mostly using a single burrow; males were foraging heavily while ranging widely to locate a number of different prey sources.

June and July: Juveniles began to forage with adult females and became more mobile as family units used different burrows every two to three days; adult males decreased movement for most of June and the first half of July in a relatively restful period before the end of July which marked the onset of breeding season.

August and September: Juveniles dispersed as adults spent less time with young and more time breeding.

October and November: Females foraged heavily, visiting a number of different prey patches; activity increased or slightly decreased compared to previous time period as they accumulated and stored fat reserves for gestation in December, January, and February. Males reduced movement dramatically and tended to concentrate foraging activity in a single productive patch.

From December to February, all age groups of both male and female badgers were very inactive and often went through extended periods of a dormant state during cold weather.

Distance traveled between consecutive daily relocations was calculated from UTM coordinates using the distance formula [Eq. (1)]:

$$DT = \sqrt{(X_2 - X_1)^2 + (Y_2 - Y_1)^2} \tag{1}$$

where DT is the distance traveled, X_1 and Y_1 are the first easting and northing UTM coordinates, and X_2 and Y_2 are the easting and northing UTM coordinates 20–30 hours after the first location. Consecutive relocations with more than 30 hours (two days) time between them were not included in the analysis.

Little is known about home-range characteristics of pre- and post-dispersal juvenile badgers. Predispersal time periods of juvenile badgers occurred in June to July, while post-dispersal areas generally began in August and continued to the end of the field season. Juveniles were not, however, considered to have dispersed until they were located outside of the mother's home range for more than two consecutive locations or at the beginning of the following field season (whichever came first).

The number of young dispersed from each adult radio-tagged female was counted by direct observations of juvenile activity outside maternal dens, or trapped while inhabiting maternal dens. Resident, adult population numbers were estimated based on home-range data of adult females. Adult females have little overlap of home ranges and aggressively defend their area (Goodrich 1994). Therefore, available habitat (total area excluding that of existing water bodies at the time) on the Arsenal was divided by average annual home-range size of radio-tagged females. Sex ratios of badger populations from southern Wyoming were then used to estimate the number of males on-site.

Age distributions were estimated directly from captures. Young-of-year animals were easily distinguished from adults by size and status of canines. Age of adult badgers was determined by counting annuli cementum of the first premolar (Messick and Hornocker 1981).

Survival of adults and juveniles was estimated using the Kaplain–Meier, non-parametric model (White and Garrot 1990) with a staggered-entry design as described by Pollock et al. (1989). Adult survival was expressed as the probability that an individual survives from the beginning to the end of the study, or 28 months. Juvenile survival was defined as the probability that an individual survived through its first year.

Landscape features of the Arsenal at the time of the study were classified into six main categories: (1) grasslands, (2) woodlands and shrubs, (3) riparian and wetlands, (4) man-made structures, (5) inactive prairie dog towns, and (6) active prairie dog towns. Chi-square analysis (Neu et al. 1974) was used to evaluate whether the area use patterns were proportional to the quantity of each category (Alldrege and Ratti 1986).

Quantifying Dieldrin Exposure

Estimates of dietary exposure relied on statistical characterization of dieldrin in abiotic and biotic media. Tissues collected from captured animals were analyzed for dieldrin concentrations. Soil was analyzed for dieldrin concentration as well. Spatial distribution of these data was combined with movement patterns of radio-tagged badgers to establish correlations between location and tissue concentrations and to model potential exposure doses.

An epizootic of plague in 1996 resulted in a near elimination of black-tailed prairie dogs (*Cynomys ludovicianus*) from the site. As the black-tailed prairie dog normally would be the main prey item, scat (fecal excreta) analysis was subsequently completed to determine the site-specific diets of badger during the study.

Blood, adipose tissue (subcutaneous and omentum), urine, and fecal samples were obtained from captured badgers while under anesthesia at the Denver Zoo Hospital. Recaptured animals were typically sampled in the field where only blood and feces were collected.

The geometric means of dieldrin soil concentrations were calculated within adaptive kernel contours using GIS ARC/INFO software and the US Army database. When multiple values were found at a single point location due to contaminant depth profiling, the average dieldrin concentrations for samples taken from surface to 30-cm depth were used.

Dieldrin and other CoC (aldrin, endrin, DDT, and DDE) residue concentrations in plasma, adipose, urine, and feces were determined by US Army contract laboratories (Environmental Systems Engineering and Denver Wildlife Research Center, Denver, CO). Minimum reporting limits for adipose tissues and feces were 0.015 $\mu g/g$ dry weight. Detection limits for plasma and urine were 1 $\mu g/L$. Dieldrin concentrations in adipose (omentum or subcutaneous) and plasma were regressed against geometric mean dieldrin soil concentrations within respective home ranges of badgers using the following approach.

Kernel home-range area estimates were generated for several contours of activity [50%, 65%, 75%, 85%, or 95% (Worton 1989)]. By definition, a contour contains an area in which the animal has a given percentage probability of being relocated (White and Garrot 1990). For example, an observer has a 95% probability of relocating an animal within the boundary of the 95% isopleth. Because badger home-range areas include both contaminated and noncontaminated soils, it was necessary to establish the most predictive contour of exposure. Therefore, correlation between biological and soil matrices was completed in two phases. First, in order to find the most predictive contour of exposure, individuals chosen for geographical, temporal, and statistical considerations (see more detailed Methods appendix) were used for an initial regression of dieldrin concentrations in plasma versus dieldrin soil concentrations within 50%, 65%, 75%, 85%, or 95% contours. After the most predictive contour was selected, these isopleths were generated from data points collected through the duration of the study (cumulative home ranges) from all individuals. Concentrations of dieldrin in plasma and adipose were then regressed with respective geometric mean dieldrin soil concentrations within their home range.

Estimating Daily Dieldrin Exposure and Tissue Concentrations

Daily dieldrin doses and tissue concentrations were estimated from toxicokinetic distribution ratios derived from laboratory dosing studies (Hoff 1998). Briefly, wild juvenile badgers were trapped and exposed to dieldrin at the University of Wyoming's Red Buttes Laboratory in Laramie, WY. Dieldrin (98% pure) was dissolved in nondenatured ethanol and injected (16.9 μL ethanol/kg body weight) into a lab mouse portion fed daily to the badgers. Two dose levels, 0.035 and 0.5 mg dieldrin/kg body weight/day (mg/kg/d; $n = 3$ and 4 badgers respectively), and a control group (ethanol only, $n = 4$ badgers) were used. A dose of 0.035 mg/kg/d was chosen after a review of existing data of dieldrin concentrations in badger prey species (black-tailed prairie dogs) within the most contaminated sections of the Arsenal; the 0.5-mg/kg/d dose level represented a dose roughly an order of magnitude greater than the average exposure in those most contaminated areas. The subchronic dietary exposure period lasted 63 days. The apparent doses were 2.5 and 35.7 mg/kg dry weight in the diet based on the measured ingestion rate of 0.014 kg food/kg/d. Cumulative doses were 2.21 and 31.5 mg dieldrin/kg body weight for the low and high doses, respectively. Plasma and tissue samples were obtained at the end of the dosing period.

Dieldrin was quantified in plasma of dosed laboratory animals and the average value (0.018 ± 0.003 mg/kg dry weight) divided by the dose (0.035 mg/kg/d) to derive the toxicokinetic distribution ratio of 0.52. Dieldrin concentrations in the plasma of Arsenal badgers were used in Eq. (2) to estimate daily dieldrin doses:

$$\text{Dose(mg/kg/d)} = [X] \div 0.52 \tag{2}$$

where $[X]$ is the measured dieldrin plasma concentration in ppm, and 0.52 is the distribution ratio ([plasma]/dose) derived in the dosing study.

Estimated doses were then divided by a daily ingestion rate (0.014 kg food/kg/d) to estimate average dietary concentrations of dieldrin (mg/kg dry weight). Average dieldrin concentrations were quantified analytically in brains (0.043 ± 0.004 mg/kg dry weight), livers (0.155 ± 0.098 mg/kg dry weight), and subcutaneous and omentum adipose (1.99 ± 0.325 and 2.94 ± 0.46 mg/kg dry weight, respectively) of badgers exposed to 0.035 mg dieldrin/kg/d in the laboratory. Tissue-to-plasma toxicokinetic distribution ratios ([tissue]/[plasma]) were calculated ([plasma] = 0.018 ± 0.003 mg/kg dry weight) to quantify the diffusion of dieldrin from the plasma into the respective tissues of dosed badgers. These ratios were used in Eq. (3) to estimate dieldrin concentrations in the brains, livers, and adipose tissues of Arsenal badgers:

$$[\text{Brain, Liver, Adipose dieldrin}] = [X] * Y \tag{3}$$

where $[X]$ was the observed dieldrin plasma concentration in ppm and Y was the average tissue-to-plasma distribution ratio in brains (2.35), livers (8.49), subcutaneous adipose (109), and omentum adipose (161), respectively.

To test the accuracy of the ratios, dieldrin concentrations in either omentum or subcutaneous adipose tissues of wild badgers were estimated from observed dieldrin concentrations in plasma samples and compared to actual measured adipose concentrations obtained during implantation surgery. Only badgers with detectable concentrations of dieldrin in both plasma and adipose were used. Average dieldrin concentrations of all samples were used when an individual had been sampled multiple times.

Quantifying Effects of Dieldrin in American Badgers on the Arsenal

Numerous endpoints to quantify badger biomarker and animal health responses to dieldrin were investigated in 1993 in controlled laboratory dosing experiments with badgers (Hoff 1998). No biomarkers collected by lethal means (brain biogenic amines, histopathology, liver and kidney cytochrome P450s, or porphyrin profiles) proved useful as indicators of exposure. However, two nonlethal biomarkers responding to controlled dosing were serum enzymes associated with liver pathology and lymphocyte counts in whole blood. Biomarkers were, therefore, limited to nonlethal collection techniques to minimize the impact of lethal sampling on the Arsenal badger population.

Mortality (inactivity) switches in radiotransmitters allowed documentation of mortality in field badgers. Necropsies were performed at the Wyoming State Veterinary Laboratory and tissues were collected for residue analyses and histopathology.

Serum clinical chemistry endpoints were albumin (ALB), total protein (TPROT), globulin (GLOB), blood urea nitrogen (BUN), creatinine (CREA), cholesterol (CHOL), glucose (GLUC), and total bilirubin (TBIL); and activities of alkaline phosphatase (ALKP), alanine aminotransferase (ALT), aspartate aminotransferase (AST), γ-glutamyl transferase (GGT), amylase (AMYL), total creatine kinase (CK), and lactic dehydrogenase (LDH) were all determined on a VET TEST 8008 (IDEXX Laboratories, Westbrook, ME) at Wyoming State Veterinary Laboratory in Laramie, WY. Serum antibody titers to canine distemper virus (CDV; morbilli virus sp.) and plague (*Yersinia pestis*) were also completed by the Wyoming State Veterinary Laboratory to determine the exposure of Arsenal badgers to these diseases. Complete blood counts were performed using a Sysmex cell counter (Model F-300, TOA Medical Electronics Co., LTD, Kobe, Japan). Whole-blood smears were stained with Buffered Differential Wrights Stain (Camco Quick Stain, Baxter Scientific Products, McGaw Park, IL) and white blood cell (WBC) differentials obtained by categorizing 100 leukocytes.

The presence and concentrations of leukocytes, nitrites, urobilinogen, protein, blood, ketones, bilirubin, glucose, WBCs, RBCs, epithelial cells, and bacteria in urine were determined using Ames Multi-Stixs (Miles Inc., Diagnostic Division, Elkhart, IN).

RESULTS AND DISCUSSION

The observations from this study are presented in three subsections. The characteristics of the population of badgers on the Arsenal are described in terms of home range, survival, densities, and dispersal. Diets of badgers are presented as a prelude to estimating exposures. The section closes with estimates of exposure and effects.

Population Characteristics

Fifty-three captures from 1025 trap-nights were recorded from March 19, 1994 to August 1, 1996. Of those 53 captures, there were 32 different individuals and 21 recaptures composed of nine adult males, eight adult females, ten juvenile males, and five juvenile females. A total of 29 peritoneal implants were placed into badgers released in the field: nine in adult females, six in adult males, four in juvenile females, and ten in juvenile males. A total of 2130 locations were estimated during 420 of the possible 688 sampling nights.

Male annual home-range areas (95% adaptive kernel contours) were larger than female annual home-range areas, consistent with other studies (Goodrich 1994, Minta 1990). Average home-range areas (see ftp site for Table S14.1) in 1994, 1995, and 1996 for adult females (mean $\pm$ std in km^2) were 3.9 ± 1.7 ($n = 5$), 9.0 ± 3.2 ($n = 5$), and 7.3 ± 1.9 ($n = 6$), respectively. Average areas for adult males (see ftp site for Table S14.2) in 1994, 1995, and 1996 (mean $\pm$ std in km^2) were 32.9 ± 30.6 ($n = 3$), 19.9 ± 11.8 ($n = 4$), and $14.3 \pm 6.9 km^2$ ($n = 3$), respectively. Average home-range areas for adult females in 1995 were 2.3 times larger than those areas in 1994.

Between the end of the 1994 and the start of the 1995 field season, the black-tailed prairie dog population on the Arsenal was nearly eliminated by an epizootic of plague (USFWS 1995). When these concentrated areas of prey were diminished, female badgers more than doubled their annual home range. No similar trends were apparent for adult males. Gender-based comparisons of these home-range area increases by females only, due to changes in prey population density, were consistent with findings by Goodrich (1994), who concluded that female badger home ranges are resource-limited, while male home-range areas are primarily influenced by the overlap of adult female home ranges.

Data combined from all possible years for each individual showed no significant seasonal differences ($p > 0.05$) in daily distance traveled for either adult female or male badgers. Though there are no statistically significant patterns, similar trends for all individuals suggest that females may travel further distances and consequently are more active, from August to November compared to April to July. Adult males are similar in that August to September are months of greatest movements while they attempt to breed with as many females as possible (Goodrich 1994). In contrast to adult females, during October to November, adult males traveled the shortest distances of all the chosen time periods.

Seasonal changes in home-range area showed trends similar to distance traveled for both adult females and males, though no significant differences were identified ($p > 0.05$). During all three years, there was a suggestion that area use increased throughout the four chosen time periods for females (see ftp site for Fig. S14.4), whereas males increased home-range areas during breeding to overlap with females and then dramatically decreased the total area right after the breeding season (see ftp site for Fig. S14.5).

Pre-dispersal areas for both males and females overlap those of their mothers, though juvenile males had a significantly (t-test $p = 0.043$) larger area (6.7 ± 1.1 km^2, $n = 3$) than juvenile females (3.0 ± 1.2 km^2, $n = 3$). A potentially confounding factor is that juvenile female data were collected in 1994 whereas juvenile male data were collected in 1996, and adult female ranges were larger in 1996 than in 1994. Larger juvenile male areas were likely an artifact of changes in adult female behavior between the two years. Post-dispersal areas of juvenile males and females had similar trends, but were not statistically significant because of low n values and large deviation within male areas. Of the 11 juveniles captured in 1994 and 1995, only two remained on-site during dispersal or survived to adulthood.

Home-range areas of American badgers are extremely variable and dependent on geographical location. Adult female badger home ranges in southern Wyoming (Goodrich 1994) were 3.5 km², whereas females in Illinois (Warner and Ver Steeg 1995) and British Columbia, Canada (Weir et al. 2003) averaged 13.05 km² and 15.6 km², respectively. Adult male home ranges in each of the studies were considerably larger. As evidenced in this study, home-range differences on the same site may also occur with reproductive season and variation in prey availability. Others (Minta 1990, Messick and Hornocker 1981) have also shown that home range area will vary, depending on prey density. Adult female home-range areas of badgers on the Arsenal during times of high prairie dog densities (3.9 km²) are very similar to those of badgers in other western states [Utah, Idaho, and Wyoming (Messick and Hornocker 1981, Lindzey 1978, and Goodrich 1994, respectively)]. As prairie dog densities decreased by 98% (USFWS 1995), adult female badger home range areas more than doubled in 1995 (8.97 km²). Area use by predators, along with their dependence on prey species, is not unique to badgers (Ehrlich and Roughgarden 1987). Shifts in prey base and availability of alternative prey species are difficult to predict but should be considered in contaminant exposure assessments.

Successful aging was completed for seven adult males and five adult females. The average ages for adult males and females at the time of capture were 4.9 ± 4.5 and 2.8 ± 1 years, respectively. These ages were within the range of age estimates in other areas of the western and midwestern United States. Average ages of adult females and males in southern Wyoming were five and four, respectively (Goodrich 1994). On another site in Wyoming, 91% of the badgers captured at greater than four years of age were males, and 55% of the badgers captured at less than four years of age were females (Minta 1990). A majority of badgers captured in Illinois were three years old regardless of the gender (Warner and Ver Steeg 1995). The estimated age for one male (15 years) is old compared to other estimates of males inhabiting the Arsenal. However, adult males in southern and western Wyoming have been captured and were found to be 13 and 14 years of age, respectively (Goodrich 1994, Minta 1990).

The number of young observed (and assumed to have dispersed) per female in 1995 (0.8, $n = 5$) was less than half that in 1994 (2.0, $n = 3$) and two-thirds of that in 1996 (1.4, $n = 5$). However, sample sizes were low and no statistical differences were noted among years ($p > 0.05$). The number of young born to badgers in southwestern Idaho was found to decrease in response to decreased prey availability (Messick and Hornocker 1981). Similar trends were seen in this study from 1994 to 1995 when prairie dog populations dramatically decreased. The average number of young dispersed over all three seasons at the Arsenal was 1.3 per female, which was the same number observed in Idaho for 3-year-old females [1.3 (Messick and Hornocker 1981)] and slightly more than in Illinois [1.0 (Warner and Ver Steeg 1995)] and Wyoming [1.0 (Minta 1990)].

Average 95% annual home-range areas for 1996 (7.3 ± 1.9 km²) were used to estimate the number of adult females that currently inhabit the Arsenal. Optimum and worst-case scenarios of prey densities for adult females were most likely observed in 1994 and 1995, respectively. Home-range areas from 1996 were therefore chosen because the badger population had adjusted to its prey base, and management techniques by the USFWS had improved the prairie dog numbers to a point that 1996 most likely represented a prey base between extremes. The area of the Arsenal is 68 km² excluding the lakes, which results in an estimate of nine adult females (range = 7–12, based on the variation of home-range area noted above). By the end of the study (August 1996), data from seven adult females were being collected. Similar analysis for adult males is difficult because home-range areas were substantially more variable. For example, in 1994 and 1995, two adult males essentially used the entire area of the Arsenal while overlapping three other males. During breeding season, transient males from off post may enter the population and skew any type of population estimate. For example, one male was captured during breeding season and, after only four relocations, was never found again. During 1996, at least five adult males were known to inhabit the Arsenal early in the season; three adult males were being tracked, one was hit by a vehicle and found dead on a highway bordering the site, and one was observed on three occasions during the season, though was never captured. In a manner similar to that of Goodrich (1994), the number of males on the Arsenal can be estimated using sex ratios (M:F, 1.3:1). If similar sex ratios exist on the Arsenal, the number of adult males would be approximately 12 (range = 9–16). Assuming 1.4 juveniles dispersed per adult female, the total number of badgers inhabiting the Arsenal in August of 1996 was 34 (range = 26–45). The density estimate including all age and gender classifications during breeding season is 0.49 badgers/km² (range = 0.38–0.65).

The estimated number of females, based on home-range areas, assumes very little overlap among individuals. Other studies support the assumption of territorial female badgers (Goodrich 1994). Efforts to capture new badgers within ranges of known adult females were unsuccessful. Estimation of adult male numbers (12, range = 9–16) based on gender ratios on the Arsenal is more problematic because of the dissimilarity of the Arsenal with other study areas. Gender ratios for adults (M:F) have varied from 1 [in Wyoming (Messick and Hornocker 1981); in Illinois (Warner and Ver Steeg 1995)] to 1.75 [in Wyoming (Minta 1990)]. When compared to other study areas, immigration and emigration of transient breeding males on and off Arsenal boundaries are probably limited to the north and northeastern sections only. The gender ratio used in the Arsenal estimate was 1.3 (Goodrich 1994) because the author used a conservative ratio based on capture data in which males are more easily captured than females. The conservative ratio further reflects the limitations in adult male movements on the Arsenal and best fits anecdotal observations on the site.

If one accepts the female population estimate as reasonable, the density of male and female badgers on the Arsenal was 0.5 badgers per km². This was lower than density estimates of badgers from other western sites [2 badgers/km² in Wyoming (Minta 1990), 2.3 badgers/km² (Messick and Hornocker 1981) and 1 badger/km² (Goodrich 1994)]. Both Goodrich (1994) and Minta (1990) described human influences on their sites as minimal. On the Arsenal, however, there are greater anthropogenic influences likely limiting movement of badgers onto the site. The most likely immigration corridor onto the site is limited to the northern boundary, because the western and southern sides are metropolitan suburbs of Denver, and an interstate highway runs parallel with the eastern border. Intensive human activity on the site for the past 50 years could easily influence dispersing juveniles and adults to leave the site. Ultimately, however, the density of this predator will be dependent on the density of prey. If data from 1994 (before the epizootic of plague affected prairie dog populations) is used for the density estimate, the result would be 1.1 badgers per km^2. Even this estimate is somewhat low compared to most other documented western areas. It is difficult to link low density estimates to contaminant effects when there are other obvious confounding influences such as urbanization bordering the site.

The estimates of survival probability of adults and juveniles were 0.68 ± 0.02 (0.49 to 1.04 = lower and upper 95% confidence limits) and 0.25 ± 0.01 (0.09 to 0.41 = lower and upper 95% confidence limits), respectively. Estimates of annual survival probability over three field seasons for adults (0.68) and juveniles (0.25) on the Arsenal were similar to estimates from Illinois [adults = 0.75, juveniles = 0.27, four year study (Warner and Ver Steeg 1995)]. Survival rate for three- to four-year-old adults in Idaho [three-year study (Messick and Hornocker 1981)] was 0.56, and juvenile survival rate in Idaho was higher at 0.46. All cited estimates for juvenile survival are probabilities of the young surviving through their first winter. Survival estimates for badgers on the Arsenal were similar to populations not inhabiting hazardous waste sites in other Midwest and Western areas.

Composition of Diet

During the 1994 season, 80% of the relocations of all badgers were found within the boundaries of active prairie dog towns. Upon visitation to natal and maternal dens, prairie dog carcasses were found on the ground outside of the entrance awaiting consumption by juveniles. On two occasions, badgers were observed carrying prairie dogs in their mouths. Other studies have shown that nearly 90% of the diet of badgers is made up of the predominant fossorial species in the respective study area (Lampe 1976, Minta 1990, Goodrich 1994). Direct observations and previous studies led to a relatively safe assumption that the primary food item for badgers on the Arsenal in 1994 was prairie dogs. However, after prairie dog populations decreased in 1994 and 1995, analyses of dietary composition completed in 1996 showed that badgers apparently shifted their prey base from mainly prairie dogs to a variety of small mammal species. A total of 23 unique prey items were identified from 21 scat samples collected in 1996 near the entrances of natal and maternal dens of transmittered females. Of the 23 prey items, 6 different prey species were keyed to family, genus, or species: voles (43.5%, *Microtus* spp.), deer mouse (21.7%, *Peromyscus maniculatus*), black-tailed prairie dog (13%), pocket gophers (8.7%, *Geomys bursarius* or *Thomomys talpoides*), kangaroo rat (8.7%, *Dipodomys ordi*), and Lagomorphs (4.3%, *Lepus califonicus, Sylvilagus auduboni, or Sylvilagus floridanus*).

Other biomonitoring studies conducted concurrently with this investigation provide some insight into relative concentrations of dieldrin within small mammal species and therefore the exposure consequences of the prey shift. Various species were collected in 1994 and 1995 within the home range of one female in Section 36 during an investigation of small mammal populations (Allen 1996). Prairie dogs were collected in 1994 from Section 36 very close to the grid used by Allen (USFWS 1995). The average concentration of dieldrin in the diet consisting of 100% prairie dogs would have been 0.069 ppm ($n = 4$). The average concentration within voles, kangaroo rats, and deer mice within the same area was 0.071 ppm (n = 18, Allen 1996). Dietary concentrations were very similar and the apparent shift in badger diet did not change badger dietary dieldrin concentrations. The trophic position of the prey items (herbivorous small mammals) did not change, which could explain similarities in dieldrin content.

Habitat Use Patterns

Landscape types used by adult female badgers ($n = 7$) were not significantly different ($p > 0.05$) than availability among the types in second-order selection (Table 14.1). However, there were trends in decreasing use of wetlands and areas with significant man-made structures. Average type use by most individuals (third-order selection) was also not significantly different ($p > 0.05$) than availability (Table 14.2). One individual, however, demonstrated (a) selection of grassland and (b) avoidance of woodland and wetland habitat (FA8, $p < 0.05$). In our study, badgers used habitat types in proportion to their availability. There was little evidence of selection or exclusion of habitat types, including areas of significant building sites, corresponding with some of the highest dieldrin soil concentrations. Spatially explicit modeling of contaminant exposure to wildlife incorporating habitat preferences for wildlife species is growing in popularity (Carlsen et al. 2004) and should be viewed as a positive evolution in wildlife risk assessments on contaminated sites. This study, however, also demonstrates that when a predator population is spatially limited within the boundaries of an "island" setting on an urban refuge, the need for space and food supersede preferences for other parameters of habitat types. The use of spatially explicit habitat modeling should therefore take care to ensure that prey availability is factored prominently among the broader contexts of the wildlife species natural history (e.g., predator, nonpredator, wildlife corridors on-and-off site) when used to assign probabilities of animals inhabiting different habitat types.

Table 14.1. Adult Female American Badger Habitat Selection (Second-Order) on the Rocky Mountain Arsenal NWR During the Spring and Summer of 1996

Habitat Type	Availability Proportion	Proportion Used Bonferroni C.I.	Relative Use Proportion
Grasslands	0.70	0.61–0.81	Equal
Woodlands	0.07	0.06–0.08	Equal
Wetlands	0.05	0.034–0.046	Less
Man-made	0.03	0.009–0.011	Less
Inactive prairie dog Towns	0.15	0.14–0.18	Equal
Active prairie dog Towns	0.01	0.009–0.011	Equal

Table 14.2. Adult Female American Badger Habitat Selection (Third-Order) on the Rocky Mountain Arsenal NWR During the Spring and Summer of 1996.[a]

Badger I.D.	Grasslands	Woodlands	Wetlands	Man-Made	Inactive Prairie Dog Town	Active Prairie Dog Town
FA1	6/93	00/00	00/00	00/00	03/07	01/00
FA4	42/33	03/05	06/05	00/00	46/60	03/00
FA8[b]	64/91	22/05	10/00	00/00	04/4.5	00/00
FA11	77/63	11/25	00/00	01/00	11/10	00/00
FA23	85/91	00/00	05/03	01/00	09/06	00/00
FA28	58/64	18/23	08/00	00/00	15/23	00/00
FA29	73/73	03/02	05/00	02/00	16/25	00/00
Mean	71/73	08/09	05/01	0.5/00	15/20	0.7/0.1
Standard Error	16/20	08/10	04/02	0.7/00	14/18	01/0.3

[a]For each type, values are percentage availability/percentage use.
[b]Avoided woodlands and wetlands.

Quantifying Exposure to Dieldrin

We used three steps to quantify exposure to dieldrin on the Arsenal. First, biological samples were taken from captured badgers. Second, concentrations of dieldrin observed in tissues were correlated to dieldrin concentrations in soil. Third, tissue distributions from laboratory studies were used to estimate daily dietary exposures and tissue burdens of Arsenal badgers.

Biological Sample Collection. The average detectable concentration of dieldrin was: in plasma, 0.0219 mg/L ($\pm$0.0237 std, $n = 23$, range $= 0.003$–0.1021 mg/L); in adipose (both omentum and subcutaneous), 1.603 mg/kg ($\pm$2.158 std, $n = 24$, range $= 0.038$–7.600 mg/kg); and in urine, 0.0505 mg/L ($\pm$0.0620 std, $n = 8$, range $= 0.0137$–0.210 mg/L). No detectable dieldrin was found in any fecal sample. The most sensitive indicator of exposure was the presence of dieldrin in adipose tissue. Nine of the 33 adipose biopsies contained dieldrin in the absence of dieldrin in concurrently collected plasma samples. The opposite occurred only twice, both when plasma concentrations were near the detection limit. Only eight of 26 urine samples yielded detectable dieldrin concentrations, and in each case, dieldrin was found in either plasma or adipose collected from the same individual.

Correlation of Dieldrin Concentrations in Tissue and Soil. Regressions of plasma dieldrin concentration as a function of geometric mean dieldrin soil concentration were significant ($p < 0.05$), but r^2 values varied for different contours (Table 14.3, Fig. 14.1). The best-fit regression line used the 65% adaptive kernel contour ($r^2 = 0.851$).

Seasonal, annual, and cumulative 65% contours were estimated for all badgers with adequate number of relocations (seasonal $= 10$, annual $= 40$, cumulative $= 40$).

Table 14.3. Plasma and Geometric Mean Soil Dieldrin Concentrations Within 50%, 65%, 75%, 85%, and 95% Adaptive Kernel Home Range Areas of Selected Badgers on the Rocky Mountain Arsenal NWR

Badger I.D.	Plasma	Dieldrin Soil Concentration: Geometric Mean (Mg/Kg Dry Weight)				
		50 %	65 %	75 %	85%	95 %
FA1	0.001[a]	0.041	0.033	0.030	0.054	0.056
FA4	0.001[a]	0.023	0.019	0.019	0.015	0.015
FA22	0.015	0.176	0.135	0.175	0.21	0.299
MA2	0.047	0.106	0.134	0.111	0.128	0.329
FA29	0.515	0.241	0.684	0.732	0.723	0.770
MA5	0.102	2.488	1.109	0.879	0.807	0.669
	r^2	0.735	0.851	0.762	0.736	0.696

[a]Detection limit for plasma used for these individuals.

Geometric mean soil concentrations of dieldrin within these contours were calculated. No significant differences ($p > 0.05$) were found among average seasonal or annual geometric mean soil concentrations of dieldrin within contours of adult males or females. Also, no differences were found between geometric mean soil concentrations of dieldrin within contours of male or female pre- and post-dispersal juveniles (t-test, $p > 0.05$). Although the shapes of 95% contours changed seasonally, temporal variation in the shapes of 65% contours of adult females and males was less dramatic (see ftp site for Figs. S14.4 and S14.5). Age, gender, and temporal variations in home-range area had no significant effect on estimates of geometric mean dieldrin soil concentrations within home ranges. Thus, the primary influence of dieldrin exposure to badgers was concentration of dieldrin in the soils of their core area of activity (65% harmonic home-range area).

Because there were no seasonal or annual differences in geometric mean soil concentrations within badger core home ranges, all relocations collected through the duration of the study for each individual were pooled and 65% adaptive kernel contours were estimated and used for additional analyses. Geometric mean soil concentrations were then calculated within these isopleths and regressed with dieldrin concentrations in plasma (Fig. 14.2A) and adipose tissues (Fig. 14.2B). Individuals used for the regression analysis represented all age and sex classifications. Average concentrations of dieldrin plasma and adipose were used from multiple samplings of individuals when they occurred. Dieldrin detection limits in plasma or adipose samples (1 μg/L and 0.015 μg/kg, respectively) were used when no detectable levels of dieldrin were found. The relationships between both plasma and adipose dieldrin concentrations and soil concentrations of dieldrin within home range contours were significant ($p < 0.0001$), with the best correlation occurring between soil and plasma ($r^2 = 0.78$) dieldrin concentrations.

Although adipose tissue dieldrin concentration was the most sensitive indicator of exposure to dieldrin (Fig. 14.2B), these levels are dynamic, concentrating with constant exposure or diluting with intermittent doses. Plasma concentrations are reflective of

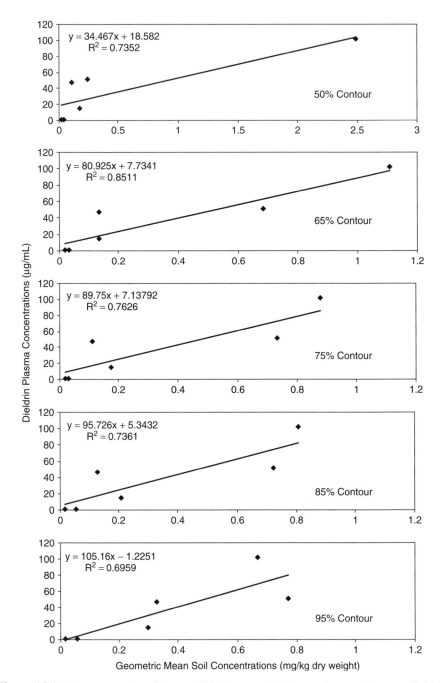

Figure 14.1. Linear regression of plasma dieldrin concentrations and geometric mean dieldrin soil concentrations from 50%, 65%, 75%, 85%, and 95% adaptive kernel contours of selected badgers from the Rocky Mountain Arsenal National Wildlife Refuge, 1994–1996.

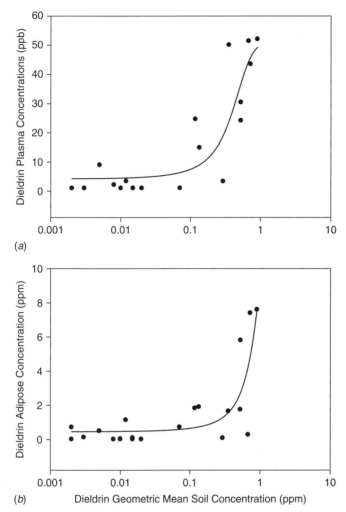

Figure 14.2. Dieldrin plasma (**A**) and adipose (**B**) concentrations versus geometric mean dieldrin soil concentrations within 65% adaptive kernel contour home ranges of American badgers inhabiting the Rocky Mountain Arsenal National Wildlife Refuge, 1994–1996.

recent exposures, as well as passive diffusion of dieldrin out of lipophilic tissues when doses in the wild are not consistent. Large carnivore species, like badgers, will concentrate or dilute dieldrin in adipose tissues as they spend time in contaminated and noncontaminated areas and also as they go through periods of fat metabolism. Therefore, plasma concentrations of dieldrin in wild badgers should be more reflective of average exposure than dieldrin adipose concentrations.

The home-range contour that was most predictive of dieldrin plasma concentrations in American badgers was the 65% isopleth. Up to threefold differences in geometric mean dieldrin soil concentrations occurred within various percent home-range contours (Table 14.3). Most risk assessments are performed with literature values of home ranges encompassing areas of 95% probability of relocation, and therefore may not most accurately estimate exposure.

Also, the shape of the curves fit to the tissue and soil relationships (Figs. 14.2A and 14.2B) are worth noting. For both plasma and adipose, little accumulation of dieldrin occurs below soil concentrations of 0.1 mg/kg dry weight. Tissue residues then sharply increased as home-range soil dieldrin concentrations approached 1.0 mg/kg dry weight. These data demonstrate that the bioavailability of dieldrin in Arsenal soils of 0.1 mg/kg dry weight and higher was sufficient enough to accumulate within the terrestrial food web. Furthermore, with the shape of the curve showing no evidence of leveling, higher levels of accumulation would be possible in more contaminated settings.

Estimating Daily Dieldrin Exposure and Tissue Concentrations. Average observed dieldrin concentrations in adipose (mean $\pm$ std; 2.28 ± 2.20 mg/kg dry weight) differed by 1.6-fold, but not significantly ($p \geq 0.18$) from those predicted (mean $\pm$ std; 3.68 ± 2.80 mg/kg dry weight) based on plasma dieldrin concentrations and plasma/tissue distribution ratios generated in laboratory dosing studies. Distribution ratios, therefore, were judged to sufficiently represent actual distribution of dieldrin in field captured badgers and were thus used to predict dietary doses and brain and liver concentrations of wild badgers having detectable levels of dieldrin in the plasma.

When plasma dieldrin concentrations were used to predict approximate dieldrin concentration in dietary items, values ranged from 0.44 to 7.52 mg/kg dry weight (Table 14.4). It should be noted that other pathways of exposure such as incidental soil ingestion and inhalational and dermal exposure may occur. Therefore, estimated exposures reflected the dietary equivalent of the combined pathways. Brain dieldrin concentrations predicted from plasma dieldrin concentrations yielded values that ranged from 0.008 to 0.137 mg/kg dry weight, and for liver the values ranged from 0.029 to 0.492 mg/kg dry weight. Average estimated brain and liver concentrations in Arsenal badgers (with detectable dieldrin plasma concentrations) were 0.115 ± 0.047 and 0.203 ± 0.155 ppm dieldrin, respectively.

Prior to the initiation of this study in 1991, a juvenile male badger was found dying from dieldrin poisoning in the South Plants area of the site. The badger's brain, liver, and adipose dieldrin concentrations were 5, 13, and 78 mg/kg dry weight, respectively, yielding a liver/brain and liver/adipose concentration ratios of 2.6 and 0.16. Gross necropsy described the badger as being extremely emaciated. Liver/brain and liver/adipose concentration ratios from our studies (Hoff 1998) for badgers dosed in the laboratory with 0.035 mg dieldrin/kg/d were 3.6 and 0.08, respectively. As the animal used fat reserves, it became more and more emaciated. Dieldrin was apparently mobilized from adipose tissue and deposited in the central nervous system, one of the only remaining high-lipid-content tissues. Although no similar residue levels were documented in badger mortalities during this study and no evidence of acute dieldrin poisoning were found, data collected from some badgers point out the importance of

Table 14.4. Estimated Daily Dose, Dietary Concentration, and Tissue Dieldrin Concentrations in Wild Badgers on the Rocky Mountain Arsenal NWR (1994–1996)

Badger I.D.	Plasma (mg/mL)	Daily Dose (mg/kg/d)	Dietary Concentration (mg/kg dry weight)	Brain Concentration (mg/kg dry weight)	Liver Concentration (mg/kg dry weight)
MA2	0.0315	0.057[a]	4.07[b]	0.074	0.266
MA5	0.0581	0.105	7.52	0.137	0.492
FA10	0.0039	0.007	0.50	0.009	0.033
FA11	0.0100	0.018	1.26	0.023	0.082
MA14	0.0125	0.023	1.61	0.029	0.106
MA21	0.0501	0.091	6.48	0.118	0.424
FA22	0.0149	0.027	1.93	0.035	0.126
FA23	0.0242	0.044	3.12	0.057	0.205
MA26	0.0034	0.006	0.44	0.008	0.029
FA29	0.0515	0.093	6.66	0.121	0.436
MA31	0.0114	0.021	1.47	0.027	0.096
MA32	0.0305	0.055	3.94	0.072	0.2d58
MA33	0.0522	0.095	6.75	0.123	0.442

[a] mg/kg/d = [plasma]/0.52.
[b] (mg/kg/d)/(0.014 kg food/kg body weight/day).

pairing geospatial area-use data with organism toxicokinetic residue data in order to document the most comprehensive picture of exposure. Geospatial use of contaminated soils was a good predictor of plasma concentrations in this study, but estimating other critical tissue burdens from laboratory data was not always as successful. Although they were not significantly different, estimated values were approximately 60% higher than observed.

For example, individuals utilizing lipid reserves in tissues alter equilibrium constants noted in nonstressed laboratory animals. A suggestion of this effect was noted in an adult female that consistently inhabited contaminated areas. This badger had a dieldrin plasma concentration of 0.0515 mg/L and an adipose dieldrin concentration of 0.274 mg/kg dry weight. Her estimated adipose dieldrin concentration, based on plasma concentrations and laboratory-derived distribution ratios, was 8.3 mg/kg. This female was captured and sampled on May 10, 1996 and had adipose concentrations much lower than they should have been if the animal was at kinetic equilibrium with dieldrin exposure. This adult female gave birth to two young in March or April and had been foraging for her and her young at that time. She was likely stressed by the need to forage for her young and may have been depending heavily on fat reserves for her own nutrition or had transferred dieldrin to her young *in utero*. This example illustrates the need for investigators to consider carefully the toxicokinetic characteristics of the chemical in question and physiological influences that field stressed animals have relative to laboratory-dosed animals when extrapolating critical tissue concentrations.

In an example of how distribution ratios were relatively accurate, one male (that consistently inhabited the same contaminated area) had estimated and observed dieldrin adipose concentrations of 6.3 and 4.1 mg/kg dry weight, respectively. In this case, the distribution ratios for the male provided relatively accurate estimates of dieldrin in adipose tissues. During the same time period, male activity was relatively low and food supply was generally plentiful.

Observed adipose dieldrin concentrations occurring below predicted levels (estimated from plasma concentrations) may indicate mobilization of adipose residues, recent consumption of highly contaminated prey items, or a significant pathway of exposure other than dietary. Badgers dosed in the laboratory with 0.035 mg dieldrin/kg/day (2.5 mg/kg dry weight dieldrin in diet), which represented an environmentally relevant exposure scenario for wild badgers on the Arsenal, had average dieldrin plasma concentrations of 0.018 mg/kg dry weight. It is unlikely that many badgers would be acutely exposed to prey items at dieldrin concentrations high enough to significantly elevate plasma levels immediately prior to being captured and their plasma sampled. Therefore, dieldrin plasma concentrations above 0.018 mg/L are most likely reflective of dieldrin mobilized from adipose. If a badger accumulates sufficient concentrations of dieldrin in its adipose tissues, periods of high activity and stress could mobilize these concentrations. Based on the behavior of the species, these time periods would most likely be the spring for adult females as they care for their young, late summer for post-dispersal juveniles as they learn to hunt for themselves, and mid-July through September for adult males as they travel extensively during the breeding season.

Quantifying Effects of Dieldrin in American Badgers on the Arsenal

Nine badgers were found dead during the course of the study, but none were determined to be caused by dieldrin poisoning. Necropsies were performed on all badgers to assess the cause of death. Three juveniles were found dead from coyote attacks, one adult and one juvenile were found dead from being hit by a vehicle, one juvenile female was confirmed to have died from Canine distemper virus (CDV), one adult female and one adult male died of unknown causes, and one death was associated with the transmitter implant surgery. Two adult male non-study animals were found dead on the Arsenal during the study, one from vehicular trauma and one from complications of CDV. When possible, tissues were taken for analytical chemical analyses.

Of the identified CoC, dieldrin was found in highest frequency of all analytes in 30 tissue samples collected from six badger carcasses (77%, 23/30 samples). Frequencies of detection for endrin, aldrin, DDT, and pp-DDE were 23%, 7%, 13%, and 10%, respectively. Concentrations of dieldrin found in the brains of dead badgers ranged from not detectable to 0.103 mg/kg dry weight. All but one of the detected concentrations fell below the low-dose dieldrin brain concentration from the laboratory dosing experiment. The dieldrin brain concentration from that one animal fell between the low and high doses administered in the laboratory study for which there were no dieldrin associated mortalities.

In US EPA's 2007 dieldrin Ecological Soil Screening Level (Eco-SSL, US EPA, 2007), a mammalian toxicity reference value (TRV) of 0.015 mg/kg/d was used to model screening level protective soil concentrations for mammals. The Eco-SSL TRV dose is equal to a concentration of 1.07 mg/kg dry weight in badger diets. Data from the toxicological literature, as well as species-specific data from this study (non-reproductive endpoints such as clinical pathology), suggest a dieldrin NOAEL (No Observed Adverse Effect Level) in badger diets of 2.5 mg/kg dry weight. Six of 32 badgers captured on the Arsenal had estimated dietary levels above this NOAEL. All of these individuals inhabited areas with the most contaminated soils in the core area of the site. It is important to reemphasize that estimated daily dieldrin dietary concentrations reflected the dietary equivalent of combined pathways of exposure, including incidental soil ingestion, dermal, and inhalation. Perhaps even more importantly, estimation of dietary dieldrin concentrations for wild badgers used plasma/dose ratios from laboratory dosing experiments. In this equation, using a food consumption rate from laboratory dosing promotes conservative estimates. Wild badgers presumably eat more food than sedentary animals in cages. If daily ingestion rates are greater in wild badgers, the estimated dietary concentrations would decrease, based on plasma/dose ratios. Actual dietary concentrations of dieldrin were probably much lower. Estimated brain and liver concentrations were well below concentrations noted in the badger dosing studies. Because exposure levels in wild badgers were generally below those determined in the laboratory study to have minimal impacts, dieldrin-associated effects were not anticipated to have occurred in Arsenal badgers documented in our study.

No significant ($p < 0.05$) correlations were found between hematological endpoints and geometric mean soil concentrations within home ranges of badgers (largest r^2 value was 0.05). Significant correlations were found between the activity of the enzymes ALT ($r^2 = 0.42$), GGT ($r^2 = 0.35$), and CHOL ($r^2 = 0.19$) and geometric mean soil concentrations within home ranges of badgers. Urinalysis methods (Ames Multi-Stixs; Miles Inc., Diagnostic Div., Elkhart IN) were primarily qualitative. Although proteinuria (≥ 2000 mg/dL) was present in some individuals ($n = 5$), only two were from animals that inhabited areas highly contaminated with dieldrin. Furthermore, proteinuria in all of these individuals was associated with nonpathological levels of hemoglobin in the urine, which were probably responsible for the elevated values.

Of all clinical pathology endpoints, serum enzymes and metabolites provided the best insight into potential response to dieldrin in free-ranging Arsenal badgers. Significant correlations were found between enzymes associated with liver pathology (ALT and GGT) and geometric mean dieldrin soil concentrations within the home ranges (65% isopleths) of badgers. Additionally, cholesterol levels had the same significant relationship. With low r^2 values associated with the relationships, diagnostic conclusions would be difficult to draw if only one of the liver enzymes was elevated. Corresponding increases in two liver enzymes, however, provide stronger evidence of liver response to chemical exposure. The rise in serum cholesterol also supports liver pathology. Fatty liver in mammals exposed to dieldrin is well documented (WHO 1989). Although not statistically significant, trends in the laboratory dieldrin-exposed badgers illustrated concurrent elevations in ALT activities, CHOL levels, and liver:body weight ratios (Hoff 1998). The evidence suggests that badgers that inhabited the most contaminated areas of the Arsenal accumulated lipid in the liver and concurrent metabolic or pathological changes resulted in elevated activities of serum enzymes ALT and GGT. Any pathological changes that might have occurred, however, were limited to functional responses, because histopathological examination did not demonstrate actual physical pathology in laboratory-exposed badgers.

Serum Antibodies to CDV and Yersinia pestis. In 1994, only 2 of 21 (9%) badgers sampled had antibody titers to CDV. In 1995, three of ten (30%) badgers sampled had antibody titers to CDV, and one badger and two coyotes were found dead due to complications to CDV. The deaths of carnivores due to CDV, along with an increase in the presence of CDV antibody titers found in the serum of badgers, suggest that a higher frequency of CDV infection occurred on the Arsenal in the fall of 1994 and spring of 1995.

Antibody titers to *Y. pestis* for 1994 mirror the timing of the disease, which significantly reduced the number of black-tailed prairie dogs on the Arsenal in 1995. For example, one badger had low titer values (1:32) on capture dates April 16, 1995 and May 12, 1995; but on July 8, 1995, it exhibited a titer of >1:512. Her offspring captured on June 28, 1995 exhibited titers of >1:512. Significant exposure to plague for this badger occurred between May 12, 1995 and June 28, 1995. These dates coincide with the time that USFWS personnel have described as the beginning of the outbreak of plague in the area which she inhabited. Badgers in other areas (e.g., Sections. 27 and 34) demonstrated plague titers of >1:512 as early as April 2, 1994.

Canine distemper virus has been shown to cause morbidity and mortality in American badgers (Thorne et al. 1982). On the Arsenal, three badgers were found dead or dying due to complications arising from CDV (two badgers, one in 1992 and one in 1993, found by USFWS personnel, and one in our study). It is, however, possible for badgers to be exposed to CDV without mortality if adequate antibody can be produced rapidly to defeat the challenge. In order to survive CDV, badgers cannot be substantially affected by immunosuppressive contaminants such as dieldrin (Bernier et al. 1987, Krzystyniak et al. 1985, Loose et al. 1981). One male badger had no antibody titer to CDV when captured in 1994. However, when recaptured on July 12, 1995, he had a 1:128 titer and was known to be alive at the end of the study in 1996. This adult male inhabited the most contaminated area on the Arsenal and had survived exposure to CDV, which would require the production of adequate levels of antibody. Other badgers inhabiting highly contaminated areas also had extremely high levels of antibody against *Y. pestis*. As with results of urinalysis, antibody titers are not useful for regression analysis because of their qualitative nature. They do, however, provide insight into the immune capabilities of wild badgers on the Arsenal. Evidence suggests that dieldrin soil concentrations found within badger home ranges on the Arsenal were not sufficient to affect the badger's ability to combat pathogenic challenge.

SUMMARY

With the exception of density, the demographics of the 1994–1996 American badger population on the Arsenal did not appear to differ from those of other populations in the

West. Low density estimates of the badger population are not likely due to exposure to dieldrin because estimated doses were below effect doses determined from controlled laboratory dosing of badgers, as well as those found for other mammals in toxicological literature. More likely, low badger densities were a result of limited immigration corridors and density of prey. Adult female home-range areas are resource limited and fluctuated with the population density of black-tailed prairie dogs. Individuals inhabiting the most contaminated areas of the site, however, may have experienced dieldrin-associated metabolic effects and liver pathology at the biochemical level.

Because of time and financial constraints, most toxicological wildlife risk assessments on hazardous waste sites are limited to comparing estimates of exposures (or possibly some residue data) to literature-derived toxicity reference values to characterize risk. This study demonstrated how contaminant-related risk assessments in wildlife populations can be improved if robust area use data are available. Detailed home range and animal movement analyses allow for linking spatially explicit attributes of a landscape such as habitat, prey, and chemical contamination with the relative contribution various stressors may have on the population.

ACKNOWLEDGMENTS

This work would not have been possible if not for the cooperation of the University of Wyoming: specifically, Dr. Beth Williams and laboratory technicians in the State Veterinary Laboratory; Dr. Hank Harlow and staff at Red Buttes Environmental Laboratory; Dr. Lee Belden and his graduate students in the School of Agriculture. Several technicians aided with data collection: Sherry Skipper, Rusty Jeffers, Matt Myers, Walt Cook, Pat Henry, Bill Henriques, and Eric Montie. GIS mapping was completed by Debbie Goeldner of Golden Geospatial Inc. in Golden, CO. Funding for the project came from USFWS and NIEHS ES04696.

REFERENCES

Alldrege JE, Ratti JT. 1986. Comparison of some statistical techniques for analysis of resource selection. *J Wildl Manage* **50**:157–165.

Allen DL. 1996. Effects of contaminants on small mammal population and community parameters on the Rocky Mountain Arsenal National Wildlife Refuge. MS thesis, Clemson University, Clemson. SC, 71 pp.

Bernier J, Hugo P, Krzystyniak K., Fournier M. 1987. Suppression of humoral immunity in inbred mice by dielrin. *Toxicol Lett* **35**:231–240.

Biota Remedial Investigation. 1989. *Final Report, Version 3.2*, Vol. I. Environmental Science and Engineering, Inc. Denver, CO. Prepared for the Program Manager, Rocky Mountain Arsenal.

Burt WH. 1943. Territoriality and home range concepts as applied to mammals. *J Mammal* **24**:346–352.

Carlsen TM, Coty JD, Kercher JR. 2004. The spatial extent of contaminants and the landscape scale: an analysis of the wildlife, conservation biology and population modeling literature. *Environ Toxic Chem* **23**:798–811.

Ehrlich PR, Roughgarden J. 1987. *The Science of Ecology*. Macmillan, New York.

Goodrich JM. 1994. North American badgers (Taxidea taxus) and black-footed ferrets (Mustela nigripes): Abundance, rarity, and conservation in a white-tailed prairie dog (*Cynomys leucurus*)-based community. Ph.D. dissertation. University of Wyoming, Laramie, WY, 101 pp.

Hegdal PL, Colvin BA. 1986. Radio Telemetry. In Cooperrider AY, Boyd RJ, Stuart HR (Eds.), *Inventory and Monitoring of Wildlife Habitat*. US Department of Interior, Bureau of Land Management, Denver, CO, pp. 679–698.

Hoff DJ. 1998. Integrated laboratory and field investigations assessing contaminant risk to american badgers (*Taxidea taxus*) on the Rocky Mountain Arsenal National Wildlife Refuge. PhD dissertation. Clemson University, Clemson, SC, 240 pp.

Johnson DH. 1980. The comparison of usage and availability measurements for evaluation resource preference. *Ecology* **61**:65–71.

Kapustka LA, Galbraith H, Luxon M, and Biddinger GR (Eds.). 2004. *Landscape Ecology and Wildlife Habitat Evaluation: Critical Information for Ecological Risk Assessment, Land-Use Management Activities, and Biodiversity Enhancement Practices*. ASTM STP 1458, American Society for Testing and Materials International, West Conshohocken, PA.

Kenwood RE. 1987. *Wildlife Radio Tagging*. Academic Press, Orlando, FL.

Krzystyniak K, Hugo P, Flipo D, Fournier M. 1985. Increased susceptibility to mouse hepatitis virus 3 of peritoneal macrophages exposed to dieldrin. *Toxicol Appl Pharmacol* **80**:397–408.

Lampe RP. 1976. Aspects of the predatory strategy of the North American badger, *Taxidea taxus*. Ph.D. dissertation. University of Minnesota, Minneapolis, MN, 102 pp.

Lindzey FG. 1978. Movement patterns of badgers in northwestern Utah. *J Wildl Manage* **42**:418–422.

Loose LD, Silkworth JB, Charbonneau T, Blumenstock F. 1981. Environmental chemical-induced macrophage dysfunction. *Environ Health Perspect* **39**:79–91.

Messick JP, Hornocker MG. 1981. Ecology of the badger in southwestern Idaho. *Wildl Monogr* **76**:1–53.

Minta SE. 1990. The badger, *Taxidea taxis (Carnivora: Mustelidae):* Spatial–temporal analysis, dimorphic territorial polygyny, population characteristics, and human influences on ecology. Ph.D. dissertation. University of California, Davis, 310 pp.

Moore TD, Spence LE, Dugnolle CE, Hepworth WG. 1974. *Identification of the Dorsal Guard Hair of Some Mammals of Wyoming*. Wyoming Game and Fish Department. Laramie, WY.

Neu CW, Byers CR, Peek J. 1974. A technique for utilization-availability data. *J Wildl Manage* **38**:541–545.

Pollock KH, Winterstein SR, Bunck CM, Curtis PD. 1989. Survival analysis in telemetry studies: the staggered entry design. *J Wildl Manage* **53**:7–15.

Stollar and Associates, Inc. 1992. *Comprehensive Monitoring Program: Biota Annual Report for 1990 and Summary Report for 1988 to 1990*. Prepared for the Program Manager, Rocky Mountain Arsenal.

Thorne ET, Kingston N, Jolley WR, Bergstrom RC. 1982. *Diseases of Wildlife in Wyoming*, 2nd edition. Wyoming Game and Fish Department, Cheyenne, WY.

US EPA (United States Environmental Protection Agency). 2007. *Ecological Soil Screening Level for Dieldrin*. OSWER Directive 9285.7-56. http://www.epa.gov/ecotox/ecossl/index.html (accessed July 23, 2009).

USFWS (United States Fish and Wildlife Service). 1995. *United States Fish and Wildlife Service Fiscal Year 1995 Annual Progress Report: Rocky Mountain Arsenal National Wildlife Refuge*. Presented to the Program Manager of the Rocky Mountain Arsenal.

Warner RE, Ver Steeg B. 1995. *Illinois Badger Studies: Final Report*. Federal Aid Project No. W-103-R-1-6 submitted to Division of Wildlife Resources and Illinois Department of Natural Resources, Springfield, IL.

Warren SD, Holbrook SW, Dale DA, Whelan NL, Elyn M, Grimm W, Jentsch A. 2007. Biodiversity and the heterogeneous disturbance regime on military training lands. *Restor Ecol* **15**:606–612.

Weir RD, Davis H, Hoodicoff C. 2003. *Conservation strategies for North American Badgers in the Thompson and Okanagan Regions*. Final Report for the Thompson–Okanagan Badger Project, 103 pp, http://www.badgers.bc.ca/TOB.htm (accessed July 23, 2009).

White GC, Garrott RA. 1990. *Analysis of Wildlife Radio-Tracking Data*. Academic Press, San Diego, CA.

WHO. 1989. *World Health Organization Report on Aldrin/Dieldrin*. Environmental Health Criteria 91, Geneva, Switzerland.

Worton BJ. 1989. Kernel methods for estimating the utilization distribution in home range studies. *Ecology* **70**:164–168.

15

ENVIRONMENTAL RISK ASSESSMENT OF PHARMACEUTICALS

Joanne Parrott, Alison McLaughlin, David Lapen, and Edward Topp

INTRODUCTION TO THE ISSUE

Pharmaceutical drugs have recently been detected around the world in municipal wastewater effluents (MWWEs) and in river waters downstream (Daughton and Ternes 1999; Kolpin et al. 2002; Metcalfe et al. 2003a, 2003b). Pharmaceuticals excreted (or discarded) by humans may enter municipal wastewaters. In addition, pharmaceuticals given to farm animals may enter water directly from contact of animals with water, or from runoff after application of manure to crop soils (Fig. 15.1).

Concentrations of pharmaceutical compounds in the aquatic environment range from the low ng/L to high µg/L (Kolpin et al. 2002; Metcalfe et al. 2003a, 2003b; Lishman et al. 2006). Concentrations in river waters and MWWEs are very much lower than the therapeutic doses for humans or animals of these pharmaceuticals compounds, but the potential for nontarget effects and long-term effects of very low environmental concentrations have raised concern among the public, scientists, and regulators.

Environmental Risk and Management from a Landscape Perspective, edited by Kapustka and Landis
Copyright © 2010 John Wiley & Sons, Inc.

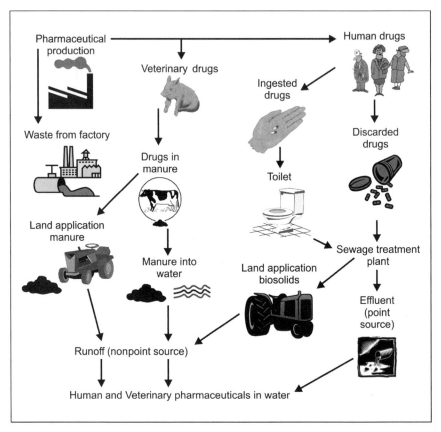

Figure 15.1. Pathway of entry of human and veterinary pharmaceuticals to the aquatic environment.

Environmental Risk Assessment (EnRA) of pharmaceuticals is an emerging practice, because typically these compounds were exempt from environmental regulation and have not been part of conventional environmental monitoring programs. For most pharmaceuticals the total volumes produced or imported are very low, making the compounds fall outside of typical large volume industrial chemical EnRA processes. Pharmaceuticals represent a special and very complex case because of the great variety of compounds within this class and because they are, by design, biologically active at quite low doses.

The lack of scientific understanding of the potential effects of very low concentrations of pharmaceuticals on nontarget organisms in the environment also complicates the process. There are some data for several classes of pharmaceuticals, but for most of the 4000+ active pharmaceuticals ingredients, environmental monitoring and biological effects data do not exist. At best we currently have methods to quantify in surface waters and in MWWEs about 300 pharmaceuticals. Although there is much data on mammalian toxicology for pharmaceuticals, we have limited data (usually acute exposure data, though sometimes chronic exposure data are available) on effects in nontarget species.

Evaluation of pharmaceuticals issues using EnRA is relatively new; improvements and changes in regulatory processes are occurring currently. In this chapter we review some of the approaches to EnRA of human and veterinary pharmaceuticals in Europe, the United States, and Canada. We also discuss effects data that suggest which pharmaceuticals we should prioritize for monitoring and effects research. Paths forward in the EnRA process and the influences of watersheds and land use are also considered, as we develop new ways to assess this complex group of compounds.

EnRA OF HUMAN AND VETERINARY PHARMACEUTICALS

There are differences in governing regulations and process, as well as in landscape characteristics in different jurisdictions. Therefore, we present European, United States, and Canadian perspectives on assessment of human pharmaceuticals in this section.

In Europe

Environmental risk assessment of pharmaceuticals in Europe falls under the European Agency for the Evaluation of Medicinal Products (EMEA) Directive (EMEA/CHMP/SWP/4447/00). Under the EMEA, most medical products with new and established active ingredients are required to undergo an EnRA at the time of registration or application for registration. The risk assessment conducted for the European Medicines Agency is the full environmental assessment, and the pharmaceutical would be exempt from any data requirements under REACH (Registration, Evaluation, Authorization, and Restriction of Chemicals, adopted December 2006).

The EnRA proceeds in a stepwise fashion starting with calculation of the predicted environmental concentration (PEC). Ferrari et al. (2004) show how several human pharmaceuticals proceed through this EcoRA process. The initial screening calculation includes values for daily dose, market penetration (percentage of population that is predicted to take the pharmaceutical, default value is 1%), wastewater per capita, and a standard factor of 10 for dilution. A PEC exceeding 0.01 μg/L triggers the requirement for environmental fate and toxicity studies and an environmental risk analysis. Data requirements for aquatic toxicity include:

- Consumption data
- log K_{ow}
- Adsorption–desorption using a batch equilibrium method (OECD 106 and OECD 121/OPPTS 835.1110)
- Ready biodegradability test (OECD 301)
- Aerobic and anaerobic transformation in aquatic sediment systems (OECD 308)

- Algae growth inhibition test (OECD 201)
- *Daphnia* reproduction test (OECD 210)
- Fish early life stage toxicity test (OECD 210)
- Activated sludge respiration inhibition test (OECD 209)

Specific data requirements are generated in a stepwise decision-tree manner, with each successive step following a previous yes/no or pass/fail criterion. If a substance is found to be not ready biodegradable in OECD 301 and the aerobic and anaerobic transformation in aquatic sediment systems test in OECD 308 indicates significant persistence of the substance in sediment, then an additional data requirement, such as toxicological testing on sediment dwelling organisms, may be required. In practice, very few pharmaceuticals appear to be ready biodegradable under OECD 301 criteria, which is likely an artifact of general pharmaceutical design. If the K_{oc} is $> 10,000 L/kg$ ($\log K_{oc} > 4$), then terrestrial environmental fate and effects analyses are undertaken. Terrestrial fate and effects studies include aerobic and anaerobic transformation in soil (OECD 307), nitrogen transformation testing (OECD 216), terrestrial plants (OECD 208), acute earthworm toxicity (OECD 207) and the Collembolla reproduction test (ISO 11267).

With reference to the OSPAR Convention (the OSlo/PARis Convention, a short title for the 1992 International Convention for the Protection of the Marine Environment of the North-East Atlantic), the EMEA guideline states that a pharmaceutical substance with a $\log K_{ow} > 4.5$ should be screened in a stepwise procedure for persistence, bioaccumulation, and toxicity according to the EU technical guidance document. Interestingly, the general criteria for good pharmaceutical design of an orally active pharmaceutical specifies a $\log K_{ow} < 5$ (Lipinski's rule of five); thus it seems quite unlikely that many pharmaceuticals will be tested in this stepwise procedure (for persistence, bioaccumulation and toxicity) because of this EMEA trigger of $\log K_{ow} > 4.5$.

The EMEA guideline states that substances that can interfere with hormonal control in organisms must be assessed regardless of the outcome of the screening calculation. Most substances other than hormones, however, would not be known *a priori* to be probable endocrine disruptors.

Environmental impact cannot be used to refuse pharmaceutical marketing authorization in the European Union Member States (Directive 2004/27/EC). When the possibility of environmental risks cannot be excluded, precautionary and safety measures may consist of product labeling, which aims at minimizing the quantity discharged into the environment by appropriate mitigation measures. The EMEA guideline recommends that package leaflets for patient information should contain the following general statement: Medicines should not be disposed of via wastewater or household waste. Ask your pharmacist how to dispose of medicines no longer required. These measures will help protect the environment.

In the United States of America

In the United States, the Food and Pharmaceutical Administration (under the Food and Drug Act; FDA) is required under the National Environmental Policy Act (NEPA) to consider the environmental impacts of approving pharmaceuticals at the time of application for registration as an integral part of its regulatory process. However, NEPA does not require that the most environmentally beneficial course of action be taken. Specific guidelines for environmental assessment of pharmaceuticals are outlined in the document Guidance for Industry: Environmental Assessment of Human Pharmaceutical and Biologics Applications (FDA 1998). A number of pharmaceuticals have categorical exemptions from the environmental assessment process, including, for example, new pharmaceuticals under investigation, or applications that will not increase use of the active moiety. The caveat is that an assessment may still be required if extraordinary circumstances indicate that a specific proposed action may affect environmental quality.

Using a tiered process, exposure in the environment is first calculated (the expected environmental concentration, equivalent to PEC) and, if a limit concentration of 1 μg/L is exceeded, environmental fate and toxicity studies are required and a risk analysis is carried out. At the first stage of this process, the initial screening calculation includes values for kilogram annual production of the active ingredient and wastewater dilution based on 1.214×10^{11} L of water per day entering municipal wastewater treatment plants (MWWTPs) (Source: 1996 Needs Survey, Report to Congress, in FDA 1998).

The guidance document outlines a tiered approach for aquatic fate and effects testing which starts with reporting of the water solubility, dissociation constant, $\log K_{ow}$, and the vapor pressure or Henry's Law Constant values. The potential for rapid breakdown is tested in a hydrolysis test (rapid if half-life is <24 hr) or aerobic biodegradation (rapid if half-life is <8 hr), or soil biodegradation test (rapid if half-life is <5 days), followed by a microbial inhibition test for those pharmaceuticals that are found not to degrade rapidly. If no rapid breakdown or degradation mechanism is identified, it is assumed that the compound will persist in the environment and, therefore, the toxicity to environmental organisms should be evaluated. If it is determined that toxicity testing is required, substances with a $\log K_{ow} < 3.5$ are subject to acute toxicity testing with one species (either algae, or daphnia, or fish). The aim of the toxicity tests is to derive an effect concentration (such as the lowest observed effective concentration, or LOEC) and a predicted no effect concentration (PNEC). The ratio of PEC to PNEC (the PEC/PNEC quotient) is assessed, and substances with values of 1 or greater must provide acute toxicity tests for another two species to complete the base set of data for algae, daphnia and fish.

Substances with a $\log K_{ow} > 3.5$ are also subject to acute toxicity testing when justification is provided, or they may proceed directly to chronic toxicity testing. Substances with a $\log K_{ow} > 3.5$ that fail the PEC/PNEC quotient with the results of acute testing should then proceed to chronic testing.

The magnitude of the safety factor or assessment factor when calculating the PNEC from the LOEC depends on the amount of ecotoxicological data available to calculate an accurate LOEC. Smaller assessment factors are used when more chronic data are available for several species. Assessment factors for the determination of the PNEC are fixed at 1000 for Tier 1 (1 species, acute test), 100 for Tier 2 (3 species, acute test), and 10 for Tier 3 (chronic testing).

In Canada

Effective September 14, 2001, substances in products regulated by the Food and Pharmaceuticals Act are legally obliged under the New Substances Notification Regulations (NSNR) of the Canadian Environmental Protection Act (CEPA) of 1999 to conduct an environmental assessment. The intent of the NSNR (for chemicals and polymers) is to ensure that new substances entering into Canadian commerce do not pose a threat to the environment and human health. As well, notification prior to manufacture or import is a requirement of CEPA 1999. Although the principles in CEPA and the NSNR lend themselves to the environmental assessment of human and veterinary pharmaceutical substances, there are questions about the applicability of the trigger quantities and data requirements with respect to new pharmaceutical substances.

The current NSNR trigger quantities for chemicals and polymers are based on kilograms manufactured or imported per year. However, the manufactured quantities of pharmaceuticals vary considerably, and it is likely that a number of pharmaceutical active ingredients of potential environmental concern may not have reached notifiable trigger quantities under the NSNR. At the highest tier of testing, physical/chemical data, ready biodegradation, and an acute fish, daphnia and algae test are mandatory. One of the criticisms of the EnRA procedure is that other tests outside of NSNR requirements may be more relevant and might improve the environmental assessments for pharmaceuticals. For example, studies of biodegradation rates in relevant environmental media and chronic aquatic toxicity endpoints may be more relevant for pharmaceuticals.

A new EnRA process for pharmaceuticals is currently being developed. Canada's Minister of Health committed to developing new, appropriate environmental regulations for substances in Food and Drug Act (F&DA) products, and a notice to this effect was published in Canada Gazette I, September 1, 2001. The Environmental Impact Initiative of the Policy Planning and International Affairs Directorate in the Health Products and Food Branch of Health Canada is the lead agency on the development of these new regulations. A multi-stakeholder Environmental Assessment Working Group (EAWG) was established in 2006 to provide broad, strategic advice on policy, operational, technical, and regulatory issues related to the development of appropriate regulations as well as the management of substances in commerce. The future recommendations of this group will be vital to the formation of new EnRAs and new trigger quantities for EnRAs for human and veterinary pharmaceuticals and other substances entering the Canadian market.

EnRA OF VETERINARY PHARMACEUTICALS

For veterinary pharmaceuticals, there has been agreement of several countries under The International Cooperation on Harmonization of Technical Requirements for Registration of Veterinary Medicinal Products (VICH). These EnRA regulations and landscape parameters that may affect veterinary pharmaceutical are presented in the following section.

VICH

The International Cooperation on Harmonization of Technical Requirements for Registration of Veterinary Medicinal Products (VICH) is a trilateral (EU–Japan–USA) program aimed at harmonizing technical data requirements for the registration of veterinary pharmaceutical products. Australia, New Zealand, and Canada have observer status at VICH meetings. The VICH has generated numerous technical documents related to data requirements for veterinary pharmaceuticals including guidance for industry on the conduct of environmental impact assessments, which includes VICH GL6 (examining potential exposure, Phase I) and VICH GL38 (examining potential toxicity, Phase 2). In the text of these documents, it is presumed that the assessment is to be conducted by the pharmaceutical manufacturer. The guidance outlined in these documents has been implemented to varying degrees by participating countries in accordance with their respective regulatory requirements and frameworks.

Overview of VICH Guidance. The VICH environmental assessment proceeds in a stepwise manner which looks first at potential exposure (Phase I) and then looks at potential toxicity (Phase II). Phase I involves the calculation of a PEC and establishes environmental concentration triggers that determine whether or not there is a need to generate toxicity data (Phase II). In Phase I, if contaminated waste or effluents are handled in such a way that they do not enter the environment; or if they are collected and incinerated, there is no need to generate data. The PEC values for veterinary pharmaceuticals used in aquaculture are compared to an aquatic concentration benchmark of 1 μg/L. The PEC values for veterinary pharmaceuticals used in terrestrial agriculture are compared to a soil concentration benchmark of 100 μg/kg. There is no aquatic PEC benchmark for veterinary pharmaceuticals used in terrestrial agriculture in VICH GL6, but there is guidance on PEC calculations for surface water in VICH GL38. Presumably, this reflects an underlying assumption that the terrestrial use of veterinary pharmaceuticals will not impact on surface waters if the initial soil concentration is estimated to be < 100 μg/kg.

Environmental Concentration Trigger: Aquatic Media. In Phase I, if the PEC in water arising from a confined facility used for aquaculture (such as a fish farm utilizing tanks or pools) results in concentrations < 1 μg/L, data need not be generated. The VICH guidance allows for additional volumes of water used during treatment in confined facilities, namely dilution, as a means of mitigating environmental concerns down to < 1 μg/L. The caveat is that if a pharmaceutical is used as an ecto- or endoparasiticide, or if the pharmaceutical is used for aquaculture, which is conducted directly in the environment (e.g., net pens in open water, which are used for the majority of salmon farming operations in Canada), then data must be generated as per Phase II, starting with degradation testing in aquatic systems and acute aquatic toxicity testing. Data for bioconcentration, bioaccumulation, and sediment organism toxicity are not necessarily required for veterinary pharmaceuticals used in open net aquaculture. This is a concern because these pharmaceuticals are discharged directly to the environment at therapeutic concentrations.

The rationale for the veterinary pharmaceuticals < 1 μg/L cutoff value was a retrospective review of the US FDA's acute ecotoxicity data (and just a few chronic studies) for 76 pharmaceutical substances that showed that 1 μg/L represented the lowest NOEC (causing no mortality). It should be noted, however, that the US FDA dataset excludes a number of the most difficult substances like synthetic estrogens (e.g., ethinylestradiol, diethylstilbestrol), fluroquinolones, and avermectins. The 1-μg/L trigger value was considered suitable by the US FDA for triggering assessments for pharmaceuticals. However, the EU Scientific Steering Committee did not consider the proposed number for the aquatic trigger scientifically valid because there were other examples of pharmaceuticals available, which showed higher toxicity. They suggested that the aquatic trigger should be lowered to 0.0004 μg/L (National Institute for Public Health and the Environment, RIVM report 601500002).

Environmental Concentration Trigger: Terrestrial Media. In Phase I, if the PEC in soil arising from the terrestrial agricultural use of a veterinary pharmaceutical is <100 μg/kg, data need not be generated. The caveat is that if the veterinary pharmaceutical is used as an ecto- or endo-parasiticide, then toxicity data must be generated as outlined in Phase II. The rationale for the terrestrial soil trigger value was a claim that levels <100 μg/kg are below the level shown to have adverse effects in the flora and fauna of terrestrial soils. However, the EU Scientific Steering Committee did not consider the proposed number for the terrestrial trigger scientifically valid (National Institute for Public Health and the Environment, RIVM report 601500002). Even if a scientifically valid environmental effects concentration could be established, there is still the complication of accurately calculating a PEC. Many of the factors used to generate soil PECs, like proportion of the year that animals are housed and manure storage time, are variable and subject to manipulation via statistical methods.

Phase II Chemistry, Fate, and Toxicity Data Requirements. Data requirements in Phase II are tiered, and they always start with a base set of physical–chemical properties tests and environmental fate studies. Physical–chemical properties tests include water solubility (OECD 105), dissociation constants (OECD 112), UV–vis absorption spectrum (OECD 101), melting point (OECD 102), vapor pressure (OECD 104), and log K_{ow} (OECD 107 or 117). Environmental fate studies include soil adsorption–desorption using a batch equilibrium method (OECD 106), soil biodegradation (OECD 307), aquatic aerobic/anaerobic degradation (OECD 308), and "optional" photolysis and hydrolysis (OECD 111) testing.

The recommended Tier A aquatic effects studies include algal growth inhibition (OECD 201 for freshwater or ISO 10253 for saltwater), daphnia immobilization (OECD 202 for freshwater or ISO 14669 for saltwater), and fish acute toxicity (OECD 203 for freshwater or alternate for saltwater). These acute toxicity tests apply to veterinary pharmaceuticals used both in aquaculture and in terrestrial agriculture, because it is acknowledged that either may impact the aquatic environment when the respective soil or water trigger value is exceeded. The recommended Tier A terrestrial effects studies include a 28-day nitrogen transformation study (OECD 216), terrestrial plant toxicity (OECD 208), and earthworm toxicity (OECD 220/222). If the substance is an ecto- or endo-parasiticide, it is also recommended that lethality studies be submitted for dung fly and dung beetle larvae.

Based on the estimation of environmental concentration, refined by consideration of metabolism and excretion factors, and using the Tier A toxicity dataset in conjunction with test-specific recommended assessment factors (safety factors or uncertainty factors), a risk quotient (PEC/PNEC) is derived. If the risk quotient indicates a potential environmental concern (if PEC/PNEC >1) or if the soil microorganism inhibition test indicates an effect $>25\%$, then testing proceeds to Tier B. The guidelines emphasize testing for the parent substance. Testing would presumably be conducted for the metabolites too, if toxicity were a known issue. However, the testing of metabolites is not an explicit recommendation.

Assessment factors or safety factors are applied to the calculation of PNEC, with higher more protective factors used when data sets are limited. Default assessment factors range from 100 for a PNEC based on acute algal toxicity to 1000 for one based on acute fish toxicity. These assessment factors may not be adequate to represent the acute to chronic ratios of pharmaceuticals (Cunningham et al. 2004). Phase II, Tier B subchronic toxicity tests are triggered by an unfavorable risk quotient (PEC/PNEC >1) using a modeled, possibly refined PEC value and a PNEC derived from acute test data generated in Tier A. Tier A data may not, however, be representative of endpoints of real environmental concern because low concentrations coupled with chronic exposure is the most probable scenario. Pulsed or acute high dose exposures in the environment are less likely.

At Tier B, a fish bioconcentration study (OECD 305) is recommended if the log K_{ow} is >4. Aquatic toxicity studies include algal growth inhibition (OECD 201), daphnia reproduction (OECD 211), fish early life stage test (OECD 210), sediment invertebrate species toxicity (OECD 218/219), and the same tests again for saltwater. Toxicity tests for sediment-dwelling species are triggered by an unfavorable (>1) risk quotient in the water column (the PEC for water is compared to the PNEC for daphnia). This is probably not the best trigger for substances that partition strongly to sediments. The recommended Tier B terrestrial effects studies are the same as those in Tier A, but expanded to potentially include a bigger test system or more species. The tests included are the 28-day nitrogen transformation study (OECD 216), terrestrial plant toxicity (OECD 208), and earthworm toxicity (OECD 220/222). Among the shortcomings of the VICH procedure are that it does not address the issues of antibiotic resistant bacteria or potential endocrine disrupting effects.

SPECIAL CONSIDERATIONS FOR ENRA OF HUMAN AND VETERINARY PHARMACEUTICALS

Pseudo-Persistence

Removal of human pharmaceuticals to biosolids and the presence of veterinary pharmaceutical residues in manure may lead to surface water contamination due to a number of factors. The apparent mobility of these residues is dependent on more than simple adsorption–desorption coefficients (e.g., log K_{oc}) for the substance. Also contributing to the potential transport of and mobility of soil incorporated pharmaceutical residues is the frequency of manure or biosolids applications, soil characteristics, tillage type, geography, and rainfall parameters. The constant low-level input of these substances into the environment has rendered many of them "pseudo-persistent" even if ultimately they are biodegradable.

Effects in Nontarget Organisms from Low Concentrations of Pharmaceuticals

Because of pseudo-persistence of these compounds, chronic exposure is the most likely exposure scenario for nontarget organisms in terrestrial and aquatic environments. As well, effects on nonlethal endpoints (e.g., growth, development, spawning) may ultimately have significant effects on the sustainability of wildlife populations. The benchmark concentration at which aquatic effects occur is not established in the case of most pharmaceuticals (Fent et al. 2006), and it is unclear how complex pharmaceutical mixtures (such as those in MWWE) may act on nontarget organisms.

Since pharmaceuticals are designed to interact with target biological receptors, there is a possibility of seeing some sort of effect on nontarget organisms. However, ecotoxicological endpoints are not necessarily qualitatively the same as in mammals. There is a surprising amount of homology between mammals and nonmammalian vertebrates. However, biological signals, structures, and pathways may be different or unknown in many invertebrate species.

Research is showing that environmental effects associated with trace levels of pharmaceuticals in the environment may be triggered at much lower concentrations than typically associated with most known industrial pollutants (Fent et al. 2006). As well, acute lethality is not likely the most appropriate benchmark for measuring the environmental impact of pharmaceuticals (Ferrari et al. 2004).

Endocrine Effects

The fundamental role of all endocrine systems is to enable a dynamic, coordinated response of a distant target tissue to signals originating from another organ and, in some instances, cues originating from outside of the body (e.g., daylight, temperature, stress). The primary objective for most endocrine systems is to maintain some form of homeostasis (e.g., maintaining even blood sugar levels, hydration levels); however, many endocrine systems play an adaptive role (e.g., maintaining a human pregnancy, frog metamorphosis, turtle egg development, or fish spawning).

Species from humans to the most ancient organisms share components of the hypothalamus–pituitary axes responsible for homeostasis. The hypothalamus–pituitary–adrenal axis is responsible for the steroid hormone component of the classic stress response, the hypothalamus–pituitary–gonadal axis is responsible for gonadal development and sexual maturation, and the hypothalamus–pituitary–thyroid axis is responsible for metabolic regulation. Often there is a surprising amount of homology between mammals and nonmammalian species (WHO 2002). In many cases, the effects of pharmaceuticals in wildlife are likely to be the same as the effects that the pharmaceuticals cause in humans, but there may be unique effects, especially if the pharmaceutical substance or target receptors play a role in fish that they do not play in mammals.

Significant species differences exist in the structure and function of endocrine and reproductive organs, and there are major species variations in toxicant metabolism. Hormonal signaling and control mechanisms in fish, birds, reptiles, amphibians, and invertebrates may depend on different chemical transmitters. Thus substances that have little impact on the hormonal stasis of humans may have dramatic endocrine effects in other species.

Several classes of pharmaceuticals have been shown to have endocrine-disrupting effects. Synthetic estrogens such as ethinyl estradiol can feminize fish, reduce fertilization rates, and obliterate populations at concentrations from 1 to 6 ng/L (Länge et al. 2001, Parrott and Blunt 2005, Kidd et al. 2007). Synthetic steroids used for growth promotion in cattle can masculinize fish (Ankley et al. 2001, 2003).

Cholesterol-lowering pharmaceuticals such as the fibric acid-derived class (which include clofibrate, ciprofibrate, fenofibrate, and gemfibrozil) may interfere with steroid hormones in nontarget organisms, because steroid hormones are all derived from cholesterol. Decreased testosterone was seen in goldfish exposed to 1.5 µg/L gemfibrozil for 14 days (Mimeault et al. 2005).

The selective serotonin reuptake inhibitors (SSRIs) are a class of antidepressants, which include fluoxetine, fluvoxamine, sertraline, and paroxetine. SSRIs inhibit the reuptake of serotonin, a neurotransmitter that is involved in various neuronal and hormonal functions. Serotonin affects mood in humans, but in bivalve mollusks serotonin is a neurohormone that induces spawning (Hamida et al. 2004). Studies have indicated that SSRIs may trigger spawning in bivalves (Fong et al. 1998) and affect fish sex steroids (testosterone, estradiol) (Brooks et al. 2003, Foran et al. 2004) at concentrations that are environmentally relevant.

Antibiotic Effects

The presence of antibiotics in the environment is a new field of investigation. Most of the research has shown that antibiotic resistant bacteria arise from human and animal fecal waste entering the environment, rather than from direct effects (selection of resistant genes in bacterial populations) of the very low levels of antibiotics in the environment.

Large quantities of enteric bacteria from human and animal fecal wastes can be released into rivers and lakes that serve as sources of water for drinking, recreation or irrigation. Antibiotic resistant fecal coliforms (e.g., *E. coli*) have been found with generally higher levels of resistance in municipal wastewaters than in pet or wildlife fecal droppings (Edge and Hill 2005). Further research will be required to ascertain the significance of antibiotic-resistant bacteria in untreated waters used for drinking (e.g., well water), recreation, or irrigation of food crops.

Veterinary: Anti-parasitic Pharmaceuticals. Anti-parasitic pharmaceuticals, such as those in the avermectin class, include ivermectin, doramectin, emamectin, and several others. These compounds are used in both farm animals and aquaculture and are associated with known environmental issues related to their extreme toxicity to invertebrates. Inhibition of the breakdown of dung in pasture has been shown to occur as a result of ivermectin-induced delayed or abnormal larval development and mortality among manure degrading insects.

The use of avermectins in aquaculture raises the issue of extreme toxicity to aquatic invertebrates and the issue of potential endocrine effects on crustaceans that might ultimately impact population health (i.e., interference with ecdysis or moulting and attendant brood loss in lobsters). The threshold of acute mortality for ivermectin has been estimated for a range of aquatic species as likely to be in the range of low ng/L. Thresholds for effects mediated through disruption of growth or reproduction occur at concentrations in the 0.001 to 0.0003 ng/L range (Garric et al. 2007).

Path Forward in the EnRA Process

There are several ways to improve and modify the existing processes to allow better EnRAs for human and veterinary pharmaceuticals. Some of these proposed improvements may be difficult to implement (doing EnRAs on classes of similarly acting substances), while others are simpler (setting lower trigger volumes for pharmaceuticals, or basing trigger volumes on the numbers of doses). Outlined below are some potential improvements and considerations that may assist in the revising of the EnRA process for pharmaceuticals.

Trigger quantities of pharmaceuticals used in the EnRA could be adjusted for potency of the pharmaceutical. The volume of pharmaceutical produced or imported could be divided by the average daily dose, to yield triggers based on dose units imported. In this manner, small volumes of very potent pharmaceuticals would be assessed under the same tier (and require the same environmental data) as large volumes of less potent pharmaceuticals. This method would have advantages because pharmaceuticals are becoming more and more specifically targeted and potent.

Environmental risk assessment of substances has conventionally been done on individual compounds rather than mixtures. This simplifies the EnRA process, but does not allow even for additivity of the mixtures of similarly acting substances in the environment to be assessed. In this light, it may not protect the environment adequately when several pharmaceuticals all with similar mechanisms of action are entering waterways.

One way of improving the single-substance EnRA is to group like-acting pharmaceuticals. We could consider grouping substances with similar chemistries and conducting umbrella (cumulative) assessments of pharmaceuticals with the same basic chemical structure or the same mechanism of toxicity. At the very least, different salts of the same active ingredient need to be considered in an additive manner, but substances could also be grouped based on mode of action. For example, the selective serotonin reuptake inhibitors (SSRIs) could form one group for assessment, avermectin-derived pharmaceuticals might form another group, fibric acid-derived blood lipid regulators another, and so on. In a comprehensive cumulative assessment, the relative pollutant contribution of individual pharmaceuticals (possibly based on the relative potency estimation from human daily dose) within a class would be assessed and, ultimately, a specific strategy for dealing with the substances as a class, or problem substances within a class, would be devised.

Consideration should be given as to the type of toxicity test and the endpoints assessed. Comparison between acute and chronic effects of six pharmaceuticals showed that standardized acute tests were not the most appropriate basis for EcoRA of these compounds (Ferrari et al. 2004). There is a lack of endocrine-specific endpoints in the current tests, and exposure durations are often too short to manifest effects of low levels of pharmaceuticals. We recommend that default aquatic ecotoxicity datasets should focus primarily on early life stage or full life cycle tests with measured endpoints such as growth, development, and reproduction.

Information in mammalian toxicity datasets can aid in focusing selection of the type and length of the test and of the test organism (Crane et al. 2006). Pharmaceuticals that are slowly excreted in mammals could trigger testing for bioconcentration and bioaccumulation potential in the environment. Pharmaceuticals that act on specific pathways or receptors may be tested in organisms having the same pathways or receptors.

While the ways of performing and EcoRA for pharmaceuticals are being developed and are continuing to improve, several areas of research have expanded the ways in which we can reduce the input of these compounds to aquatic environments. Landscape factors and land-use practices can influence the exposure concentrations of nontarget organisms in the environment, because they greatly influence the environmental fate of human and veterinary pharmaceuticals.

Factors Affecting the Distribution of Human Pharmaceuticals in the Environment

Outflows from municipal wastewater treatment plants (MWWTPs) and the land application of sewage sludge (biosolids) represent two significant potential environmental exposure routes for human pharmaceuticals. The relative significance of either pathway for any pharmaceutical will depend on the mass of that chemical that reaches the MWWTP, its persistence during the sewage treatment process, the fraction of remaining residues that are released in the MWWTP outflow, and the fraction of residues that partition into particulate organic material that is subsequently recovered as biosolids.

The degree and type of wastewater treatment (i.e., primary, secondary, tertiary) will vary the effectiveness of pharmaceutical degradation and thus the availability of residues for release into the environment (Carballa et al. 2004). Hydraulic retention time and solids retention time are two key factors that can influence the degree of breakdown of human pharmaceuticals, with longer retention times generally resulting in greater breakdown (Ternes et al. 2004). Organisms in smaller rivers in heavily populated areas using large MWWTPs would be exposed to the highest aquatic concentrations of pharmaceuticals, particularly during periods of low river flow (Benotti and Brownawell 2007). Boulder Creek, CO, is one location where female-skewed sex ratios (4 female: 1 male) are seen in wild fish captured from this municipal effluent-dominated stream (Woodling et al. 2006).

Sewage biosolids are commonly disposed of by application onto the surface of land, burying deep beneath the soil surface (landfilling) or incineration. Incineration will destroy any pharmaceutical, whereas adjacent surface water or groundwater could become contaminated with pharmaceuticals transported from sites receiving surface-applied or landfilled material. The probability of adjacent water becoming contaminated with land-applied pharmaceuticals can be managed through regulations that specify suitable soil and land characteristics for application and that specify timing of application to soils. Soil characteristics include antecedent moisture and tillage, while land characteristics include distance to watercourse, depth to water table, slope, and tile drainage.

When biosolids are applied to agricultural land, application rates (volume or mass applied per unit hectare) are typically limited to meet and not exceed crop nutrient (phosphorous or nitrogen) requirements. The physical characteristics of the biosolids, the moisture content for example (dewatered biosolids are typically 30% moisture content; liquid biosolids are typically 97% water), and the application method (e.g., surface broadcasting, subsurface injecting) can be managed to limit the risk of movement of the pharmaceuticals following application. When equivalent commercial rates of dewatered biosolids or liquid biosolids were experimentally applied on a tile drained field, much more of the carbamazepine and triclosan carried in the slurry were mobilized by rain and exported in tile drainage water than that carried in dewatered biosolids (Fig. 15.2). The depth of application (surface, subsurface) of the slurry had relatively little effect on the mass of pharmaceuticals exported.

These mass balance estimates can be used to (a) inform EcoRAs with respect to exposure from land application of biosolids and (b) define management practices that limit potential exposure. They also clearly indicate that the physical characteristics of the biosolids can dramatically influence transport potential.

Factors Affecting the Distribution of Veterinary Pharmaceuticals in the Environment

Large amounts of antibiotics are used for growth enhancement, prophylaxis, and therapy in commercial livestock and poultry production (Kools et al. 2008). Agriculture also uses other pharmaceuticals, notably large amounts of ecto- and endo-parasiticides. In many commercial production systems, animals and poultry are raised confined in barns and their waste is collected and stored for several months before being used as fertilizer on agricultural land when climate and crop conditions are suitable. The fate of veterinary pharmaceuticals in stored manure will vary according to the length and method of storage and treatment to which it may be subjected. In contrast, animals grazing on pasture or in feedlots will excrete directly onto soil.

The risk of pharmaceuticals being transported to water following contact with soil will depend on their persistence, mobility, and proximity to water. Similar to the measures used to limit the movement of human pharmaceuticals in biosolids applied to fields, manure application can be managed to minimize risk. Factors such as application method and rate, soil tillage and incorporation, and offset distances from surface or groundwater are considered to limit the movement of veterinary pharmaceuticals

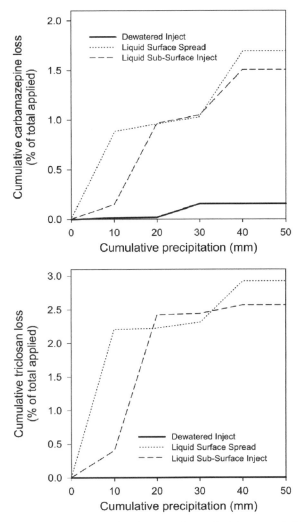

Figure 15.2. Export via tile drainage of carbamazepine and triclosan carried in dewatered or liquid biosolids expressed as a percentage of the total mass of each pharmaceutical applied. The liquid biosolids were applied to the soil surface, or injected at a depth of about 10 cm.

into waterways (Lapen et al. 2008). Desiccation cracks and worm channels in soil can enhance the flow of runoff from manure-amended soils containing antibiotics into tile drains and surface waters. However, mitigation measures such as mixing (e.g., tilling or disking) the soil prior to the application of animal waste containing antibiotics can limit movement of antibiotics in runoff (Kay et al. 2004).

The risk of off-site transport will vary according to a number of soil factors, notably texture; landscape factors, notably slope; and climate factors, notably intensity of rainfall. A number of exposure simulation models have been developed that employ scenarios incorporating these factors (e.g., Schneider et al. 2007). In this manner, exposure scenarios for pharmaceuticals can be modeled, based on livestock density and dosing of pharmaceuticals for particularly vulnerable regions (e.g., sandy soil, sloped land, high rainfall areas).

CONCLUSIONS

Pharmaceutical drugs have recently been detected around the world in municipal wastewater effluents and in river waters downstream. Concentrations of pharmaceutical compounds in the aquatic environment range from the low ng/L to high μg/L. Concentrations in river waters and MWWEs are very much lower than the therapeutic doses for humans or animals of these pharmaceuticals compounds, but the potential for nontarget effects and long-term effects of very low environmental concentrations have raised concern among the public, scientists, and regulators.

Environmental Risk Assessment of pharmaceuticals is an emerging practice, because typically these compounds were exempt from environmental regulation and have not been part of conventional environmental monitoring programs. Pharmaceuticals represent a special and very complex case because of the great variety of compounds within this class and because they are, by design, biologically active at quite low doses.

The EnRA of pharmaceuticals is a relatively new field, and improvements and changes in regulatory processes are occurring currently in Canada. In Europe and in the United States, EcoRA processes have been set through EMEA and VICH. Paths forward in the EnRA process in Canada include (a) use of long-term bioassays using aquatic and terrestrial organisms, (b) setting trigger quantities for each stage of EnRA based on amount in commerce and the daily dose of the pharmaceutical, and (c) grouping classes of similarly acting or similarly structured compounds for risk assessment.

Land use is another factor affecting the exposure of nontarget organisms in the environment. Environmental concentrations of human and veterinary pharmaceuticals in rivers arise from direct input from sewage treatment plants and indirect input from manure or biosolids in soils. The effectiveness of sewage treatment plants for breaking down human pharmaceuticals depends largely on sludge retention time. Application method of sewage biosolids or animal manure to soil, buffer zones around streams, and precipitation, slope of land, and soil permeability all affect the potential transport of human and veterinary pharmaceuticals to surface waters.

The EnRA of pharmaceuticals is a relatively new requirement and procedure. As the research progresses on effects of low concentrations on nontarget organisms, improvements will be made in assessing this complex and diverse group of compounds.

REFERENCES

Ankley GT, Jensen KM, Kahl MD, Korte JJ, Makynen EA. 2001. Description and evaluation of a short-term reproduction test with the fathead minnow (*Pimephales promelas*). *Environ Toxicol Chem* **20**:1276–1290.

Ankley GT, Jensen KM, Makynen EA, Kahl MD, Korte JJ, Hornung MW, Henry TR, Denny JS, Leino RL, Wilson VS, Cardon MC, Hartig PC, Gray LE. 2003. Effects of the androgenic growth promoter 17-α trenbolone on fecundity and reproductive endocrinology of the fathead minnow (*Pimephales promelas*). *Environ Toxicol Chem* **22**:1350–1360.

Benotti MJ, Brownawell BJ. 2007. Distributions of pharmaceuticals in an urban estuary during both dry- and wet-weather conditions. *Environ Sci Technol* **41**:5795–5802.

Brooks BW, Foran CM, Richards S, Weston JJ, Turner PK, Stanley JK, Solomon K, Slattery M, La Point TW. 2003. Aquatic ecotoxicology of fluoxetine. *Toxicol Lett* **142**:169–183.

Carballa M, Omil F, Lema JM, Llompart M, Garcia-Jares C, Rodriguez I, Gomez M, Ternes T. 2004. Behavior of pharmaceuticals, cosmetics and hormones in a sewage treatment plant. *Water Res* **38**:2918–2926.

Crane M, Watts C, Boucard T. 2006. Chronic aquatic environmental risks from exposure to human pharmaceuticals. *Sci Tot Environ* **367**:23–41.

Cunningham VL, Buzby M, Hutchinson T, Mastrocco F, Parke N, Roden N. 2004. Effects of human pharmaceuticals on aquatic life: Next steps. *Environ Sci Technol* **40**:3456–3462.

Daughton CG, Ternes TA. 1999. Pharmaceuticals and personal care products in the environment: Agents of subtle change? *Environ Health Perspect* **107**:907–938.

Edge TA, Hill S. 2005. Occurrence of antibiotic resistance in *Escherichia coli* from surface waters and fecal pollution sources near Hamilton, Ontario. *Can J Microbiol* **51**:501–505.

FDA. 1998. Guidance for Industry: Environmental Assessment of Human Drug and Biologics Applications, U.S. Department of Health and Human Services, Food and Drug Administration, Center for Drug Evaluation and Research (CDER), Center for Biologics Evaluation and Research (CBER), July 1998, CMC 6, Revision 1, http://www.fda.gov/downloads/Drugs/GuidanceComplianceRegulatoryInformation/Guidances/ucm070561.pdf (accessed 25 January 2010).

Fent K, Weston AA, Caminada D. 2006. Ecotoxicology of human pharmaceuticals. *Aquat Toxicol* **76**:122–159.

Ferrari B, Mons R, Vollat B, Fraysse B, Paxéus N, Lo Giudice R, Pollio A, Garric, J. 2004. Environmental risk assessment of six human pharmaceuticals: Are the current environmental risk assessment procedures sufficient for the protection of the aquatic environment? *Environ Toxicol Chem* **23**:1344–1354.

Fong PP, Humuninski PT, D'Urso LM. 1998. Induction and potentiation of parturition in fingernail clams (*Sphaerium striatinum*) by selective serotonin reuptake inhibitors (SSRIs). *J Exp Zool* **280**:260–264.

Foran CM, Weston J, Slattery M, Brooks BW, Huggett DB. 2004. Reproductive assessment of Japanese medaka (*Oryzias latipes*) following a four-week fluoxetine (SSRI) exposure. *Arch Environ Contam Toxicol* **46**:511–517.

Garric J, Vollat B, Duis K, Péry A, Junker T, Ramil M, Fink G, Ternes TA. 2007. Effects of the parasiticide ivermectin on the cladoceran *Daphnia magna* and the green alga *Pseudokirchneriella subcapitata*. *Chemosphere* **69**:903–910.

Hamida L, Medhioub M-N, Cochard JC, Le Pennec M. 2004. Evaluation of the effects of serotonin (5-HT) on oocyte competence in *Ruditapes decussatus* (Bivalvia, Veneridae). *Aquaculture* **239**:413–420.

Kay P, Blackwell PA, Boxall AB. 2004. Fate of veterinary antibiotics in a macroporous tile drained clay soil. *Environ Toxicol Chem* **23**:1136–1144.

Kidd KA, Blanchfield PJ, Mills KH, Palace VP, Evans RE, Lazorchak JM, Flick RW. 2007. Collapse of a fish population after exposure to a synthetic estrogen. *Proc National Acad Sci* **104**:8897–8901.

Kolpin DW, Furlong ET, Meyer MT, Thurman EM, Zaugg SD, Barber LB, Buxton HT. 2002. Pharmaceuticals, hormones, and other organic wastewater contaminants in U.S. streams, 1999–2000: A National reconnaissance. *Environ Sci Technol* **36**:1202–1211.

Kools SAE, Moltmann JF, Knacker T. 2008. Estimating the use of veterinary medicines in the European union. *Regul Toxicol Pharmacol* **50**:59–65.

Länge R, Hutchinson TH, Croudace CP, Siegmund F, Schweinfurth H, Hampe P, Panter GH, Sumpter JP. 2001. Effects of the synthetic estrogen 17α-ethinylestradiol on the life-cycle of the fathead minnow (*Pimephales promelas*). *Environ Toxicol Chem* **20**:1216–1227.

Lapen DR, Topp E, Metcalfe CD, Li H, Edwards M, Gottschall N, Bolton P, Curnoe W, Payne M, Beck A. 2008. Pharmaceutical and personal care products in tile drainage following land application of municipal biosolids. *Sci Tot Environ* **399**:50–65.

Lishman L, Smyth SA, Sarafin K, Kleywegt S, Toito J, Peart T, Lee B, Servos M, Beland M, Seto P. 2006. Occurrence and reductions of pharmaceuticals and personal care products and estrogens by municipal wastewater treatment plants in Ontario, Canada. *Sci Tot Environ* **367**:544–558.

Metcalfe CD, Koenig BG, Bennie DT, Servos M, Ternes, TA, Hirsch, R. 2003a. Occurrence of neutral and acidic drugs in the effluents of Canadian sewage treatment plants. *Environ Toxicol Chem* **22**:2872–2880.

Metcalfe CD, Miao X-F, Koenig BG, Struger J. 2003b. Distribution of acidic and neutral drugs in surface water near sewage treatment plants in the lower Great Lakes, Canada. *Environ Toxicol Chem* **22**:2881–2889.

Mimeault C, Woodhouse AJ, Miaob X-S, Metcalfe CD, Moon TW, Trudeau VL. 2005. The human lipid regulator, gemfibrozil bioconcentrates and reduces testosterone in the goldfish, *Carassius auratus*. *Aquatic Toxicol* **73**:44–54.

Parrott JL, Blunt BR. 2005. Life-cycle exposure of fathead minnows (*Pimephales promelas*) to an ethinylestradiol concentration below 1ng/L reduces egg fertilization success and demasculinizes males. *Environ Toxicol* **20**:131–141.

Schneider MK, Stamm C, Fenner K. 2007. Selecting scenarios to assess exposure of surface waters to veterinary medicines in Europe. *Environ Sci Technol* **41**:4669–4676.

Ternes TA, Joss, A, Siegrist, H. 2004. Scrutinizing pharmaceuticals and personal care products in wastewater treatment. *Environ Sci Technol* **Oct 15**:392A–399A.

WHO (World Health Organization) International Program on Chemical Safety. 2002. In Damstra T, Barlow S, Bergman A, Kavlock R, Van Der Kraak G (Eds.), *Global Assessment of the State-of-the-Science of Endocrine Disruptors*. WHO/PCS/EDC/02.2, 180 pp.

Woodling JD, Lopez EM, Maldonado TA, Norris DO, and Vajda AM. 2006. Intersex and other reproductive disruption of fish in wastewater effluent dominated Colorado streams. *Comp Biochem Physiol Part C: Toxicol & Pharmacol* **144**:10–15.

16

ECONOMIC ANALYSIS OF ECOLOGICAL GOODS AND SERVICES

Ronald J. McCormick, James Pittman, and Timothy F. H. Allen

This planet has—or rather had—a problem, which was this: Most of the people living on it were unhappy for pretty much of the time. Many solutions were suggested for this problem, but most of these were largely concerned with the movements of small green pieces of paper, which is odd because on the whole it wasn't the small green pieces of paper that were unhappy.

Douglas Adams, *The Hitchhikers Guide to the Galaxy*

THE ECOLOGY OF ECONOMICS FROM A COMPLEX SYSTEMS PERSPECTIVE

The writers of this chapter represent the fields of economic ecology (McCormick), ecological economics (Pittman), and complex systems (Allen). Our shared premise, presented as a unifying thread of thought throughout, is that there is not enough ecology in economics, not enough economics in ecology, and not enough systems analysis in either. We do not intend a treatise on the fundamentals of ecological economics, because many works already exist on the subject (e.g., Costanza et al. 1997, Daly and Farley 2003). As well, we do not offer hard and fast rules for setting a value for ecosystem goods and services or for developing indicators of sustainability.

In this chapter we show how neoclassical economics isolates itself deep in a bubble, well away from ecological theory, and how ecological economics has risen to the surface of that bubble, seeking a way out. Current avenues out of the neoclassical bubble focus on issues of entropy and scale. With the addition of some relatively new formalizations of older ecological thought, we show a way of gently opening a window in the economics bubble. By smoothly blending economic stocks and flows into an overarching ecological framework, we can realize the full potential of using ecosystem services valuation to guide the sustainable use of landscapes by human societies (Ozkaynak et al. 2003). From such an integrated framework, how one determines the value of goods and services derived from an ecosystem will depend entirely on how one determines what values are to be sustained.

Human Society and Global Landscapes

There has long existed a tension between a society, its economy, and their supporting landscapes. With some exceptions during the last 15,000 years of societal evolution, our planet's landscapes have held a controlling position. Up to the start of the industrial age, the biome you lived in set the bounding context within which your social–economic system operated. As Ruddiman (2005) presents, with the beginnings of land clearing for agriculture roughly 8000 years ago, the dynamic tension between human society and its supporting landscapes started to fundamentally shift in character. Rapid emergence of the fossil-fuel-driven economy in the last two centuries subsumed the agricultural economy into a new, global system (Allen et al. 2001).

By capturing and using fossil sun, humanity increased its global population base by producing more food, and it began the final phase of wresting control of the earth's ecosystems and climate away from the planet. Wresting, as a metaphor, perhaps puts too strong a connotation of achieved control on the point. The effect of dissipating the majority of fossil fuels stored in the planet in less than three centuries didn't actually result in our control of global biomes, it merely made humans a larger player in the global climate game. Planetary land and water systems simply continued to process compounds as always, tucking away here and there elements and compounds when and where possible. Eventually, compounds just remained in the atmosphere as the maximum rate of long-term storage in ecosystem structures was exceeded. The complete restructuring of the global atmosphere to a state not seen for at least 600,000 years will likely remain for all time as humanity's largest architectural construct (Weisman 2007).

Neoclassical economics disregards environmental externalities, choosing not to envision the possibility of a global storage limit. Environmental economics considers, but does not effectively deal with, the finite absorbance rate and storage capacity of global ecosystems. Ecological economics emerged with a wiser vision [for a good synopsis, see Krishnan et al. (1995)], and still today it seeks a fundamental theory from which to conceptualize, test, and improve our understanding of a society, its economy, and their supporting landscapes (Hawken 1993, Hawken et al. 1999, Daily and Ellison 2003). There remains much work to do, and in this chapter we offer ecological economists an opening to revise their point of view, to place ecological theory at the fore of their science, and to become economic ecologists (O'Neill 1996) in the process.

Ecology Inside Economics

What is the scale of the economy that fits inside the ecology? This is a preeminent question in ecological economics today and is referred to as the "sustainable scale." Daly (1992) defines sustainable scale in economics as "...nothing other than an intergenerational distribution of the resource base that is fair to the future." Some obvious follow-on questions include:

- Who decides what is fair?
- How far away do we consider "the future" to be?
- Can a maximal scale be set?
- Is there an optimal scale?
- Can monitoring systems be put in place to rapidly rescale the economy in response to unexpected changes in the ecosystem?

Ecological economics likely can address the questions of maximal, minimal, or optimal scale using current valuation theory and methods (discussed in detail in Chapter 17). However, on the question of rapid responses to shifting ecological conditions, pertinent economic theory—be it classical, neoclassical, environmental, or ecological—does not have built into it the fundamentals of complex systems analysis necessary to plan and implement an appropriate, timely, and scaled response.

Any assessment of a complex system, particularly one requiring the integration of social, economic, and ecological characteristics, starts by defining one level of analysis as a simple starting point, but goes on to address complexity by invoking more levels (Ahl and Allen 1996). Setting a level of analysis forces the explicit definition of what one is talking about. Specifically, what criteria were used to decide what elements to include or exclude, and how the scale at which those elements will be examined was selected. Selection criteria serve to define an ecological type—that is, the level of organization to be studied. Scale decomposes into two elements: grain, the smallest unit measurable; and extent, the spatial and temporal range over which units are observed (recall definitions in Chapter 4). Scale also carries with it the duality of the size of the material system observed, as opposed to the scaling implied in the protocol of observation such as the choice between binoculars or a microscope.

Level of analysis comes from combining a *level of observation* (scale) with a *level of organization* (ecological type) for the defined problem. Business tools of Checkland (1999) have already worked in their own sector, but Allen and Hoekstra (1992) have indicated how to bring these methods into ecology. Ecological types include landscape, ecosystem, population, community, organism, and biome. Each ecological type has a list of attributes and criteria for definition. Types depend on what is identified in the foreground, bearing in mind the context (a person may be a father, a professor, or a primate, depending on the class to which they are assigned).

Types are defined independent of scale. Ecosystems are defined by the flow of energy, materials, or information, which can occur at any scale, such as within a leaf, across an entire forest, or over the width and breadth of a continent. Similarly, a landscape is defined as a bounded unit of land, which can range from very small to very large in spatial extent. Saying your level of observation is at "the landscape scale" is at best vague, and it often leads to confusion and misinterpretation of what is intended for the analysis (Allen 1998).

Without a defined level of analysis, an ant on a tree is just an organism on an organism from the perspective of traditional biology. Select the ant as the focal ecological type and the tree becomes a landscape from the ant's perspective. As well, it is the ant's perspective that sets the scale at which observations should be made, bounding the questions that systems ecologists might ask to those relevant to the ant, not the observer. Conversely, if the relationship is one of stewardship on the part of ants that protects an acacia tree, the system becomes a coevolved mutualism. Complex systems theory provides a powerful model from which to develop a conceptual model of a social–ecological system of interest. While the conceptual model includes multiple perspectives for analysis at multiple scales, it helps investigators to maintain a clear vision of the system by being very explicit about the many possible levels of analysis.

The theory and tactics of orthodox economics, even those of ecological economics, are not designed to address multiple levels of analysis within the same system. Heterodox and post-autistic economics that properly acknowledge environment understand the need to do this, but are still somewhat limited by the need to use data and tactics available from traditional economic analyses. Economic ecology represents a strategic shift in focus for how a system is defined, and how socially relevant values might be established for wicked problems that can then lead to "clumsy solutions" (Shapiro 1988, Verweij et al. 2006).

Tactics Versus Strategy

Robert Rosen (1981), in his Presidential Address to the Society for General Systems Research, presented the case that a systems theoretic perspective provides for strategic innovation more readily than traditional paradigmatic approaches, which are essentially pursuing variations of known tactics. As an example, Rosen presented the story of "The Purloined Letter" by Edgar Allen Poe. The police prefect, in searching a house for said letter, had moved from coarse visual examinations of the residence to fine-scaled probes of walls and couches, which still did not reveal the letter's location. The prefect was using long-accepted police tactics in his search, but his basic strategy, developed over many years of police work and quite successful in most situations, was failing him. The letter, which had been re-addressed and left in plain sight, was hidden in a manner that was outside the purview of the prefect's search strategy. For those readers who don't know the rest of the story, Dupin, the private detective consulted by the frustrated police, found the letter almost immediately upon entering the house by using a new and insightful search strategy. Rosen's retelling of the story brought focus to the ineffectiveness of exhaustive yet strategically misguided techniques previously used in pursuing answers to current problems. He urged scientists not only to reassess the tactics used, but also reassess the strategic assumptions inherent in accepted data acquisition or technique development tactics.

With Herman Daly and Joshua Farley, co-authors of *Ecological Economics* (Daly and Farley 2003), in the audience, one of the authors (McCormick) presented the fundamentals of setting up a clear level of analysis during one of "the scale question" interactive sessions at the 2005 US Society for Ecological Economics meeting in Tacoma, Washington. Daly responded after the presentation that ecological economists had seen this before, and it did not apply to their situation regarding the question of sustainable scale. His response, and the general agreement of other economists present at the time, points to the current preeminence of economics as the focus for addressing problems on social–ecological landscapes. Ecological economics takes the direct view that the economy is a wholly owned subsidiary of the ecology, but still persists in analyzing those hierarchically nested systems using the entrenched tactics of economics.

This forces a limited view of ecology as well as economics, in that the economy has many scales, and there are highly nonlinear changes in ecological effects resulting from small, linear changes in the scale of the economy. Any discussion of sustainability or sustainable scale must include different components, different rates, and different human values (Carter 1979) with each level of analysis (type and scale) addressed. As well, ecological systems are nested, interlinked hierarchical systems that contract and expand in a nonlinear manner when pressed with a slight linear increase in societal demand for goods and services (Carpenter et al. 1999, Gunderson and Holling 2002). This concept proves particularly relevant when assessing neoclassical economic models, which do not include the element of time in a manner sufficiently able to respond to rapid ecosystem changes.

Evolving from neoclassical to environmental to ecological, economic theory and analysis has remained essentially the same, though with slightly wider eyes reducing assumed externalities at each step. However, those eyes are covered by the same filter as classical economics, namely that there is a single actor, with perfect access to knowledge, acting in a definable, predictable, and rational manner, and the economics of multiple actions can be aggregated using equilibrium models and Gaussian distributions. Several authors point out the serious shortcomings inherent in these model assumptions (Mandelbrot 1963, Keen 2002, Taleb 2007).

Neoclassical economics today appears to have less scientific rigor than at any point in the past due to a pursuit of mathematical rigor, via adoption of models from physics, over the refinement and testing of scientific theory (Mirowski 1990). Econometric models developed without a sound and comprehensive theory backing them are just a lot of math, blind and deaf to the world at large (Keen 2001). As laid out by the modern era forefathers, with Adam Smith's "An Inquiry into the Nature and Causes of the Wealth of Nations" (1776) being one of the most well known, economics was approached more from the standpoint of philosophy than mechanistic science. Smith's economics leads one to a develop a narrative for how human society uses the global landscape, as well as offering guidance in understanding the motivations behind the local actions of government, business, and individuals. Samuelson's (1948) original work, and the 16 subsequent editions, in setting out a mathematical approach to the discipline, put economic theory onto an ever-narrowing path of complicating abstractions of the marginal utility maximizing individual (Samuelson 1997). Long before it took on the present-day role of sole agent in determining global consumption, Veblen (1899) referred to this model individual as a "homogeneous globule of desire."

Economics Inside Ecology

Individuals acting locally, regionally, or globally create the economy; mathematical models do not. Neoclassical models lack predictive relevance because economies emerge from six billion decisions being made every day on where to maximize utility. The models also do not, nay, cannot, account for the existence of and effects from "projectors" in society. Projectors, one or several individuals pressing people to "invest here" or "buy this stock now," irrationally and disproportionately influence economic data and market indicators. The long-term economic cascade from the 2008 collateralized debt obligations meltdown in the United States, the internet stock bubble in the late 1990s, and the global economic crisis caused by the small group of "geniuses" at Long-Term Capital Management (Lowenstein 2001) represent projector aftereffects.

Adam Smith and Jeremy Bentham debated the effects of prodigal and projector actors on economics more than 250 years ago (Paganelli 2003). Ecological economics still deals with those issues today, because the types of gambles that projectors take, and convince others to take as well, have typically resulted in the wasting of ecological resources (e.g., liquidating vast volumes of standing timber to cover shortfalls in junk bond market speculation). It only takes a rumor that a company has a problem to very rapidly devalue the stock, often to the point of nonrecovery. Finding a sustainable scale in such a system presents no small challenge to ecological economists.

Societal economic direction cannot be controlled by simplistic notions of interest rate changes and tax cuts. At best, the economy that emerges out of the planetary population's choices can be nudged or encouraged to move in a desired direction, but there is no guiding theory or model that can confirm this. This makes ecological economic problems appear wicked, and it points to the difficulty of finding efficient solutions. Trusting market forces to find the "efficient" solution is far too adroit for a truly complex problem. The market is a thermodynamic expression of happenings, and it can be effective for some purposes. However, missing from the perspective of the market are coded elements that are not rate-dependent and thus act in a realm outside thermodynamic processes. Complex, wicked problems don't have straightforward solutions, because they include inefficiencies that embody an outcome not coded for in some preferred outcome developed in part from a non-neutral process. Complex problems require consideration of multiple, usually contradicting, viewpoints and assessment of multiple stakeholder problem-solution narratives (McCormick et al. 2004). Planned inefficiency, acceptable contradiction, and multiple views of the same system represent the building blocks for developing a clumsy ecological policy solution to a wicked social–economic problem—that is, shifting from efficient economic tactics to complex, interwoven, and adaptable ecological strategies.

Entropy, Exergy, and Valuation

Costanza et al. (1997) note there are two low-entropy sources of energy available to society, sunlight and terrestrial (e.g., hydro, wind, wave, geothermal, fossil). That these two sources are the only ones available, and that the majority of society currently uses terrestrial sources, is interesting. But, the low-entropy classification does not lead one to a better understanding of how and why society uses these energy sources, or if the type of energy available ultimately dictates the type of economy possible.

Costanza et al. (1997) also note that while sunlight is abundant, it is time- and rate-limited; conversely, terrestrial energy is limited in total abundance, but relatively unlimited in availability. These additional considerations start to get closer to our current understanding of how and why ecological systems use energy. Economists have wrestled with the apparent contradiction that biological and economic systems create order, seemingly in direct defiance of the second law of thermodynamics [for just one example, see Daly (1997)]. There is no such defiance because matter–energy gradients in the global context are harvested by biosocial entities and are then degraded faster inside social–ecological structures than they would have been in the context from which they were taken. The sum of the biosocial structure and the context in which it resides taken together degrades as a whole in exactly the normal way under the second law. Kay and Schneider (1992) reinterpreted the second law of thermodynamics to show that biological systems will organize to dissipate the work available from an incoming exergy gradient. As the substance inside a biosocial structure is pushed away from equilibrium, it creates dissipative structures to resist being pushed further. The order in biosocial structures is a consequence of that resistance, and thus it is quite the reverse of denying the second law; biosocial structures actually depend on the second law and on the degradation that it predicts.

Comparing and contrasting the mathematics behind entropy, emergy, and exergy are beyond the scope of this chapter. However, some understanding of exergy as an indicator of value is important, so we refer the reader to "Exergy analysis of ecosystems" by Fraser and Kay (2003) and the work of Hau and Bakshi (2004). Much as Ozkaynak et al. (2003) point to a need for qualitative aspects to ecological economics, Fraser and Kay (2003) describe the need to consider energy quality in the study of ecosystems. Even with discussions of energy quality, their exergy analysis of ecosystems is quite quantitative, and calculating exergy loss in ecosystem operations should become a useful valuation tool.

An ecosystem is an assembly of energy converting (dissipating) structures (plants, animals, fungus, etc.) that produce goods and services available to human economies. Entropy production is a first law of thermodynamics accounting for destruction of exergy, a measure of energy quality per the revised second law (Kay and Schneider 1992). Ecological economics focuses on low-entropy energy sources and high-entropy consumer products. Entropy only addresses the continuous, inexorable increase in system disorder. Since entropy never stops increasing in magnitude, using it as a valuation tool presents definitional, logistical, and logical complications. Exergy, by accounting for order in a system, and specifically energy available for organizing a social–ecological landscape, can be quantified for a given system starting state prior to dissipation by the system, and the efficiency of that dissipative action can be calculated.

High-entropy/low-entropy quantification arguments within the ecological economics community are an indication of (a) the physics envy existing within modern economics and (b) the tactical stranglehold that mathematical rigor has on the subject. A strategy of evaluating how energy quality relates to the complexity and structural maturity of an ecosystem moves one away from only dealing with magnitudes of energy input and output and gets at the effect of human management actions on the ability of ecosystems to dissipate energy. A greater ability to dissipate energy indicates a greater ability to produce ecosystem goods and services available to human societies. As the theory and science evolve, exergy analysis offers the potential for a strategic breakthrough in the valuation of ecosystem goods and services.

Sustainability

Allen et al. (2001), in analyzing together the historical evolution of ants and humans, Kay's dissipative gradients, and economic epochs starting with our ancestral hunter/gatherer tribes, developed high-gain/low-gain theory (Fig. 16.1). Underlying systems thinking is the premise that systems behave as a whole and that such

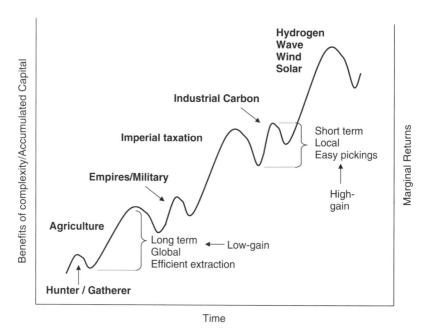

Figure 16.1. Depiction of high-gain and low-gain historical epochs. High-gain periods end when the resource supply runs out. Low-gain periods end when demand exceeds growth. This sequence does not imply inevitability in a grand narrative, but it simply highlights the path to where we are now.

behavior cannot be explained solely in terms that simply aggregate the individual elements. This premise does not sit comfortably with reductionist methodologies that decompose the whole to evaluate the parts in isolation. There are insights to be gained from such decomposition of the whole, but in the end it cannot explain the manifest emergence of the whole. For instance, one cannot delve into water looking at molecules and hope to find a wet one. Wetness is possessed by water as an aggregate, not by the individual particles. There comes a point when the significance of something at the level of the whole is lost in the reduction. If you are reading this sentence in print and you do not understand it, then knowing the chemistry of the ink will not help (Pattee 1978).

We are now in a position to improve the way that the study of biological systems informs understanding of social systems (see Chapter 8). There have been two tracks taken in the rapprochement between biological and social science. One is some version of applying evolutionary ideas across the divide. The other invokes complexity science and emergence as a model. There has been relatively little attention paid to combining the two approaches, which is unfortunate, because therein lies a path to a synergistic program. Such a combination would take full advantage of understanding established separately in the respective scientific disciplines. Evolution turns on the selection of structures, such that past experience becomes coded, so that fortunate accidents are stabilized in modified structures that come out of the evolutionary process. Entirely separate are processes addressed in complexity science, where new structures appear under emergence. Whereas the mainstream of complexity science makes much of emergence, it rarely moves on to include how coded information acts as a twin source of order in a process that uses the principles of evolution.

The distinction between selection of structures and emergence was captured in Schrödinger's 1944 book *What Is Life*, even before the nature of the genetic code was known. He referred to "order from order," which is a general expression of the selection of structures, where coded information is passed from cell to cell and organism to organism. Schrödinger's second comment on organization he called "order from disorder." This was a particularly impressive insight, in that it predated Prigogine's work on the thermodynamics of far from equilibrium systems. There thermodynamic processes drive emergence of new structures *de novo*, not merely modified structures. The potentiality for emergence exists ahead of time, but is only manifested when a gradient is applied to some material. Note that order and structure are not selected from preexisting structures in "order from disorder," but come only at the end of the process when new structures emerge.

While there have been mistakes made in applying biological insight to social systems, the worst being social Darwinism, it is possible to move biological models into social settings in a valid way. For instance, equations that describe how gold creates order in societies would be basically the same as those where heat flows in thermodynamic processes. The way energy works to order biology can be understood in terms that follow the laws of thermodynamics. Concentrated wealth, such as gold or food, is diffused; and in that process, work is done and order is created. In economics, the market works in thermodynamic terms. By contrast, regulations in a society are coded, and these impose order by means of their static quality of persist-ing over time, operating as constraints. Constraints in these terms are related not to thermodynamics, but to processes that select structures in a way that echoes evolution in biological systems. The thermodynamics of biological and social functions are not in and of themselves efficient and otherwise, since thermodynamics does not have a preferred state. Efficiency and other goal-centered notions introduce some plan that does expect some outcome. Combining static coded effects with thermodynamic flux gives a balanced understanding of both biological and social situations.

All biological and social systems exist using both thermodynamic flux and coded controls in a way that lifts both types of systems away from simple physicality. In mere physical systems, where coding is absent, levels of analysis are fairly clear. For instance, quarks, atoms, and molecules are never confused, because of the orders of magnitude in scale that separates these levels. The reason why biological and social systems are so elaborate is that emergence is stabilized by coding for it, making for new structures that persist over time. These reliable structures are then candidates for forming populations that themselves can undergo flux in a new process of emer-gence at a scale quite close to that of the original emergence. Manfred Eigen (Eigen and Winkler-Oswatitsch 1992) calls the new, more elaborate patterns "hypercycles." By coding for levels in hypercycles, levels in biosocial structures are densely packed, inviting confusion for the observer and analyst and requiring exquisite attention to level of analysis as an issue in itself. In the systems we have studied, it appears that com-plexity is not a material issue, but one of level of analysis. Polanyi (1968) addressed the levels in these terms as he discussed "life's irreducible structure." Change the level of analysis a hair's breadth in addressing a physical system, and one gets only a little more or less of the same thing. Change the level of analysis in a biosocial system, and what appeared before as a thermodynamic issue becomes predictable only in terms of coded information in the system.

We have found that, at a given level of analysis, systems are best described and predicted in terms of either thermodynamic emergence or stable coded structures, but not both at the same time. One has in the foreground a process demonstrating movement through fluxes, or one has that same process constrained by some limit. If the constraints are central to the pattern, it makes for something static, albeit with dynamics going on inside the limits. The difference is not in the materiality of the situation, but in the level of analysis that brings either the flux or the statics into focus. Pattee (1978) explains that the rapprochement of static codes versus dynamical flux introduces a complementarity of the same sort as the wave particle duality in physics. The distinction between wave-like things and particle-like things is not a material issue, because it is electrons all the way. It is in fact a level of analysis issue. In Allen's previous work he has addressed many systems in these dual terms and has come to see the two sorts of order and prediction. If the system is bounded so that it is seen as controlled by energy flux, we call it a high-gain system. If that same system is seen as bounded by efficiencies and controls, then we call it a low-gain system (Figure 16.1).

We emphasize again that this distinction is not a material issue, but rather relates to how we bound the entities we are studying. If the energy and force for order is seen as outside the system, it behaves as high gain. If the system is seen as bounded larger, so that those same forces for order are inside the system, then it behaves as low gain. Allen et al. (2003) speak of external complexification, where the system in seen as high gain, as opposed to internal complexification, where it is low gain. When Tainter (1988) argued that complexification was employed as a problem solving device of societies, he was generally speaking of a low-gain conception of the societies in question. The distinction here is between defining systems in either energetic terms or as coded, constrained entities. The power of addressing systems in these terms is that a large number of other characteristics line up on one side or the other with consistency.

We have been able to see these patterns of high and low gain consistently. High gain appears as an entity that uses a ready-made resource, perhaps oil in the industrial world. Low gain sees systems that do not have a ready-made resource available, and thus must go to some lower-quality version of that same resource as it exists diffusely in a wider context. Through increases in efficiency, low-gain systems can use low-grade material and degrade it further so as to be able to use it, low quality notwithstanding. It is possible to see even oil production as a low-gain activity. As easy-to-access sources of oil have dried up, ever more elaborate devices, such as deep under sea drilling, are brought online to capture more difficult sources of oil. Oil still gets burned in a high-gain fashion, but now the system is defined so that the elaborate devices for capturing it are inside the system, making oil extraction an organized low-gain activity. These changes in level of analysis matter because each conception allows us to see its own set of predictions about stability, longevity, size, organization, and management possibilities.

Tainter et al. (2003) have posited 12 characteristic comparisons of low-gain and high-gain systems as predictive hypotheses (Table 16.1). We were able to argue above using oil, because what we said was intuitively reasonable. Ecosystem valuation techniques are needed for counting calories, cost and fluxes, so as to make concrete and scientific the insights we have heretofore intuited. Tainter et al. (2003, 2006) have addressed colonial insects and historical societies in these same high- and low-gain terms. Allen (unpublished manuscript) has looked at land-use patterns in a social landscape in contemporary Southern Wisconsin, invoking the differences between road and railroad transportation. Tainter has pointed to information coming from outside two different societies that has led to remarkably similar changes. From self-reliant local organizations, both systems have made local people underemployed. The reasons in both systems turn on constraints coming from the larger social units that invoke values about wild, beautiful places that are said to need to be made sustainable. All of this now needs measurement and calibration so that our insights about particular systems can be made general and applicable as society takes responsibility for management that it is already imposed by default.

Toward an Economic Ecology

Adam Smith (1776) discussed at length the social and moral issues concerning usury, and he also discussed how landlords or loaners did nothing to "earn" their rents.

Table 16.1. Characteristics of High- and Low-Gain Extraction Systems

High-Gain Systems	Low-Gain Systems
1. Use steep energy gradient	Use shallow energy gradient
2. High-quality resource will likely initiate a high-gain phase	Entering a low-gain phase subject to chance
3. A new type of resource use	Expands an existing resource use
4. Resources abundant	Resources scarce
5. Local and concentrated	Extensive
6. Dissipative and inefficient	More energy degradation, higher efficiency
7. Minimal demands on system	High demands on system
8. Impressive in energy capture	Impressive in organization and structure
9. Brief duration	Lasts significantly longer
10. New levels at top of hierarchy	New levels in middle of hierarchy
11. Self-organized	Organized in reference to environment
12. Manage through context or will self-repair	Manage by manipulating parts

Source: Adapted from Tainter et al. (2003).

Marx's basic theory had similar issues, in that no one could allocate their labor requirement onto another layer of society. These ideas are reflective of a low-gain society, where capitalism is decidedly high gain. As Gates (1998) notes, the global version of capitalism creates more capital, not more capitalists. Thus, as currently implemented, the efficient and impersonal market-based economy lacks the legitimacy of Smith's (1759) ethical and moral basis for economics.

We have argued that economics has lost its theoretical foundation and has wandered far away from its base cause, the efficient and sustainable management of the household (home, city, state, province, territory, country, world). Chapter 17 looks at the current conduct of economic valuation of ecological resources, the how and why of value creation. Chapter 18 returns to and expands on the idea of an economics to address multilevel, complex systems, essentially an economic ecology that helps us to manage social-ecological landscapes in an effective, just, and sustainable manner.

REFERENCES

Ahl V, Allen TFH. 1996. *Hierarchy Theory: A Vision, Vocabulary and Epistemology*. Columbia University Press, New York.

Allen TFH. 1998. The landscape level is dead: Persuading the family to take it off the respirator. In Peterson D, Parker VT (Eds.), *Scale Issues in Ecology*. Columbia University Press, New York, pp. 35–54.

Allen TFH, Hoekstra TW. 1992. *Toward a Unified Ecology*. Columbia University Press, New York.

Allen TFH, Tainter JA, Pires JC, Hoekstra TW. 2001. Dragnet ecology—"Just the facts, ma'am": The privilege of science in a postmodern world. *Bioscience* **51**:475–485.

Allen TFH, Tainter JA, Hoekstra TW. 2003. *Supply-Side Sustainability*. Columbia University Press, New York.

Carpenter S, Ludwig D, Brock W. 1999. Management of eutrophication for lakes subject to potentially irreversible change. *Ecol Appl* **9**:751–771.

Carter, J. 1979. The "Crisis of Confidence" Speech. Online at http://www.pbs.org/wgbh/amex/carter/filmmore/ps_crisis.html. Last accessed June 5, 2009.

Checkland P. 1999. *Systems Thinking, Systems Practice*. John Wiley & Sons, Chichester, West Sussex, England.

Costanza R, Cumberland J, Daly H, Goodland R, Norgaard R. 1997. *An Introduction to Ecological Economics*. CRC Press, Boca Raton, FL.

Daily GC, Ellison K. 2003. *The New Economy of Nature: The Quest To Make Conservation Profitable*. Island Press, Washington, DC.

Daly H, Farley J. 2003. *Ecological economics: principles and applications*. Island Press, Washington, DC.

Daly HE. 1992. Allocation, distribution and scale: Towards an economics that is efficient, just, and sustainable. *Ecol Econ* **6**:185–193.

Daly HE. 1997. Georgescu-Roegen vs. Solow/Stiglitz. *Ecol Econ* **22**:261–266.

Eigen M, Winkler-Oswatitsch R (translated by Woolley P). 1992. *Steps Towards Life: A Perspective on Evolution*. Oxford University Press, Oxford, UK.

Fraser RA, Kay JJ. 2003. Exergy analysis of ecosystems: Establishing a role for thermal remote sensing. In Quattrochi DA, Luvall JC (Eds.), *Thermal remote sensing in land surface processes*. CRC Press, Boca Raton, FL, pp. 283–360.

Gates JR. 1998. *The Ownership Solution: Toward a Shared Capitalism for the Twenty-First Century*. Perseus Books, New York.

Gunderson L, Holling CS. 2002. *Panarchy: Understanding Transformations in Human and Natural Systems*. Island Press, Washington DC.

Hau JL, Bakshi BR. 2004. Expanding exergy analysis to account for ecosystem products and services. *Environ Sci Technol* **38**:3768–3777.

Hawken P. 1993. *The Ecology of Commerce*. HarperCollins, New York.

Hawken P, Lovins A, Lovins LH. 1999. *Natural Capitalism*. Little, Brown and Company, New York.

Kay J, Schneider E. 1992. Thermodynamics and measures of ecosystem integrity. In McKenzie DH, Hyatt DE, McDonald VJ (Eds.), *Ecological Indicators*, Vol. **1**. Proceedings of the International Symposium on Ecological Indicators. Elsevier, Fort Lauderdale, FL, pp. 159–182.

Keen S. 2001. Economists have no ears. *Post-autistic Economics Newsletter Issue* No. 7, July, Article 4. http://www.btinternet.com/~pae_news/review/issue7.htm. Last accessed January 25, 2008.

Keen S. 2002. *Debunking Economics: The Naked Emperor of the Social Sciences*. Zed Publishing, London.

Krishnan R, Harris JM, Goodwin NR (Eds.), 1995. *A Survey of Ecological Economics*. Island Press, Washington, DC.

Lowenstein R. 2001. *When Genius Failed: The Rise and Fall of Long-Term Capital Management*. Random House Trade Paperbacks, New York.

Mandelbrot B. 1963. The variation of certain speculative prices. *J Business* **36**:394–419.

McCormick RJ, Zellmer AJ, Allen TFH. 2004. Type, scale, and adaptive narrative: Keeping models of salmon, toxicology and risk alive to the world. In Kapustka LA, Galbraith H, Luxon M, Biddinger GR (Eds.), *Landscape Ecology and Wildlife Habitat Evaluation: Critical Information for Ecological Risk Assessment, Land-Use Management Activities, and Biodiversity Enhancement Practices*. ASTM STP 1458. ASTM-International, West Conshohocken, PA, pp. 69–83.

Meadows D. 1998. *Indicators and Information Systems for Sustainable Development*. The Sustainability Institute, Hartland Four Corners, VT.

Mirowski P. 1990. From Mandelbrot to Chaos in Economic Theory. *Southern Econ J* **57**:289–307.

O'Neill RV. 1996. Perspectives on economics and ecology. *Ecol Appl* **6**:1031–1033.

Ozkaynak B, Devine P, Rigby D. 2003. Whither ecological economics? *Intern J Environ Pollut* **18**:317–335.

Paganelli MP. 2003. In medio stat virtus: An alternative view of usury in Adam Smith's thinking. *History Political Econ* **35**:21–48.

Pattee HH. 1978. The complementarity principle in biological and social structures. *J Soc Biol Structures* **1**:191–200.

Polanyi M. 1968. Life's irreducible structure. *Science* **160**:1308–1313.

Rosen R. 1981. The challenges of system theory. Presidential Address, 26th Annual Conference of Society for General Systems Research, Toronto, Canada. *General Syst Bull* **XI**(2): 2–5.

Ruddiman W. 2005. *Plows, Plagues, and Petroleum: How Humans Took Control of Climate*. Princeton University Press, Princeton, NJ.

Samuelson PA. 1997. *Economics: The Original 1948 Edition*. McGraw-Hill, New York.

Shapiro MH. 1988. Introduction: Judicial selection and the design of clumsy institutions. *Southern Calif Law Rev* **61**:1555–1569.

Smith A. 1759. *The Theory of Moral Sentiments*. Edinburgh. Online at http://www.adamsmith.org/smith/tms-intro.htm. Last accessed June 5, 2009.

Smith A. 1776. *An Inquiry into the Nature and Causes of the Wealth of Nations*. Edinburgh. Online at http://www.adamsmith.org/smith/won-intro.htm. Last accessed February 1, 2008.

Tainter JA. 1988. *The Collapse of Complex Societies*. Cambridge University Press, Cambridge, UK.

Tainter JA, Allen TFH, Little A, Hoekstra TW. 2003. Resource transitions and energy gain: Contexts of organization. *Conserv Ecol* **7**(3): 4. [online] URL: http://www.consecol.org/vol7/iss3/art4.

Tainter JA, Allen TFH, Hoekstra TW. 2006. Energy transformations and post-normal science. *Energy* **31**:44–58.

Taleb NN. 2007. *The Black Swan: The Impact of the Highly Improbable*. Random House, New York.

Veblen T. 1899. *The Theory of the Leisure Class*. Oxford World's Classics (1997), Oxford University Press, Oxford, UK.

Verweij M, Douglas M, Ellis R, Engel C, Hendriks F, Lohmann S, Ney S, Rayner S, Thompson M. 2006. Clumsy solutions for a complex world: The case of climate change. *Public Administration* **84**:817–843.

Weisman A. 2007. *The World Without Us*. Thomas Dunne Books, New York.

17

ECOSYSTEM SERVICE VALUATION CONCEPTS AND METHODS

James Pittman and Ronald J. McCormick

> If we keep our pride;
> Though paradise is lost;
> We will pay the price;
> But we will not count the cost.
>
> Neil Peart, *Bravado*

Economics as a science has historically focused on research and analysis to inform and guide the allocation of scarce resources to meet human needs. Early methodological developments in economics place primary emphasis on efficiency in expenses from utilization of human labor, as well as investment and maintenance of built or man-made capital (building, vehicles, equipment, etc.), given that these resources were relatively scarce in comparison to the abundance of natural resources. Centuries later, natural resources have generally become more scarce as a result of increasing consumption and pollution, to such an extent that it has been difficult to adequately address and manage given the limitations of conventional economic theory and methods. Pioneering ecological economist Herman Daly frames this as the transformation from an "empty-world" perspective, in which the scale of the economy is quite small with unlimited room to grow in relation to natural limits, to our current "full world"

Environmental Risk and Management from a Landscape Perspective, edited by Kapustka and Landis
Copyright © 2010 John Wiley & Sons, Inc.

reality in which resources are scarce and waste assimilation overburdened as the scale of the economy has grown closer to the total carrying capacity of natural systems. In response to this trend, ecological economics represents the most recent development in scientific theory and methods to provide a new level of sophistication in optimizing the maximum potential utility from capital investments that will cultivate resilience in natural and human resource use.

Because it is no longer sufficient for economics to focus on efficiency through resource allocation, we must also optimize the generative processes of resource production. Given that the health and well-being of humans both individually and collectively depend on natural resources, the flow of value and benefit from this source is the ultimate means generating economic utility. Economic utility is maximized when any decision on resource allocation yields a return on investment greater than the investment itself; investment banking demonstrates this dynamic as interest rates determine the net rate of return on invested principle. In the case of natural resource productivity, the rate of return on investment serves to guide resource efficiency in a manner that will cultivate principal investments in natural capital in order to maximize the production of ecosystem services. Efficient and resilient ecosystems are essential to support a human society, and subsequently to sustain an economy: The planet, the people, and the prosperity of both are inextricably intertwined and interconnected.

In order to address contemporary challenges, modern economic theory and methods require a worldview that balances distributed allocation of investments in built or man-made capital as well as in social or cultural capital and natural capital. This subtle shift in perspective is fundamentally important to the ecological economic worldview since it allows for any potential to raise the importance of investments in natural and social or cultural capital to a level that is as critical as, if not more critical than, investments in man-made built capital or even in financial capital. This can be seen most clearly in the tendency for investments in natural capital to appreciate in value over time as resilient, diverse forests, rivers, and wetlands demonstrate natural processes of ecosystem succession, species adaptation, and evolution. In contrast to the appreciation of natural capital over time, built capital depreciates and requires continual investment to maintain.

The framework for a transformation in economic worldview is perhaps most clearly summarized by ecological economic distinctions between weak and strong definitions of sustainability:

- A goal of maintaining the total stock of all forms of capital regardless of conversions of natural capital to built capital would constitute weak sustainability.
- Strong sustainability requires precautionary conservation of critical natural capital as holding primary importance as the productive means giving rise to other forms of capital (Pearce and Atkinson 1993).

Ecosystem service economics provides methods and tools for putting this theory in practice, so that public and private policy makers can recognize the value of critical natural capital from an economic perspective, in terms of both qualitative assessment and quantitative valuation.

Public and private decision-makers rely on financial and economic tools that offer limited use when valuing ecosystems and the services they provide. This is precisely why public officials and policy managers as well as business executives and managers are more and more frequently considering indicators of environmental performance. Such an integrative approach helps to clarify the extent to which quality of life is dependent on ecosystem resilience and integrity and influenced by resource scarcity and pollution. It also reveals a need for further research into ethical and social justice ramifications of how equitably humans are sharing the opportunities and burdens of environmental resources, constraints, and effects.

Development of more robust tools for project appraisal, program evaluation, and policy analysis is driven by a need for nonfinancial performance measures. This need is most obvious when evaluating social and environmental effects that are not readily monetized—that is, adequately quantified in financial terms using market data. These effects, often called "negative externalities," constitute a failure of financial analysis and economic markets to adequately address critical issues of environmental and social sustainability. Examples of externalities include, but are not limited to, industrial pollution, accidental toxic contamination, environmental risk, hazard mitigation, consumption of ecological resources at levels higher than regeneration rates, or any other form of cost avoidance with potential to limit optimal performance in areas of environmental, health, or safety management.

Negative externalities ultimately amount to production or development that imposes, usually without explicit approval, economic costs and risks on social and ecological systems. Admittedly, there are cases, however rare, of positive externalities, benefits received without cost, to a social or ecological system. Conventional financial and economic analysis tools tend to not consider externalities due to market prices or financial accounting not representing the underlying economic value. The existing prevalent use of these economic tools for project and policy analysis proves a need for innovative economic tools that include negative externalities for use in public and private decision-making.

The economic valuation of ecosystem services, inclusive of ecosystem goods, provides a context for assessing the economic impact of ecological and social externalities imposed on a system. Externalities that cause degradation, fragmentation, or destruction of ecosystem components result in the loss of economic value by making the ecosystem less efficient in providing goods and services. With this concept of ecosystem services, the fields of economics, ecology, and other scientific disciplines have established a broader framework for understanding the economic value that ecosystems provide (Dasguptha 1982, Daily 1997, Costanza et al. 1997, Heal 2000). The emerging theory and application base of ecosystem service analysis is consistent with and complementary to more popular concepts like natural capitalism (Lovins et al. 1999). The concept of ecosystem services was most prominently popularized by the publication of the Millennium Ecosystem Assessment (2005). This publication assesses the current state of functional decline in global ecosystems on which human life is dependent, as well as in possible future scenarios and policy recommendations for intervention.

The complex dynamics of ecosystems result from the interaction of system structural components (e.g., plants, animals, bacteria) with system processes (e.g.,

Figure 17.1. Relationship of ecosystems to the goods and services produced (Batker et al. 2005a, 2005b).

photosynthesis, precipitation, nutrient transport, gene flow). Ecosystem processes produce structures, which, in turn, house new processes, which create new structures. It is this materially, energetically, and informationally open hierarchy (Kay et al. 1999) of process–structure–process that human society designates as having an economically useful function (e.g., produces wood, filters water, and stores water). With explicit recognition of functional value, an economy emerges from the societal demand for specific goods and services produced by the ecosystem (Fig. 17.1).

Ongoing development of ecosystem service concepts regularly spawns new and practical valuation tools for use in making management decisions. Whether qualitative or quantitative, ecosystem service valuation methods provide one of the first lenses for understanding the scale of social–economic effects on the total ecological productivity of the planet. Specifically, value transfer methods were used to analyze local sites using aggregated economic data from other research locations; this provided one of the first clear pictures of human impacts on natural systems at a global scale (Costanza et al. 1997).

More recent innovations include the development of a comprehensive classification of service types, such as: regulation functions; habitat or supporting functions; production or provisioning functions; and information or cultural functions (de Groot et al. 2002, Farber et al. 2006). Current ecosystem service analyses focus on integrating economic theory and ecological science to assess uncertainties related to nonlinear changes and critical thresholds in ecosystem integrity (Farber et al. 2002). Initial work related consideration of critical thresholds to issues of scarcity in ecosystem services (Batabyal et al. 2003). However, the tendency of these factors to result in oversimplifications or underestimates in valuation, as discussed in detail below, presents a clear and present need for additional research and analysis.

Theoretical and applied methods of valuing ecosystem service based on nonmarket data exist in the similar sciences of environmental and ecological economics. These sciences differ, though, on issues related to the intrinsic or existence value of ecosystems. More importantly, the two approaches differ on the ethical question of whether valuation might actually speed the liquidation of ecosystem structures via designation as natural capital assets available for sale on emerging ecosystem markets.

It is increasingly clear that intact ecosystems hold enormous value as societal assets—value ultimately much greater than that potentially realized by fully liquidizing the present goods and services for trading on the open market. The economic value of direct, consumptive resource use is often quite minimal in comparison to the total combined value of nonconsumptive uses. These two broad categories of value comprise the total economic value of ecosystem goods and services. Yet, due to imprecision and uncertainty of nonmarket quantitative valuation methods, any such calculations will likely significantly underestimate total economic value.

In order to understand the differences between these elements comprising total economic value, it is useful to begin with a conventional, utilitarian approach. Each individual ecosystem service may correspond to one or more types of value. These include direct and consumptive use value, direct and nonconsumptive use value, indirect use value, and non-use values including option use value, bequest value, and intrinsic (existence) value. Economists and ecologists (Daily 1997) have jointly explicated categories of use values in much more detail than covered here. Readers interested in further comparison, contrast, and discussion of issues regarding the potential for double-counting in ecological valuation are directed to this reference.

Direct consumptive use values are the most obvious value-holding utility held by ecosystems. This form of use value is seen most clearly in those resources (timber, fish, coal, water, etc.) extracted from ecosystems and used in a way that renders the resources inaccessible for future use. Ecosystem processes result in the production of tangible materials for consumption, and as such these are commonly categorized as marketable ecosystem goods. Given that direct consumptive use is facilitated by market exchange of resources between buyers and sellers, these use values are easily captured using market-based data on the price of goods.

Direct nonconsumptive use values are in contrast based on uses for which there is direct human benefit received from an ecosystem, but each use does not decrease the extent to which another individual might use the same resource. Examples of direct nonconsumptive use values include recreational, educational, and aesthetic use of ecosystems by humans. Direct nonconsumptive uses generally are not bought or sold on a market, and thus the use value is not captured by market price data.

Indirect use values include benefits from an ecosystem delivered to a human in a form that does not require individual awareness of or choice to receive the benefit. Indirect use values accrue to humans via the normal functioning of stable, intact ecosystems. Examples include climate regulation, water and air purification, disturbance moderation, and food production for nonhuman species. Other services provided by ecosystems in the categories of regulation and habitat function are discussed in detail below.

Non-use values of ecosystem services are complex in nature, and they present many difficulties in both basic comprehension and empirical analysis. Option use values are comprised of direct use values not used prior to or at the time of analysis. For example, producing timber for sale from a forest would be a direct consumptive use. However, if those same trees were left standing as part of a forest landscape valuation, the foregone opportunity for direct use constitutes an option use. Bequest value is similar in that it represents a foregone opportunity for direct use, with the value expressly passed on to a subsequent generation. Bequest value also includes indirect use values that are preserved for future generations. Intrinsic, or existence, values are those values inherent in ecosystems above and beyond any other category of human use. These values consider forests, rivers, and other ecosystems as valuable in and of themselves; thus, most non-use values extend beyond the utilitarian worldview and into the realm of environmental ethics.

Ecosystem service analyses recognize that complex, functionally intact ecosystems tend to increase in value over time (naturally appreciating as assets), whereas human-built capital tends to depreciate in value with each year (Batker et al. 2005a). Ecosystems self-organize, and the inherent processes and structures remain open to the introduction of material, energy, and information into the system (Kay et al. 1999). This generally leads to increasing system complexity and order through evolution, adaptation, and succession. Human-built systems are designed (externally organized), requiring continual maintenance, and are not open to new materials not planned for in the original design. For example, the built-capital mechanical system of an automobile, regardless of how well-built, tends to degrade over time due to entropy. Standard financial valuation methods readily calculate the gradual entropic degradation as depreciation, which occurs despite the energy and resource inputs of fuel, oil, parts, and labor, and the value fully depreciates at the unforeseen introduction of other solid objects, such as another automobile, into its structure. In contrast, the natural capital living system of a forest tends to increase in complexity and biodiversity as reproduction and succession occurs. This happens despite entropy as a result of continual inputs of energy from the sun, water, and nutrients from the atmosphere, along with novel genetic material from mutation or external sources.

The characteristics of self-organization, complexification, and adaptation are what produce emergent resilience in natural and human systems (Gunderson and Holling 2002). The resilience of a system changes with access to varying levels of resources or exposure to varying levels of risk. Increasing access to resources and decreasing exposure to risk should subsequently result in an increase in any investment. Thus, investment in improving the resilience of ecosystems as natural capital assets will provide for a more efficient and cost-effective return on investment as compared to investment in human-built capital. Fragmentation, conversion, or direct loss of intact ecosystems ultimately results in a loss of value; economically significant value is not sufficiently captured by market price signals. Market prices are set almost entirely by costs related to production, distribution, sale, and direct consumption of goods and services. As is evident in the full range of use and non-use values, ecosystem services provide a much broader spectrum of value than those typically considered by economists.

Note that there remains no consensus on a classification system for ecosystem services and that the research literature published to date has at least two, if not more, different systems. One classification system (Table 17.1) includes 23 types of ecosystem services (deGroot et al. 2002), while another (Table 17.2) uses 18 types (Farber et al. 2006), with some overlap between the two lists. Both systems were likely adapted from an earlier source (Costanza et al. 1997).

REGULATION FUNCTION ECOSYSTEM SERVICES

Regulation functions are one of the most widely studied function types in the economic valuation literature. Regulation function ecosystem services (RFES) includes processes and functions supporting the maintenance of essential life support systems. This includes atmospheric composition, climate regulation, disturbance moderation, water regulation, soil formation and retention, nutrient transport, waste treatment, pollination, and biological control. RFES represents indirect use values in that no direct

Table 17.1. Ecosystem Processes, Functions, Goods and Services (Batker et al. 2005)

Functions	Infrastructure and Processes	Examples of Goods and Service
Regulation Functions	**Maintenance of Essential Ecological Processes and Life Support Systems**	
1 Gas regulation	Role of ecosystems in bio-geochemical cycles	Provides clean, breathable air, disease prevention, and a habitable planet
2 Climate regulation	Influence of land cover and biological mediated processes on climate	Maintenance of a favorable climate promotes human health, crop productivity, recreation, and other services
3 Disturbance prevention	Influence of ecosystem structure on dampening environmental disturbances	Prevents and mitigates natural hazards and natural events generally associated with storms and other severe weather
4 Water regulation	Role of landcover in regulating runoff and river discharge	Provides natural irrigation, drainage, channel flow regulation, and navigable transportation
5 Water supply	Filtering, retention and storage of fresh water (e.g., in aquifers and snowpack)	Provision of water for consumptive use; includes both quality and quantity
6 Soil retention	Role of vegetation root matrix and soil biota in soil retention	Maintains arable land and prevents damage from erosion, and promotes agricultural productivity
7 Soil formation	Weathering of rock, accumulation of organic matter	Promotes agricultural productivity, and the integrity of natural ecosystems
8 Nutrient regulation	Role of biota in storage and recycling of nutrients	Promotes health and productive soils, and gas, climate, and water regulations
9 Waste treatment	Role of vegetation and biota in the removal or breakdown of xenic nutrients and compounds	Pollution control/detoxification, filtering of dust particles through canopy services
10 Pollination	Role of biota in the movement of floral gametes	Pollination of wild plant species and harvested crops
11 Biological control	Population control through trophic–dynamic relations	Provides pest and disease control, reduces crop damage

(continued)

Table 17.1. (*Continued*)

Functions	Infrastructure and Processes	Examples of Goods and Service
Habitat Functions	**Providing Habitat (Suitable Living Space) for Wild Plant and Animal Species**	
12 Refugium function	Suitable living space for wild plants and animals	Maintenance of biological and genetic diversity (thus the basis for most other functions)
13 Nursery function	Suitable reproduction habitat	Maintenance of commercially harvested species
Production Functions	**Provision of Natural Resources**	
14 Food	Conversion of solar energy into edible plants and animals	Hunting, gathering of fish, game, fruits, etc.; small-scale subsistence farming and aquaculture
15 Raw materials	Conversion of solar energy into biomass for human construction and other uses	Building and manufacturing; fuel and energy; fodder and fertilizer
16 Genetic resources	Genetic material and evolution in wild plants and animals	Improve crop resistance to pathogens and pests
17 Medicinal resources	Variety in (bio)chemical substances in, and other medicinal uses of, natural biota	Drugs, pharmaceuticals, chemical models, tools, test and essay organisms
18 Ornamental resources	Variety of biota in natural ecosystems with (potential) ornamental use	Resources for fashion, handicraft, jewelry, pets, worship, decoration and souvenirs
Information Functions	**Providing Opportunities for Cognitive Development**	
19 Aesthetic information	Attractive landscape features	Enjoyment of scenery
20 Recreation	Variety in landscapes with (potential) recreational uses	Travel to natural ecosystems for eco-tourism, outdoor sports, etc.
21 Cultural and artistic information	Variety in natural features with cultural and artistic value	Use of nature as motive in books, film, painting, folklore, national symbols, architecture, advertising, etc.
22 Spiritual and historic information	Variety in natural features with spiritual and historic value	Use of nature for religious or historic purposes (i.e., heritage value of natural ecosystems and features)
23 Science and education	Variety in nature with scientific and educational value	Use of natural systems for school excursions, etc., use of nature for scientific research

Table 17.2. Ecosystem Service Categories, Use Values, Valuation Potential, Valuation Methods, and Transfer Potential

Service Category	Use Value	Valuation Potential	Valuation Method	Transfer Potential
Gas regulation	IU, NU	Medium	CV, AC, RC	High
Climate regulation	IU, NU	Low	CV	High
Disturbance prevention	IU, NU	High	AC	Medium
Biological regulation	IU, NU	Medium	AC, PF	High
Water regulation	IU, NU	High	MP, AC, RC, HP, P, CV	Medium
Soil retention	IU, NU	Medium	AC, RC, HP	Medium
Waste regulation	IU, NU	High	RC, AC, CV	Med/High
Nutrient regulation	IU, NU	Medium	AC, CV	Medium
Water supply	DU, OU, DU	High	AC, RC, M, TC	Medium
Food	DU, OU, DU	High	MP, PF	High
Raw Materials	DU, OU, DU	High	MP, PF	High
Genetic resources	DU, OU, DU	Low	MP, AC	Low
Medicinal resources	DU, OU, DU	High	AC, RC, PF	High
Ornamental resources	DU, OU, DU	High	AC, RC, HP	Medium
Recreation	DU, OU, DU	High	TC, CV, RA	Low
Aesthetics	DU, IU, OU, NU	High	HP, CV, TC, RA	Low
Science and education	DU, IU, OU, NU	Low	RA	High
Spiritual and historic	DU, IU, OU, NU	Low	CV, RA	Low

Use value types: direct use (DU), indirect use (IU), option use (OU), non-use (NU).
Valuation methods: avoided costs (AC), contingent valuation (CV), hedonic pricing (HP), market pricing (MP), production function (PF), replacement cost (RC), travel cost (TC), ranking (RA).
Source: Adapted from Farber et al. (2006).

human consumption of natural resources occurs. Nonetheless, RFES represent tremendous value in providing a human-habitable climate, with water and nutrient flows that sustain the diversity of life that society depends upon. Economists have established that the RFES of stormwater management and flood protection provided by wetland and riparian systems are of vast economic value (Farber 1987, Thibodeau and Ostro 1981).

One current area of research focuses on climate regulation, specifically the process of carbon sequestration in biotic organisms. This constitutes an essential part of the regulation function, and valuation methods include the use of a combination of market prices and avoided cost values. For tradable pollution permits useable for offsetting emissions, emerging carbon trading markets currently price carbon sequestration, as a climate regulation service, at about $3.65 per tCO_2e (ton of carbon dioxide equivalent emissions) (CCX 2007). This falls well short of the best approximation of the total social–economic benefit of climate regulation services from ecosystems. Current literature estimates range from $14 per tCO_2e (Tol 2005) to $29 per tCO_2e (Stern et al. 2006).

Valuation research on water purification services provided by ecosystems reveals very interesting conclusions through use of replacement cost methods. New York

City decided to undertake a $1.5 billion investment in watershed conservation for natural water purification in order to avoid an initial cost of $8 billion to build a filtration plant (Krieger 2001). Similarly, the city of Portland, Oregon currently spends $920,000 annually to restore and protect the Bull Run watershed, which naturally filters Portland's drinking water supply, avoiding the construction of a $200 million water filtration plant (Krieger 2001). Both cities avoid additional costs that would accrue with required plant maintenance and normal operating expenses. In both cases, ecosystem processes residing within the watershed landscape structure provide functional water filtration for a lower initial cost, and they are nearly free of operating and maintenance expenses.

HABITAT FUNCTION ECOSYSTEM SERVICES

Habitat function ecosystem services (HFES) include processes and functions supporting habitats suitable for wild plants and animals, producing great value by supplying nutrients, essential elements, and other resources to biological systems. Services included in this category include refugia and nursery functions. HFES provide indirect use and non-use value in that no direct human consumption occurs. As well, functional and intact habitat integrity directly supports regulation and production functions that provide direct value to human economic systems.

Habitat functions are relatively difficult to estimate using valuation methods. Due to differences in valuation category systems, HFES are distinguished from other service types by some theorists (de Groot et al. 2002), while other practitioners use systems that do not distinguish these as unique service types within a distinct function category (Farber et al. 2006, Costanza et al. 1997). In the application of ecosystem service valuation, there is a profound dearth in, and thus great need for, primary valuation research into HFES as indicators of ecosystem integrity (Swedeen and Pittman 2007).

PRODUCTION FUNCTION ECOSYSTEM SERVICES

Production function ecosystem services (PFES) include processes and functions supporting the provision of natural resources. Services include production of food and raw materials as well as genetic, medicinal, or ornamental resources. The resources produced are frequently referred to as ecosystem "goods." These ecosystem functions and services result in direct economic benefits by producing items with explicit market value. PFES provide the most direct consumptive use value of all categories of ecosystem service. Market-based valuation methods (e.g., timber, fish, or agricultural product pricing) are commonly used to estimate PFES that result in the production of ecosystem goods directly consumed by society.

Combining production function data with a fish population model resulted in PFES value estimates of C$0.93 to C$2.63 per hectare of drainage basin, which equates to about C$1322 to C$7010 per km of in-stream salmon habitat in British Columbia

(Knowler et al. 2002). Similar analyses regarding polychlorinated biphenyl (PCB) contamination in the Chesapeake Bay produced a valuation estimate of $2.3–$7.7 million in damage costs to the striped bass fishery. These estimates can serve as proxy valuations for habitat function ecosystem services. However, if applied in the context of a comprehensive ecosystem service valuation analysis, issues of double-counting would prevent use of the PFES values for estimates of HFES values.

INFORMATION FUNCTION ECOSYSTEM SERVICES

Information function ecosystem services (IFES) includes processes and functions supporting the provision of opportunities for human cognitive development. Services include generation of recreational opportunities, information pertaining to aesthetic, cultural, artistic, spiritual and historic resources, and information contributing to science and education. Society uses information functions directly, though only in a nonconsumptive manner. Thus, for many IFES other than recreational or aesthetic uses, it becomes necessary to use nonmarket valuation methods to complement market-based valuation methods.

IFES also produce indirect use, option use, or bequest value. A study in the Portland, Oregon area found that residential property values increased $436 for every 1000 feet closer that a property was to a wetland (Mahan et al. 2000). Additional research has also assessed how other environmental amenities enhance property values (Crompton 2001, Anderson and Cordell 1988, Dorfman et al. 1996).

The functional categories of ecosystem services tend to correspond with various categories of usage: In some cases the services are directly used for consumptive purposes while in other cases the services are not used directly, relegated only to optional use or otherwise more peripheral in terms of utility. A number of economic valuation methods can be applied to the estimation of total economic value provided by ecosystem services depending on the type of usage classification that corresponds with each (Table 17.2). The following discussion of ecosystem service valuation methods was adapted from consistent commonalities from a literature review of multiple sources (Costanza et al. 1997, Farber et al. 2002, Batker et al. 2005a). Readers are encouraged to review all of these sources for specific information on methods, discussion of the pros and cons of using a particular method, and sources of primary research examples, case studies, and data. The information contained in these references is simply too vast to summarize completely herein.

Market price valuation methods used to analyze services, particularly those resulting in the production of goods, can directly obtain a value for a good or service from an analysis of people's willingness to pay (WTP). For example, market price methods are used in the valuation of food and natural resources (e.g., water, timber, minerals). With the emergence of markets for ecosystem services, further discussed below, market price methods can be applied as proxy value estimates of replacement cost, as in the forest-based carbon sequestration example mentioned earlier.

Avoided cost valuation analyzes services that society benefits from due to avoidance of costs otherwise incurred if those services were not available from natural ecosystems. For example, avoided cost methods are used to value the disturbance moderation services of flood control. Provided by river and wetland systems, these services prevent or minimize the cost of residential and commercial property flood damage. Avoided cost methods are also used to value natural wastewater treatment services provided by wetlands and soils, insofar as the service results avoided health-care costs by limiting the exposure of humans to contaminants.

Replacement cost valuation analyzes benefits to society from ecosystem services that, if not available, but still necessary, would require replacement through investment in technological solutions or human-built capital. For example, replacement cost methods are used to value water treatment services provided by wetlands, because these ecosystem services replace the need for and cost of technological filtration systems (as discussed above in the regulation function ecosystem services section).

Production function and factor income valuation methods are used to analyze services that benefit society due to production of value in the form of marketable goods (production function) or realized income (factor income). For example, the production function method is used to value improvement of water quality based on market data showing increases in fish harvests; the factor income method would establish value based on increased income in the fishing industry.

Travel cost valuation methods are used to analyze services that benefit society via a site-based amenity. Absence of a desired service from a nearby ecosystem would require travel to an alternate site to receive the same benefit, and the cost incurred would establish a value for the local amenity. For example, travel cost methods could be applied for the valuation of recreation services in areas that attract visitors from a distance, revealing the value of recreational areas in close proximity based on a willingness to pay for travel to more distant recreational areas.

Hedonic pricing valuation methods reflect the societal value of a service based on a willingness to pay prices for goods or services in a market indirectly associated with a particular ecosystem service. For example, the hedonic pricing method, applied to aesthetic or recreational services of ecosystems, would establish value by comparing appraisal prices of residential homes near a landscape type (e.g., forest, wetland, lake) to prices of comparative properties located further from the same type of landscape.

Contingent valuation methods estimate demand for an ecosystem service through direct stakeholder queries. The structure of the queries can include consideration of hypothetical scenarios involving alternative management approaches. For example, contingent valuation methods would gather data on stated preference or WTP for increased access to recreational open space or cleaner air. Where hypothetical scenarios include more than one service or ecological condition, the method is often referred to as conjoint analysis.

Ranking or participatory valuation methods establish the relative value of ecosystem services through use of quantitative rubrics rather than monetized valuation. These methods tend to include more extensive interpersonal discourse or information review by stakeholders in order to capture more subjective perceptions of value. Such methods are quite useful in cases where seemingly objective quantitative valuation would fail to capture the full perception of value assessed using rating or ranking models of hypothetical scenarios or choices presented to stakeholders as in contingent valuation.

Value transfer methods build supplementarily on these other methods in that value transfer is used to draw conclusions about a study site based on data from a different research site with similar ecosystem characteristics. This method is known as benefits transfer or cost transfer depending on the type of values being used, though in the case of ecosystem service valuation the benefit transfer is more common. Value transfer is increasingly applied to economic analysis of ecosystem services since it is less costly and time-consuming than primary research methods, but it is also a controversial method due to the same ease. Value transfer is essential to developing comprehensive dollars-per-unit-area analyses for multiple ecosystem services, given the extensive data aggregation required when assessing differently scaled landscapes.

Leading developments in ecosystem service valuation techniques include the use of powerful information technology systems in assessing complex landscapes. System dynamics models of ecosystem processes and functions are used to increase the precision and certainty of valuation methods. Two such tools are the Global Unified Meta-model of the Biosphere (GUMBO), a global-scale simulation model of economic, social, and ecological dynamics (Boumans et al. 2002), and the Multi-scale Integrated Model of Earth Systems (MIMES), a regional or local-scale simulation model of these dynamics currently in development for dynamic generation of policy scenarios (Gund Institute 2007). The latter tool highlights (a) the applicability of ecosystem service valuation for broad landscapes and (b) the need for collaborative data gathering, management, and integration to support the ongoing development of powerful models and information technology tools.

Data repositories, such as the Ecosystem Service Database (Villa et al. 2002), strengthen the rigor of ecosystem service valuation with powerful geo-spatial query and aggregation tools for cross-scale analyses. This database is a key part of the Assessment and Research Infrastructure for Ecosystem Services (ARIES) project funded by the National Science Foundation (Ecoinformatics Collaboratory 2007). Using these tools, local decision-makers can easily generate static analyses using ecosystem service valuation. Current applications include proof-of-concept case studies focusing on Central America and the Pacific Northwest.

Practical application of ecosystem service analysis and valuation to environmental management of large landscapes has increased since publication of the Millennium Ecosystem Assessment (2005), particularly in the Puget Sound region of the Pacific Northwest. Applications include policy analysis, project appraisal and program evaluation related to land-use planning on Maury Island (King County Department of Natural Resources and Parks 2004), conservation and restoration of salmon habitat in the Green/Duwamish River watershed (Batker et al. 2005a), and forest management on public lands in the Tolt River watershed (Batker et al. 2005b). On a finer scale, ecosystem service valuation has been applied to restoration of riparian buffers near agricultural lands in the King Conservation District (Batker et al. 2006) as well as levee setback or removal with forest and wetland restoration along the Cedar River (Swedeen and Pittman 2007). Ecosystem service valuation has even been applied at the scale of individual trees to show economic value provided in an urban context (McPherson et al. 2002).

Interest in the methodologies extends beyond the Pacific Northwest region, as evidenced by the many informational websites and publications on the concept of ecosystem services and applied valuation methods. One such resource is the Ecosystem Valuation web site (King et al. 2006), developed in partnership with the United States Department of Agriculture National Resource Conservation Service (USDA NRCS) and the National Oceanic and Atmospheric Association (NOAA). Informational websites with less extensive information have been posted by national science associations and other federal agency offices (e.g., ESA 2007, USDA 2007). Funding from the US EPA has also supported nonprofit research on potential standardization of ecosystem services concepts and analysis (Boyd 2006).

Additional applications of ecosystem service valuation methods have focused on using various Long-term Ecological Research Sites (LTER) as case studies (Farber et al. 2006). As well, a recently released report by the New Jersey Department of Environmental Protection (2007) includes the first-ever comprehensive analysis of ecosystem services at the state level. Clearly, interest in ecosystem service concepts and methods is steadily increasing.

Ecosystem service valuation methods and advanced technical modeling of natural systems have also received peer critique, as is common in any science. Academic researchers have noted concern with the validity of valuation and value transfer methods and recommend alternative participatory and multicriteria assessment methods (Spash and Vatn 2006); however, these methods are similarly critiqued as relatively new and controversial tools for economic analysis of policies and projects. Moreover, the criticism of ecosystem service valuation method for estimating the value of Earth's primary ecosystems exceeds available income or ability to pay for such services (Bockstael et al. 2000, Spash and Vatn 2006) reveals the urgency of using such analysis tools. Ecological economists espousing the use of ecosystem service valuation methods contend quite explicitly that traditional finance and economics relying on market-based data and conventional valuation tools do not reveal the very real limits to growth (Costanza et al. 1997). Indeed, income to pay for development or consumption that exceeds global productivity ultimately constitutes a growing generational environmental debt (Azar and Holmberg 1995), with issues of intergenerational social equity beyond inequities of risk distribution in the present.

Although ecosystem service valuation methods have only recently been combined with advanced technical modeling of ecosystem functions and processes, these modeling tools also have noteworthy limitations. Critics assert that verification and validation of natural systems models are prone to uncertainty and relativity that will always challenge the predictive value of such tools (Oreskes et al. 1994). These assertions have some validity, although these concerns apply to any quantitative modeling methods—including all conventional economic modeling or financial projections of costs and benefit that disregard ecosystem service values. In the context of ecosystem service management, valuation provides an opportunity to increase the rigor of economic and financial tools currently in common, if not ubiquitous, use.

Ecosystem services valuation offers a significant improvement over traditional benefit–cost economic analysis methods. Benefit–cost methods commonly represent social and environmental externalities, such as the value of ecosystem services, as having no economic value. The growing number of projects applying ecosystem service valuation in addition to the emergence of many markets supporting payments for ecosystem services both stand as clear indications that decision-makers are beginning to understand and plan for these critical issues in project appraisal and policy analysis. Even noting the limitations of these practical methods, there remains a great opportunity for rigorous, transparent applications to provide improvement on current tools and methods. Known issues of uncertainty and relativity simply emphasize the importance of using ecosystem services valuation and modeling as descriptive information tools rather than with an absolute, predictive determinism. Methods that are participatory and include multiple criteria, though often demanding greater investments in research time and money, are certainly also complementary tools for consideration of nonfinancial project or policy impacts.

Ecosystem service economics and nonmarket valuation methods are increasingly being used in a real-world context to inform public policy development in areas of (a) conservation and restoration ecology and (b) resource and hazard management. Ecosystem services valuation offers a means for assessing the long-term effects of policy decisions on natural capital assets that influence economic markets, community well-being, and environmental integrity. The importance of ecosystem service economics can be seen in the broad range of service categories not currently managed in market-based or regulatory systems. Perhaps more importantly, the importance of valuation can be seen in the extent to which the novelty of this approach indicates that the value of these ecosystem services has not been adequately represented in policy and planning to date. This fact, combined with the extensive gaps in available valuation data for comprehensive ecosystem service valuation analysis, presents circumstances indicating that valuation estimates at any level of analysis are indeed underestimates of the total economic value produced in the form of ecosystem services. This presumption of underestimation highlights the more precautionary approach advocated in ecological economics, in contrast to environmental economics.

There are a variety of reasons leading to conclusions that ecosystem service concepts and valuation methods ultimately present an underestimate of the total economic value of ecosystems and effective environmental management and performance on any scale. One notable underestimate would be in decision cases that include energy resource utilization because these are not adequately represented with respect to scarcity or nonrenewable resource consumption rates, but also externalities of various energy resources such as the long-term assimilation impacts of nuclear energy. For this valuation, avoided costs might be generated from health and safety risk analysis based on the economic value of human life, a highly controversial valuation science that is in common use in county, state, federal, and international policy (Ackerman and Heinzerling 2004). Valuation methods also tend not to include consideration of the relative scarcity of, or externalities from, use of renewable resources such as wind, solar, and intertidal, although hydro-electric power is represented to some extent through valuation of water supply services.

Similarly, intrinsic or existence values might also apply in cases wherein ecosystem degradation, fragmentation, or contamination has potential to subsequently impact human health, safety, or life. This is likely, given that economic modeling and analysis of the value of human health and life, though quite commonly used, are of questionable scientific rigor if not indeed unethical (Ackerman and Heinzerling 2004). Regardless, ecosystem service valuation methods most likely do not include sufficient consideration of these unmitigated risks as a result of poor environmental performance that results in exposure to toxic releases cumulative effects of pollution emission. One other important consideration is a lack of any representation for market externalities in the relative abundance or scarcity of geological resources such as industrial metals or fossil fuels for which market price is not accurate or available. The latter of these market externalities may still be quantified for consideration of relative scarcity, consumption, or impact using ecological footprint methods or integration into aforementioned system dynamics models such as GUMBO or MIMES.

PAYMENT FOR ECOSYSTEM SERVICES

The majority of this section has focused on nonmarket valuation methods for estimating the economic value of ecosystem services. This is due to the fundamentally precautionary perspective of ecological economics in presuming that market values, or indeed quantitative valuation methods overall, will always underestimate the total economic value of ecosystem services—particularly in the case of critical natural capital. That notwithstanding, recent years have provided opportunities for extensive growth in emerging markets supporting the payment for ecosystem services.

Private landowners, businesses, governmental agencies, and other stakeholders are increasingly exposed to emerging markets for buying and selling ecosystem service credits. These market-based systems allow for many investments in natural capital ranging from forest carbon sequestration, wetlands preservation, water quality, and biodiversity as a nonconsumptive way to preserve the integrity and resilience of ecosystems in complement to existing water and resource commodity markets. One of the best single sources for extensive information on these markets supporting payments for ecosystem services is the Katoomba Group's Ecosystem Marketplace website (http://www.ecosystemmarketplace.com/viewed July 20, 2009).

There are still skeptics of ecosystem service markets, with concern that the approach amounts to the reduction of nature to privately held commodities. While this discussion is too complex and intricate to fully present in this chapter, suffice to say that it affirms the importance of complementary regulatory approaches to payments for ecosystem services where appropriate, strict regulatory controls on markets, economic analysis of nonmarket value from ecosystem services, and, most importantly, precautionary principles in planning for ecosystem service management.

For added clarification, consider an example: If confronted with the hypothetical choice between giving up all material belongings or the supply of life-supporting oxygen, any rational economic actor would choose the latter; thus it can be said they are willing to accept an economic loss equal to the total value of all material belongings in order to retain the service of oxygen production.

In conclusion, the total economic value of ecosystem goods and services lies not only in the market value of commercial goods extracted from natural systems and the value of those goods and services which may hold market value now or at some time in the future, but also in the intrinsic value held by the complex interconnections of resilience and biodiversity that are integral to resilient, efficient ecosystems supporting life on Earth. Relatively recent innovations in the sciences of environmental and ecological economics provide a means for estimating one, multiple, or comprehensive aggregations of ecosystem service values. The scientific rigor and ethical subtleties of these applied methods are actively discussed and debated by academic researchers in higher education as well as practitioners in the nonprofit, business, and government sectors. Unfortunately, while economic concepts and methods are being created and applied for estimating the total value of ecosystems to inform public decision-making, there is still a critical need for these scientific developments to be integrated to financial accounting, analysis, and reporting methods that guide private decision-making (Pittman and Wilhelm 2007). Simply put, financial analysis of the value produced by ecosystem services based on market data, either existing or from emerging ecosystem markets, does not accurately represent the value of social and environmental externalities integral to the total economic value of an ecosystem. These innovative concepts and methodological tools will be essential to public and private managers and decision-makers seeking to fully understand risk in terms of environmental and, to some extent, health and safety performance.

REFERENCES

Ackerman F, Heinzerling L. 2004. *Priceless: On Knowing the Price of Everything and the Value of Nothing*. London Press, New York.

Anderson L, Cordell H. 1988. Influence of trees on residential property values in Athens, Georgia: A survey based on actual sales prices. *Landscape Urban Planning* **15**:153–164.

Azar C, Holmberg J. 1995. *Defining the Generational Environmental Debt. in Socio-ecological Principles and Indicators for Sustainability*. Chalmers University of Technology, Göteborg.

Batabyal A, Kahn J, O'Neill R. 2003. On the scarcity value of ecosystem services. *J Environ Econ Manage* **46**:334–352.

Batker D, Barclay E, Boumans R, Hathaway T, Burgess E, Shaw D, Liu S. 2005a. *Ecosystem Services Enhanced by Salmon Habitat Conservation in the Green/Duwamish and Central Puget Sound Watershed*. Earth Economics, Seattle, WA.

Batker D, Pittman J, Burgess E. 2005b. *Ecosystem Services Valuation Study for the Tolt River Watershed: A Draft General Technical Report to the United States Department of Agriculture Forest Service*. Earth Economics, Seattle, WA.

Batker D, Pittman J, de la Torres I. 2006. *Special Benefit from Ecosystem Services: An Economic Analysis of King Conservation District Activities*. Earth Economics, Seattle, WA.

Bockstael NE, Freeman AM, Kopp RJ, Portney PR, Smith VK. 2000. On measuring economic values for nature. *Environ Sci Technol* **34**:1384–1389.

Boumans R, Costanza R, Farley J, Wilson M, Portela R, Rotmans J, Villa F, Grasso M. 2002. Modeling the dynamics of the integrated earth system and the value of global ecosystem services using the GUMBO model. *Ecol Econ* **41**:529–560.

Boyd J. 2006. *Practical Measurement of Ecosystem Services: Can We Standardize the Way We Count Nature's Benefits*. Prepared by Resources for the Future for the Environmental Protection Agency National Center for Environmental Economics.

CCX (Chicago Climate Exchange). 2007. *Market Data Overview* http://www.chicagoclimatex. com (accessed August 5, 2007).

Costanza R, d'Arge R, de Groot R, Farber S, Grasso M, Hannon B, Limburg K, Naeem S, O'Neill R, Paruelo J, Raskin R, Sutton P, van den Belt M. 1997. The value of the world's ecosystem services and natural capital. *Nature* **387**:253–260.

Crompton J. 2001. *The impact of parks and open space on property values and the property tax base*. National Recreation and Park Association, Ashburn, VA: http://rptsweb.tamu.edu/ faculty/pubs/property%20value.pdf (accessed June 2004).

Daily G (Ed.). 1997. *Nature's Services: Societal Dependence on Natural Ecosystems*. Island Press, Washington, DC.

Dasguptha P. 1982. *The Control of Resources*. Harvard University Press, Cambridge, MA.

de Groot R, Wilson M, Boumans R. 2002. A typology for the classification, description, and valuation of ecosystem functions, goods, and services. *Ecol Econ* **41**:393–408.

Dorfman J, Keeler A, Kriesel W. 1996. Valuing risk-reducing interventions with hedonic models: The case of erosion protection. *J Agric Resource Econ* **21**:109–119.

Ecoinformatics Collaboratory. 2007. *Assessment and Research Infrastructure for Ecosystem Services (ARIES) Project Description*. http://ecoinformatics.uvm.edu/projects/aries.html (accessed June 15, 2007).

ESA (Ecological Society of America). 2007. *Communicating Ecosystem Services Website*. http://www.esa.org/ecoservices (accessed June 15, 2007).

Farber S. 1987. The Value of Coastal Wetlands for Protection of Property Against Hurricane Wind Damage. *J Environ Econ Manage* **14**:143–151.

Farber S, Costanza R, Wilson M. 2002. Economic and ecological concepts for valuing ecosystem services. *Ecol Econ* **41**:375–392.

Farber S, Costanza R, Childers D, Erickson J, Gross K, Grove M, Hopkinson C, Kahn J, Pincetl S, Troy A, Warren P, Wilson M. 2006. Linking ecology and economics for ecosystem management. *Bioscience* **56**:117–129.

Gund Institute. 2007. *Multi-scale Integrated Models of Ecosystem Services*. http://www.uvm.edu/ giee/mimes (accessed June 15, 2007).

Gunderson L, Holling C. 2002. *Panarchy: Understanding Transformations in Human and Natural Systems*. Island Press, Washington, DC.

Heal G. 2000. *Nature and the Marketplace: Capturing the Value of Ecosystem Services*. Island Press, Washington, DC.

Kay JJ, Regier HA, Boyle M, Francis G. 1999. An ecosystem approach for sustainability: Addressing the challenge of complexity. *Futures* **31**:721–742.

King County Department of Natural Resources and Parks. 2004. *Ecological Economic Evaluation: Maury Island, King County, Washington*. King County Department of Natural Resources and Parks, Land and Water Division.

King D, MazZotta M, Markowitz K. 2006. *Ecosystem Valuation Website*. Developed by the USDA National Resource Conservation Service and National Oceanic and Atmospheric Association. http://www.ecosystemvaluation.org/(accessed 20 July 2006).

Knowler DJ, MacGregor BW, Bradford MJ, Peterman RM. 2002. Valuing freshwater salmon habitat on the west coast of Canada. *J Environ Manage* **69**:261–273.

Krieger D. 2001. *Economic Value of Forest Ecosystem Services: A Review*. The Wilderness Society, Washington, DC.

Lovins A, Lovins H, Hawkin P. 1999. A road map for natural capitalism. *Harvard Business Rev* **May–June**: 145–158.

Mahan B, Polasky S, Adams R. 2000. Valuing urban wetlands: A property price approach. *Land Econ* **76**:100–113.

McPherson EG, Maco SE, Simpson JR, Peper PJ, Xiao Q, Van Der Zanden AM, Bell N. 2002. *Western Washington and Oregon Community Tree Guide: Benefits, Costs, and Strategic Planting*. International Society of Arboriculture, Pacific Northwest Chapter, Silverton, OR.

Millennium Ecosystem Assessment. 2005. *Ecosystems and Human Well-Being—Current States and Trends: Findings of the Conditions and Trends Working Group*. Island Press, Washington, DC.

New Jersey Department of Environmental Protection. 2007. *Valuing New Jersey's Natural Capital: An Assessment of the Economic Value of the State's Natural Resources*. State of New Jersey, Trenton, NJ http://www.state.nj.us/dep/dsr/naturalcap/nat-cap-1.pdf accessed July 2, 2009.

Oreskes N, Shrader-Freshette K, Belitz K, 1994. Verification, validation and confirmation of numerical models in the earth sciences. *Science* **263**:641–646.

Pearce D, Atkinson G. 1993. Capital Theory and the Measurement of Sustainable Development: An Indicator of Weak Sustainability. *Ecol Econ* **8**:103–108.

Pittman J, Wilhelm K. 2007. Creating appropriate economic and financial indicators of sustainability. In Litten L (Ed.), *New Directions in Institutional Research: The Sustainability Challenge*. Jossey-Bass, San Francisco, pp. 55–70.

Spash C, Vatn A. 2006. Transferring environmental value estimates: Issues and alternatives. *Ecol Econ* **60**:379–388.

Stern N, Peters S, Bakhshi V, Bowen A, Cameron C, Catovsky S, Crane D, Cruickshank S, Dietz S, Edmonson N, Garbett S-L, Hamid L, Hoffman G, Ingram D, Jones B, Patmore N, Radcliffe H, Sathiyarajah R, Stock M, Taylor C, Vernon T, Wanjie H, Zenghelis D. 2006. *Stern Review: The Economics of Climate Change*. HM Treasury, London.

Swedeen P, Pittman J. 2007. *Economic Assessment of Cedar River Flood Hazard Management Plan*. Earth Economics, Seattle WA.

Thibodeau F, Ostro B. 1981. An economic analysis of wetland protection. *J Environ Manage* **12**:19–30.

Tol RSJ. 2005. The marginal damage costs of carbon dioxide emissions: An assessment of the uncertainties. *Energy Policy* **33**:2064–2074.

United States Department of Agriculture (USDA). 2007. *Valuing Ecosystem Services Website*. http://www.fs.fed.us/ecosystemservices (accessed June 15, 2007).

Villa F, Wilson M, de Groot R, Farber S, Costanza R, Boumans R. 2002. Designing an integrated knowledge base to support ecosystem services valuation. *Ecol Econ* **41**:445–456.

18

METRICS AND INDICES FOR SUSTAINABLE SOCIAL–ECOLOGICAL LANDSCAPES

Ronald J. McCormick

> Most people think the future is the ends and the present is the means. In fact, the present is the ends and the future the means.
>
> Fritz Roethlisberger

This discussion started (Chapter 16) by introducing a question from ecological economics, "What is the scale of the economy that fits inside the ecology?" We presented a general deconstruction of neoclassical economics and explored how scale appears merely implicit in economic models, whereas explicit definition of scale is required in any model of a complex system. Providing explicit definitions of scale and ecological type sets a clear level of analysis. In Chapter 16, we described how recent advances in systems theory (high-gain/low-gain analysis) allows for the blending of market thermodynamics and social regulation via the field of economic ecology, which naturally leads to issues of sustainability. As society moves to a low-gain perspective on managing for sustainable social–ecological landscapes, the assumptions and limitations inherent in available economic valuation techniques become quite apparent.

The primary limitation of most valuation techniques lies in how one sets the boundaries of a system of interest. Benefit–cost analysis has limited temporal relevance due to the inherent uncertainty in any financial forecast data. The majority of financial prognostications put forth in early 2008 proved to be completely and utterly useless by the autumn of that year. At the root of the subprime mortgage meltdown was the idea that one could take securities developed under one (local) level of analysis and (presumed minimal) risk scenario and expand those same risks upscale to a global market. Moving upscale changed the initial level of analysis that the system was working within, but most analysts, placing total trust in mathematical models, missed that fact. Indeed, even Alan Greenspan, in his testimony to the U.S. Congress on October 23, 2008, said "I found a flaw in the model that I perceived is the critical functioning structure that defines how the world works." The flaw was that his model assumed self-correction in the securities market would take place at an individual institution level, thus protecting "local" investors. Self-correction did occur, but not at the level in the system where it was needed, it occurred at a global, and nearly catastrophic, level.

Though the subprime debacle is attributable to no single action, a key indicator of lender risk, interest rate, was reduced in variance as part of the process of offering more loans (Rajan et al. 2008). A person with a poor credit rating received a loan at an interest rate very similar to what someone with a good credit rating was offered. The effect of this was that once a single degree of separation existed between the person receiving the loan and the person or institution holding the note, there was no way to compare relative risk between mortgage packages (interest rate provided insufficient signal). Lack of transparency of actual risk at a low level in the process allowed unequally risky loans to appear as equally risky loans at a different level in the system.

Human systems, especially socioeconomic systems with designed allowable interactions, are fraught with these kinds of unrecognized risks. Landscapes supporting ecosystem processes and intact communities are self-organizing (no set interaction rules) and open to the flow of energy, materials, and information (total transparency). If a plant or animal signals that "I taste bad" via bright red markings, then most animals will heed the warning and steer clear. However, as the communication of information is open, some animal might see and accept the risk and have a snack. Whether accepting that risk was wise or unwise is answered very quickly via feedback from the consumer's body, not years later via a party not involved in the original snack decision. Timely and relevant feedback and open communication make ecosystems function effectively. Businesses that mimic these characteristics usually prosper (*sensu* Pascale et al. 2001). Sustainable social–ecological landscapes need to adopt these strategies in order to manage risk and flourish in the face of change.

Aldo Leopold urged people to "think like a mountain," a mantra often used by the deep ecology movement. However, while environmentalists were thinking like a mountain, the rest of society built roads, logged off the trees, extracted any valuable ores, and harvested the edible wildlife, all while the mountain's glaciers were melting and relentlessly eroding it, becoming silt in some valley. Perhaps in this, the 21st century, we need to "think like an ecosystem," to know where the matter and energy we use came from, where it is going, and how effectively we use it while we have control. This is the heart of economic ecology and is the root of sustainability.

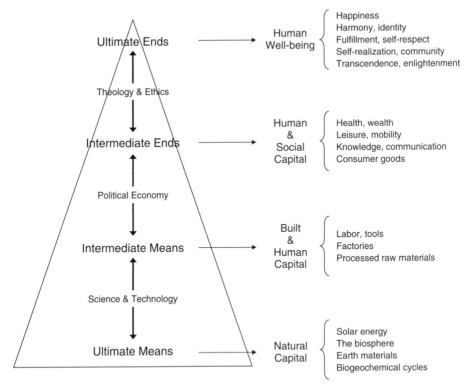

Figure 18.1. Connections inherent in any social–ecological landscape. [Modified from Meadows (1988).]

ULTIMATE MEANS, ULTIMATE ENDS

Referring to a social–ecological landscape is a short-hand way of referring to a society's ultimate ends, the goals and values of that society, and the constraints placed on those goals by the natural capital available on a landscape, the ultimate means of ecological goods and services. Herman Daly, in his 1973 book *Toward a Steady-State Economy* [as quoted in Meadows (1998)], outlined a hierarchy where the economy and daily human endeavors (intermediate means and ends) were placed in context between natural capital and human well-being (Fig. 18.1). This framework remains highly relevant today, some 35 years later.

Meadows (1998) used Daly's framework to guide the selection of indicators for sustainable development. Her explanation of what indicators are and why they are so difficult to develop closes with the notion that (a) we need more than just indicators, (b) we need information systems that are organized and scaled along a hierarchy from fine-grained to coarse-grained, and (c) information is open to and modifiable at all levels. This echoes my call to "think like an ecosystem" and reflects a low-gain view of how a sustainable world might be organized. Meadows went on to show how systems thinking is vital to indicator development, and how conceptual modeling and dynamic modeling are necessary to fully develop and evaluate any multilevel indicator system.

Hierarchy theory (Ahl and Allen 1996) and systems analysis focuses on boundaries, specifically the porosity and rate of exchange at boundaries. The question "What is the scale of the economy?" is a simple model that seeks to know how much the boundary of the economy can infringe on the whole of the ecology. Indicators that only track the central tendencies of a system will not provide timely data on a system's variance, and a system can fail well before a signal is seen in those indicators. Indicators need to focus on the boundaries of the system, the place where information, matter, and energy are exchanged at a rate. Again, Meadows (1998) reflected these notions in her statement "The three most basic aggregate measures of sustainable development are the *sufficiency* with which ultimate ends are realized for all people, the *efficiency* with which ultimate means are translated into ultimate ends, and the *sustainability* of use of ultimate means."

Obviously, a thorough reading of Meadows' work is recommended, because I have only presented the key points and there is much more depth to be explored. It is to me one of the most useful and cogent presentations on indicators and information systems. My intent in presenting Meadows' work is to offer not just another generic set of indicators that may not work for your landscape, but instead a conceptual model of how to select and structure your own unique and usable indicator set. To effectively develop a set of indicators, you will also need a conceptual model of your social–ecological landscape, outlining the hierarchy of interacting elements.

Conceptual Modeling in Ecological Risk Management

Monetized valuation systems are not sufficiently scalable and flexible to help us track what we extract, degrade, and return to a landscape. Indicators are needed at multiple levels in the system in order to fully assess the rate we extract goods and services from a landscape, as well as the rate at which we expel waste back into that landscape. There are myriad indicators that could be developed to answer these questions. The Ecological Footprint (Wackernagel and Rees 1996) is one very good example of a scaleable, multilevel indicator. The issue then is to decide which indicators actually tell us what we need to know when we need to know it. The field of risk assessment and management offers a process by which key social issues and ecological stressors can be brought to the surface and analyzed in context. The holistic risk framework is presented elsewhere in this book (Chapter 8). Here I focus on the initial step of problem formulation, specifically developing a conceptual model of the landscape and ecosystem processes of interest.

A conceptual model (CM) depicts the ecological and socioeconomic relationships among components of the landscape and potential stressors from human activities, whether direct or indirect. Stressors can relate to biological, chemical, and physical interactions that currently exist as well as those anticipated to occur in the future, including those specifically related to planned human activities. A CM is therefore a characterization of interactions occurring across spatial and temporal scales relevant to decisions to be made about a project. Temporal and spatial extent, time and space, the

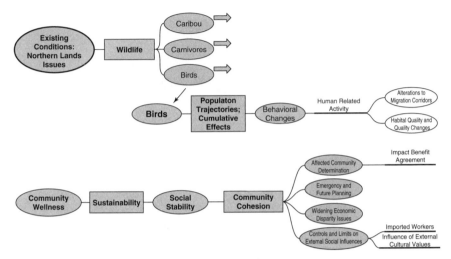

Figure 18.2. Example linkage diagrams. One pathway from the top diagram can be inter-preted as indicating that "alterations to migration corridors, related to human activities in the area, causes behavioral changes in individual birds, which can cumulatively affect population trajectories in birds, and is a subissue of general wildlife issues that are existing conditions within Northern Lands." One pathway from the bottom diagram can be interpreted as indicat-ing that "imported workers create external social influences that affect community cohesion and overall social stability, a subissue of the larger sustainability issues related to overall community wellness."

when and where of what we are interested in, are key elements of any level of analysis. Defining a level of analysis is part of the process of creating a conceptual model, the first step in developing risk-based management scenarios for complex landscapes.

A conceptual model should be developed as part of any holistic risk assessment project plan. It should be designed to be dynamic, yet readily accessible to multiple stakeholders. For this chapter, I developed an example CM for a generic development project in northern Canada. While the CM is generalized for presentation in this book, it does reflect actual stakeholder issues publicly expressed during the review and approval process for projects in the region. The CM (which can be found on the book's ftp site) has 64 pages of linkage diagrams. Details about how to read the diagrams and navigate through the CM are included in the first few pages. Each linkage diagram (Fig. 18.2) frames a perceived connection from an anticipated activity/source resulting from project operations to a specific environmental element (air, water, soil, vegetation, wildlife, humans, etc.). Linkage issues are potentially attributable to sources—that is, community members, elders, project developers, scientists, government officials, and other nongovernmental organization stakeholders.

The sample CM presented here is visually interactive and provides 16 levels of analysis covering all phases and timeframes for a project. A project CM should contain cross-references to each issue identified by stakeholders as well as those identified by government officials and technical experts. Every issue identified is incorporated into

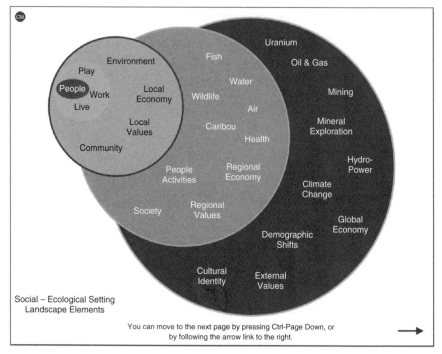

Figure 18.3. Nested relationships for residents of a northern social–ecological landscape.

one or several of the linkage diagrams. The CM thus serves as a foundation for organizing a project assessment by various disciplines including human and ecological risk.

Means and Ends on a Specific Landscape

People are closely linked to their ecological setting, their landscape. Page 1 of the example CM (Fig. 18.3) shows the many ways people interact with place to reflect local values and produce a local economy. The key terms used are as those expressed by stakeholders to describe elements of the landscape within which they live, work, and play. This high-level conceptual diagram shows people nested within a local community and its environs. Each community is a self-contained and local unit (indicated by the dark solid boundary line around the people-community-environment circle), while also belonging to and being nested within a larger, regional landscape.

Regional social, economic, and cultural values emerge from the interactions of many local communities and a diverse landscape; these values drive human activities to produce a regional economy. A tacit societal agreement emerges from the many communities and guides management actions within the regional landscape with respect to caribou, fish, water, human health, and other issues. Even so, the regional social structure is connected to the global landscape with varying degrees of external influences (shown by the lighter-colored shared boundary).

Globally, economic drivers increase the desire for access to local resources. The resulting infusion of external values and different cultural norms directly affects local peoples and their landscapes. Along with this interaction of global and local values, human immigration and climate change place added pressures on regional societies and ecosystems.

The conceptual model identifies and expands on the interconnections between global drivers and local conditions, with an emphasis on boundaries and interaction rates. For example, as shown on the Social–Ecological Landscape Elements page, local communities are nested within their regional environment. They are tied closely to this environment and rely on it. However, human activities originating from outside the regional landscape weaken or eliminate local–regional bonds, effectively pulling the local people–community–economy into the global circle by altering the regional landscape. The CM details these cross-scale interactions for the spatial extent of northern lands affected by the proposed project. The social–ecological relationships within this landscape are illustrated within the CM over time from the present to long after project closure.

From Conceptual Model to Indicators to Scenarios

Following Meadows' (1998) list of criteria for what makes a good indicator, one could develop informative metrics for any and all of the endpoints (far right topic, Fig. 18.2) of each linkage diagram in the conceptual model. Employment statistics for the entirety of the far northern territories would be useful for one level of analysis presented in the CM (Existing Conditions: Northern Lands), as would within-territory numbers (Existing Conditions: Regional), as well as individual community values (Existing Conditions: Local). The criteria used to collect these values would change with project phase (temporal scale) as well. Similarly, caribou numbers could be collected for all Northern Lands as an aggregate population number, while hours of hunting effort per caribou harvested might serve to assess the status of local caribou numbers in every affected community.

The key point when developing indicators is to tailor each to each level of analysis and to make sure that the metrics used reflect the desires and values of affected stakeholders. The trouble with this key point is that desires and values of stakeholders change with each level of analysis, especially with time. Thus, again as pointed out by Meadows, every indicator should have a sunset date. At some reasonable point in time, every indicator must be fully reviewed by the relevant stakeholders, and a decision must be made as to whether to keep collecting those data or to start collecting different data to better reflect current needs.

Multilevel conceptual models, dozens of levels of analysis, dozens of dozens of possible indicators spread across space and time; suddenly the idea of tracking the sustainable use of a landscape appears exceptionally complicated. Much of the current discourse regarding sustainability metrics appears complicated for just this reason, becoming mired in the details of things that can be measured, but don't necessarily tell us anything. Essentially they are metrics (e.g., Indicators of Sustainable Community

1998 by Sustainable Seattle) for *becoming less unsustainable* than ones for tracking our sustainable use of ecosystem goods and services.

Every social–ecological landscape is unique in its complexity. The path to sustained use of each landscape will vary widely around a set of common ideas. I have presented my view of what those common ideas might entail, and I have shown how to integrate theory into practice. The last issue in dealing with each unique complexity lies in how to interpret the signal from your indicator set. Humans are hard-wired to only be able to value something in comparison to something else (Ariely 2008). An indicator absent a context lacks value for making decisions about using an ecosystem service. Most indicators are evaluated with respect to historic values, which can be informative for short-term (year-to-year) goals. How a given value relates to possible future landscape states is just as, or perhaps even more, important when thinking like an ecosystem.

As far as we know, no one can predict the future. At best, we can only generate possible scenarios using general, plausible trends. The business world does this quite regularly (e.g., Shell International Ltd. 2005), and they regularly review and update their scenarios. Van Der Heijden (2005) and Schwartz (1996) are only two of many books written on how to develop scenarios for your particular social–ecological landscape. By describing three or four possible future states for the economic, social ecological, and political context that you will be evaluating your landscape indicators in, you can assess whether an up or down trend is acceptable.

Referring back to the conceptual model example, at present, using only past and current values, a drop in employment in the mining sector is considered to be poor economic and social news. Maintaining the current high-gain system where the entire region is reliant on a single-source employer will ensure that reduced employment is always bad news. Re-envisioning an integrated local and regional landscape that has in place alternative employment and other programs to absorb inevitable downturns in a single industry is one very plausible scenario, and a scenario where rising single-sector unemployment is not considered to be poor economic and social news.

Consider two plausible scenarios having the same indicator trend, but with two different evaluations. Social and monetary valuation systems, intermediate means and ends, will shift and change with time. Through continuous dialogue and stakeholder input, a community can decide which possible scenario within the social–ecological landscape it desires as an ultimate end, and it can then set about developing conceptual models and indicators to address the source of the ultimate means to that end. What I have shown here is that by using risk-based decision analysis tools, one can guide stakeholders in a process toward holistic conceptualization of the problems being addressed, minimizing the chances of a "solution" becoming the "next and larger problem."

REFERENCES

Ahl V, Allen TFH. 1996. *Hierarchy Theory: A Vision, Vocabulary and Epistemology*. Columbia University Press, New York.

Ariely D. 2008. *Predictably Irrational: The Hidden Forces That Shape Our Decisions*. Harper-Collins, New York.

Meadows D. 1998. *Indicators and Information Systems for Sustainable Development*. The Sustainability Institute, Hartland Four Corners, VT.

Pascale R, Millemann M, Gioja L. 2001. *Surfing the Edge of Chaos: The Laws of Nature and the New Laws of Business*. Three Rivers Press, New York.

Rajan U, Seru A, Vig V. 2008. *The Failure of Models that Predict Failure: Distance, Incentives and Defaults*. Available at Social Science Research Network: http://ssrn.com/abstract= 1296982. Last accessed November 6, 2008.

Schwartz P. 1996. *The Art of the Long View: Planning for the Future in an Uncertain World*. Currency Doubleday, New York.

Shell International Limited. 2005. *Shell Global Scenarios to 2025*. Royal Dutch/Shell Group.

Van Der Heijden K. 2005. *Scenarios: The Art of Strategic Conversation*. John Wiley & Sons, West Sussex, England.

Wackernagel M, Rees W. 1996. *Our Ecological Footprint: Reducing Human Impact on the Earth*. New Society Publishers, Gabriola Island, BC.

EPILOGUE

The collection of chapters in this book argue individually and collectively that it is possible to do risk assessments at very large scales for a variety of environmental, ecological, and human health goals. We often hear colleagues fretting over obstacles they encounter as they attempt to perform risk assessments. Some of the obstacles, as we have argued elsewhere, are inherent to dealing with complex systems; others are contrived problems that we have allowed to stymie our efforts. Too many of our friends and colleagues passively accept regulatory policies as dogma, as the only way that one can conduct an risk assessment, whether human health or ecologically oriented.

We want to underscore that in each of the chapters of this book, the authors have reached beyond the comfort of the known, beyond narrowly interpreted regulatory guidance so that they could be truer to the technical realities of the disciplines as they see them. Ironically, much of the technical guidance, especially in the ecological arena, tends to be criticized as being too vague—that is, not having sufficient detail to know what is being requested. Yet, innovative approaches to address the challenges of a particular project often are killed by claims that "guidance doesn't allow such innovation." Our experiences have been that the guidance has been constructed, in many cases brilliantly even if unwittingly, such that innovation actually is encouraged. Far from being prescriptive, much guidance opens the door for better applications of the sciences.

We hope you have found the book and the chapters to be thought-provoking, even inspiring. We will consider the effort a success if in the future we find more projects in which the assessors demonstrate that they have consciously addressed the difficult challenges of defining spatial and temporal scales in ways that match the scope of the relevant ecological processes associated with their project.

Please give us feedback on the content and general themes portrayed in this book. We look forward to hearing from you.

Environmental Risk and Management from a Landscape Perspective, edited by Kapustka and Landis
Copyright © 2010 John Wiley & Sons, Inc.

INDEX

Environmental Risk and Management from a Landscape Perspective, edited by Kapustka and Landis
Copyright © 2010 John Wiley & Sons, Inc.